THE EVOLUTION OF BUSINESS IN THE CYBER AGE

Digital Transformation, Threats, and Security

THE EVOLUTION OF BUSINESS IN THE CYBER AGE

Digital Transformation, Threats, and Security

Edited by

Divya Gupta Chowdhry

Rahul Verma

Manisha Mathur

AAP | APPLE ACADEMIC PRESS

Apple Academic Press Inc.
4164 Lakeshore Road
Burlington ON L7L 1A4
Canada

Apple Academic Press Inc.
1265 Goldenrod Circle NE
Palm Bay, Florida 32905
USA

International Standard Book Number-13: 978-1-77188-810-3 (Hardcover)
International Standard Book Number-13: 978-0-42927-648-4 (eBook)

Library and Archives Canada Cataloguing in Publication

Title: The evolution of business in the cyber age : digital transformation, threats, and security / edited by Divya Gupta Chowdhry, Rahul Verma, Manisha Mathur.

Names: Chowdhry, Divya Gupta, editor. | Verma, Rahul, 1989- editor. | Mathur, Manisha, editor.

Description: Includes bibliographical references and index.

Identifiers: Canadiana (print) 20190215615 | Canadiana (ebook) 20190215658 | ISBN 9781771888103 (hardcover) | ISBN 9780429276484 (ebook)

Subjects: LCSH: Business enterprises—Computer networks—Security measures. | LCSH: Confidential business information—Security measures. | LCSH: Computer security. | LCSH: Information technology—Management.

Classification: LCC HD30.38 E96 2020 | DDC 658.4/78—dc23

CIP data on file with US Library of Congress

Apple Academic Press also publishes its books in a variety of electronic formats. Some content that appears in print may not be available in electronic format. For information about Apple Academic Press products, visit our website at **www.appleacademicpress.com** and the CRC Press website at **www.crcpress.com**

About the Editors

Divya Gupta Chowdhry
Director, Jagran Institute of Management, India

Divya Gupta Chowdhry, PhD, is Director at the Jagran Institute of Management, established under the benevolence and support of Dainik Jagran Group, India. She has rich experience in the corporate sector, teaching experience in academics, and more than 14 years of experience in research. She has presented more than 35 papers in various international, national conferences, and seminars in India and abroad. She has published more than 30 research papers to date. She has edited and contributed various chapters in several books. Several research scholars have submitted synopses for PhD degrees under her supervision. She has also chaired many national and international conferences. She is also an active member of governing body, BOS, advisory board, R&D boards of various colleges and institutes, editor of various peer-reviewed prestigious journals, and a paper setter in the area of management. She is a gold medalist throughout her academic career.

Rahul Verma
Lecturer, Department of Training and Technical Education, Bhai Parmanand Institute of Business Studies, India

Rahul Verma, MCom (International Business) & PGDBA (Finance & Marketing), is a Lecturer in management with the Department of Training and Technical Education, Government of the National Capital Territory of Delhi, India, and is pursuing a PhD in Commerce from Mewar University, India. His research interests include international business, marketing, and finance. He has attended and presented papers at several national and international conferences and seminars. He has published more than 19 research papers with prestigious publishers and has edited a book, *Radical Reorganization of Existing Work Structures through Digitalization,* published by IGI Global in 2017. He is also on editorial boards of several peer-reviewed journals.

Manisha Mathur
Information Assistant, National Law University, India

Manisha Mathur, MSc (Information Technology) & MA (Psychology), is an Information Assistant with the National Law University, India. She is currently doing research on "Cyber Crimes against Women in Rajasthan" as a research scholar at Mewar University, India. She has presented several papers at national and international conferences and seminars and has published several papers in prestigious national and international journals. She has also delivered many guest lectures at various institutions. Her chapter on "Digitization's Impact on Work Culture" was published in the book *Radical Reorganization of Existing Work Structures through Digitalization.* She has a keen interest in research in various fields, including social work.

Contents

Contributors

S. Anitha
ACS College of Engineering, Biomedical Engineering, #207, ACS College of Engineering, Kambipura, Mysore Road, Bangalore, Karnataka – 560079, India, E-mail: dranithasammilan@gmail.com

Sugandha Soni Chopra
Rama University, Management, EWS-4111, AWAS Vikas-3, Panki Road, Kalyanpur, Kanpur, U.P. – 208017 / 676-A, Near Canara Bank, Gandhi Nagar, Jammu – 180004, India, E-mail: sugandhasoni101984@gmail.com

Debashis Ghosal
Crossover Leadership Partner, 104, Coral Block, Sri Sairam Manor, 8-3-678, Pragathi Nagar, Yousufguda PO, Hyderabad, Telangana – 500045, India, E-mail: debg2006@gmail.com

Dinesh Goyal
Professor, Suresh Gyan Vihar University, Center for Cloud Infrastructure & Security, Suresh Gyan Vihar University, Mahal, Jagatpura, Jaipur, Rajasthan – 302017, India, E-mail: dinesh8dg@gmail.com

Aman Jantan
University of Science, School of Computer Science, Penang, Pulau Pinang – 560079, Malaysia, E-mail: aman@usm.my

S. Karkuzhali
Kalasalingam Academy of Research and Education, Computer Science and Engineering, Anand Nagar, Krishnankoil, Tamilnadu – 626126, India, E-mail: vijikarkuzhali@gmail.com

Muhammad Ubale Kiru
University of Science, School of Computer Science, Penang, Pulau Pinang – 560079, Malaysia, E-mail: muhd.kiru@student.usm.my

Jyoti Kukreja
Jagannath International Management School, Management, MOR, Pocket-105, New Delhi – 110019, India, E-mail: jyoti.kukreja@jagannath.org

Hemanth G. Kumar
ACS College of Engineering, Biomedical Engineering, #207, ACS College of Engineering, Kambipura, Mysore Road, Bangalore, Karnataka – 560079, India, E-mail: hemumanju@gmail.com

Vannie Naidoo
University of KwaZulu – Natal, South Africa, E-mail: Naidoova@ukzn.ac.za

Anand Nayyar
Duy Tan University, Graduate School, Viet Nam, Room No: 102, 254 Nguyen Lan Vinh Street, Da Nang, Vietnam, NA, Da Nang – 550000, Vietnam, E-mail: anandnayyar@duytan.edu.vn

P. Palos-Sanchez
University of Seville, Department of Business Administration and Marketing, Avda. Ramón y Cajal 1, Seville – 41018, Spain, E-mail: ppalos@us.es

E. Sweetline Priya
Madras Christian College, Department of Computer Application (BCA), F1 B Block, Shahin Villas, Rajiv Nagar 4th Street, Kovilampakkam, Chennai, Tamilnadu – 600117, India, E-mail: sweetlinepriya.edwin@gmail.com

R. S. Rajput
Research Scholar, Suresh Gyan Vihar University, Center for Cloud Infrastructure & Security, Suresh Gyan Vihar University, Mahal, Jagatpura, Jaipur, Rajasthan – 302017, India, E-mail: rajpoot.rs@gmail.com

Rudra Rameshwar
Thapar Institute of Engineering and Technology, L.M Thapar School of Management, Thapar School of Management, Thapar University, Patiala, NA, Patiala, Punjab – 147004, India, E-mail: rudrarameshwar@gmail.com

A. Reyes-Menendez
King Juan Carlos University, Business Economics, P° Artilleros s/n, Madrid – 28032, Spain, E-mail: ana.reyes@urjc.es

Shulagna Sarkar
Institute of Public Enterprise, Plot No. 98, Road No. 3, Surya Enclave, Tirumalgiri, Secunderabad, Telangana – 500013, India, E-mail: shulagnasarkar@gmail.com

J. R. Saura
King Juan Carlos University, Business Economics, P° Artilleros s/n, Madrid – 28032, Spain, E-mail: joseramon.saura@urjc.es

S. Senthilkumar
Ayya Nadar Janaki Ammal College, Chemistry, Srivilliputtur Main Road, Sivakasi, Tamilnadu – 626124, India, E-mail: senthilkumar@anjaconline.org

B. C. Sinha
Rakshpal Bahadur College of Engineering and Technology, Management, Knowledge Park, Part III, Greater Noida, U.P. – 201310, India, E-mail: sinhachandra99@gmail.com

Arun Solanki
Gautam Buddha University, School of Information and Communication Technology, School of ICT, Noida, NA, Noida, U.P. – 201312, India, E-mail: ymca.arun@gmail.com

Shweta Tripathi
Devi Ahilya Vishwavidyalaya, Indore, Madhya Pradesh, India, E-mail: shwetatrip02@gmail.com

Rahul Verma
Department of Training and Technical Education, Bhai Parmanand Institute of Business Studies, India

Abbreviations

ABE	attribute-based encryption
ACC	American College of Cardiology
AI	artificial intelligence
AICPA	American Institute of Certified Public Accountants
AIDC	automatic identification and data capture
AM	application master
AMD-V	AMD-virtualization
AMQP	advanced message queuing protocol
APIs	application programming interfaces
AWS	Amazon web services
BL	blended learning
BLE	Bluetooth low energy
BoF	buffer overflow
BPI	business performance innovation
BYOD	bring your own device
CA	certificate authority
CARP	channel aware routing protocol
CBSE	Central Board for Secondary Education
CDR	call detail record
CDU	channel distribution usage
CEA	Cotton Egypt Association
CERT	computer emergency response team
CHIP	cognitive human interface personality
CISO	Chief Information Security Officers
CoAP	constrained application protocol
CON	confirmable message
CPS	cyber-physical systems
CRC	checksum and cyclic redundancy check
DAC	discretionary access control
DACL	discretionary access control list
DAGs	directed acyclic graphs
DAO	destination advertisement object
DCPS	data-centric publish-subscribe

DDM	download dropper method
DDoS	distributed denial of service
DDS	data distribution service
DECT	digital enhanced cordless telecommunications
DER	distributed energy resources
DI	direct instruction
DIS	DODAG information solicitation
DLRL	data-local reconstruction layer
DNS	domain name system
DODAG	destination-oriented directed acyclic graph
DoS	denial of service
DSOs	distribution system operators
EC2	elastic compute cloud
EHEA	European Higher Education Area
EMI	English-medium instructors
EMIS	Education Management Information Systems
EPC	electronic product code
EPS	endpoint security
ETL	extract, transform, and load
FC	flipped classroom
FCM	flipped classroom model
FFD	full functional device
GDPR	general data privacy regulations
GST	goods and services tax
HBR	Harvard Business Review
HCM	human capital management
HDFS	Hadoop distributed file system
HIPAA	Health Insurance Portability and Accountability Act
HR	human resources
HTTP-DDoS	hypertext transfer protocol-distributed denial of service
HVAC	heating, ventilation, and air conditioning
IaaS	infrastructure as a service
ICT	information and communication technology
IEEE	electrical and electronics engineers
IETF	internet engineering task force
IGL	inquiry guided learning
IIoT	industrial internet of things
ILAs	interactive learning activities

IMC	integrated marketing communication
IOC	indicators of compromise
IoT	internet of things
IoT-A	internet of things-architecture
IP	internet protocol
IPDR	internet protocol detail record
ISMS	information security management system
IT	information technology
ITRC	identity theft resource center
ITU	international telecommunication unit
JiTT	just-in-time teaching
KM	knowledge management
LbD	learning by doing
LeD	learning dialogs
LxI	learning experience interaction
MAB	multi-armed bandits
MAC	mandatory access control
MBR	master boot record
MFA	multi-factor authentication
MFC	model-based flipped classroom
MOODLE	modular object-oriented dynamic learning environment
MQTT	message queuing telemetry transport
MTC	machine type communication
N-AIM	National-Artificial Intelligence Mission
NCCP	National Cybersecurity Policy
NCRB	national crime records bureau
NERC	North American Electric Reliability Corporation-Infrastructure Protection
NICE	National Cybersecurity Education Initiative
NIST	National Institute of Standards and Technology
NJCCIC	New Jersey Cybersecurity and Communications Integration Cell
NON	nonconfirmable message
NOSQL	not only SQL
NSA	national security agency
OC	option count
OMG	object management group
OSI	open systems interconnection

OTF	on-the-fly
OTrP	open trust protocol
PaaS	platform as a service
PB	petabytes
PBL	problem-based learning
PINs	personal identification numbers
PKI	public key infrastructure
PLTL	peer-led team learning
PMO	Prime Minister's Office
PRAW	periodic restricted access window
RaaS	ransomware as a service
RBAC	role-based access control
RDBMS	relational database management system
RDS	rural development strategy
RFD	restricted functional device
RFID	radio-frequency identification
RGB	returnable glass bottles
RM	resource manager
ROI	return on investment
ROLL	routing over low-power and lossy links
RPL	routing protocol for low
SaaS	Software as a Service
SEO	search engine optimization
SII	simply industrial internet
SMAC	social, mobility, analytics, and cloud
SMEs	small and medium enterprises
SMQTT	secure MQTT
STEM	science, technology, environment, and management
SVG	scalable graphics file
TB	terabytes
TBL	task-based learning
TDMA	time division multiple access
TEE	trusted execution environment
TIE	The Indus Entrepreneurs
TIM	traffic indication map
TSCH	time-slotted channel hopping
UAC	user account control
UAVs	unmanned aerial vehicles

UPI	unified payment interface
UTAUT	Unified Theory of Acceptance and Use of Technology
VUCA	volatile, uncertain, complex, and ambiguous
WHO	world health organization
WSN	wireless sensor network
XML-DDoS	extensible markup language distributed denial of service
XMPP	extensible messaging and presence protocol
XSS	cross-site scripting
YARN	yet another resource negotiator

Foreword

Digital business is accelerating and changing the business environment. Doing business in a digital age must evolve to meet the challenge. In cyberspace, businesses have a ballpark figure worth of the global economy of about $445 billion, and businesses have lost productivity and time as well as sensitive financial and personal information due to cybercriminal activities because conducting criminal activity in cyberspace has gotten easier. Business recovery from cyberattacks may perhaps be complex and more costly than imagined, and it is necessary to get a deeper look at the business impact of cyberattacks. Therefore, businesses cannot ignore the threat of cyberattacks, and thus, businesses need to implement their own crackdown on cyberattacks as the fact remains that we cannot ignore technology.

Security in cyberspace is a greatly multifaceted cross-disciplinary area that includes a computer system, software engineering, infrastructure of network, the infrastructure of management, policymaking, law-making, enforcement, and crimes investigation. The Internet today is a widespread business information infrastructure, and therefore, cybersecurity is a major area of concern. Consequently, a multifaceted approach to solutions needs to be presented.

This book provides a valuable window on the digital business environment and covers the necessary components from detecting cyber threats to securing mobile communication devices. In this new age of global interdependence and interconnectivity, it is necessary to provide security practitioners, students and professionals with state-of-the-art knowledge on the frontiers of the digital business environment. This book is a good step in that direction.

— **Vijay Kumar Vaidya**
Vice-Chancellor,
Mewar University, Rajasthan, India

Acknowledgment

The editors would like to acknowledge the help of all the people involved in this project and, more specifically, to the authors and reviewers who took part in the review process. Without their support, this book would not have become a reality.

First, the editors would like to thank each one of the authors for their contributions. Our sincere gratitude goes to the chapter's authors, who contributed their time and expertise to this book.

Second, the editors wish to acknowledge the valuable contributions of the reviewers regarding the improvement of quality, coherence, and content presentation of chapters. Most of the authors also served as referees; we highly appreciate their double task.

— Divya Gupta Chowdhry
Rahul Verma
Manisha Mathur

Preface

Cyberspace, or the virtual world, came into existence as the result of the evolution and progression of digital communication technologies. Cyberspace became a vital vehicle of communication via the Internet. The Internet opened the door to a treasure trove of information. Information technology (IT) deals with information systems, data storage, data access, data analyzing, data modification, intelligent decision-making, and so many field related areas of data and information. However, the emergence of technologies expands the exploitation of technology. Most people have been the victim of cybercrime in cyberspace, and cybercriminals and threats are on the rise.

The aim of this book is to study and analyze cyberattacks that are increasing every day. Cyber threats are illicit doings or unlawful acts, and cyberattacks are through digital means or the Internet in which computers or digital devices are either tools or targets or both. Therefore, we need to acknowledge that cyberattacks and the challenges they present are part of the future of cyberspace. It is important to know how technologies can be exploited by invaders, which is drastically detrimental to our society, businesses, finances, etc. Protecting yourself and your business and providing a secure environment from these threats has become a major area of concern. The public is not aware of the latest cyber-secure technologies; there is an important need to develop an awareness of cyber technologies and threats. Cyber threats and cybersecurity as a subject of technical, legal, and social study are not only advancing but mind-blowing too, as it is comprised of new concepts and challenges.

In this digital era, cyberattacks are the reckless nurturing zones of crime. Cybercrime masterminds are exploiting the vulnerabilities of cyber technologies and creating a hazardous milieu in cyberspace. Our modern digital society needs to identify and understand cyberattacks, cybercriminals, and cyber technologies.

The question arises of how to protect a business or organization from cyberattacks? In the light of all this, authoring a book on cyberattacks, cybersecurity, and cyber challenges is really a hard nut to crack.

Protection and a secure business environment should be provided on cyberspace, which is a very difficult task due to the weakness of the infrastructures of technologies and laws. Today's need is to provide secure cyber business environments for users on various sectors, and for law enforcement to play a key role in providing strong evidence. This is also a primary objective of this book.

With this book in hand, readers will be able to develop a clear concept and have a better understanding of cyber technologies, cyberattacks, and cyber business environment.

The scope of the book takes into account the major cyberattacks being undertaken in the business world. This book explains in detail the possible data breaches and crimes. This book even explains the facts of cyberattacks prevailing at present. It includes the basic business threats in cyberspace and how businesses can protect themselves in the cyber world.

ORGANIZATION OF THE BOOK

The book is organized into two sections, namely "Digital Business Transformation," consisting of seven chapters; and "Cyber Threats and Cybersecurity," consisting of six chapters. A brief description of each of the chapters follows:

- Chapter 1 introduces the concept of digitalization in the context of people and the workplace. The chapter aims to highlight the attempt of digitalization of the workplace and also elaborates on the concepts associated with artificial intelligence (AI) revolution. The authors set the scene for discussing the factors that determine the scope of AI in replacing human interface in organizations. In particular, the chapter identifies various cases in successful digitalization implementation across the globe.
- Chapter 2 establishes that how entrepreneurs of small and medium enterprises (SMEs) with insufficient capabilities and limited resources drove digital transformation in their companies. The author of this chapter expands our understanding of both digital entrepreneurship and digital transformation. This chapter also presents the new capacity to gain an accurate into how digital platform service providers can help SMEs appearance and strive to gain.

- Chapter 3 analyses and compares the advantages of online learning and traditional classroom learning environments, and presents the notion of 'blended learning (BL),' which is a new learning environment. The authors identify that the role of teachers is increasingly important in the new cultural and educational environments that are being created and to be created with the help of information communication and technology (ICT).
- Chapter 4 in its endeavor examines the role of E-HRM as a tool in increasing the competitive edge of Indian organizations, and the response has been sought after from employees of eight organizations using the structured questionnaire as an instrument. Based on a survey, the author reveals that all attributes of financial contribution are not above the test value and there exists a difference in level for some attributes in context to private vis-a-vis public and manufacturing vis-a-vis services.
- Chapter 5 takes the philosophical orientation and debates that when change usually occurs, things fall apart, but the culture to connect for any marketing strategist is pertinent. The author in this piece of writing includes a case let and a case study on the marketing strategies for successful global industry pie. The chapter sets the scene for a discussion on the point that technology is not a choice but more of a necessity nowadays. The author contends that technology 'makes' or 'breaks' marketers.
- Chapter 6 outlines an in-depth overview of Internet of Things (IoT) like technical details and enabling technologies followed by an in-depth technical overview of protocols enabling IoT—a device-to-device communication. The authors address the cybersecurity issues surrounding IoT to enable researchers and developers to speed-up research in these areas, along with an overview of various simulation tools available for carrying out research on IoT.
- Chapter 7 presents a brief introduction to big data and big data analytics, the various categories of big data, the characteristics that determine the big data, how different technologies are involved in big data analytics, and a brief introduction on the tools involved handling big data. The author also identifies the applications of big data in various fields like health care, banking, marketing, politics, meteorological department, finance, and so on.

- Chapter 8 systematically reviews various threats in different fields of business and the various existing methods to overcome the said threats. Based on a survey focused on users' practices in internet banking, the authors propose a survey model conducted on internet banking in various countries. The authors proposed model bridges the gap by increasing the responsibility of banks to reduce cybersecurity risks for users.
- Chapter 9 focuses on the world's most frightening cybersecurity threat known as ransomware. The overall aim of the chapter is vested upon uncovering the underlying mysteries behind the sudden growth and popularity of ransomware and how to fight and stop it. The authors seek to bring together the collective professionals' views and recommendations on how to set up the strategic defense-in-depth for fighting against ransomware.
- Chapter 10 sets the scene for discussions presented by various authors dealing with the role of technology in teaching and learning, the role of teachers in making aware their students towards cybercrime and cybersecurity, and empowerment of teachers thereof.
- Chapter 11 aims to identify the challenges the Digital Business Environment has to face and to propose viable solutions that would allow avoiding cyberattacks and result in a reinforcement of cybersecurity in the IoT. To this end, in order to deepen the impact that cybersecurity has on the IoT for individuals, society, nations, and global humanity, the authors systematically review the literature and perform an exploratory analysis of the results on this ubiquitous topic.
- Chapter 12 sets the scene for a discussion about the factors leading to cloud being vulnerable to threats and attacks and also on the methods and solutions to make secure cloud system. In particular, the chapter identifies that cloud security can be implemented into four sections, namely cloud security behind the virtualization, securing data over the cloud, cloud channel security, and securing cloud system management.
- Chapter 13 concludes and explores crucial issues associated with cyber trends, namely cybersecurity, cybercriminals, cybercrimes, cyberethics, cybersecurity techniques, social media, and cybersecurity and challenges faced in implementing cybersecurity. The authors contend that cybersecurity is vital in the Fourth Industrial

Revolution, which can cause a major technology disruption and affect the lives in ways we never thought possible.

—Divya Gupta Chowdhry
Rahul Verma
Manisha Mathur

Editorial Advisory Board

PART I
Digital Business Transformation

CHAPTER 1

Digital Workplace: The Human Interface

DEBASHIS GHOSAL[1] and SHULAGNA SARKAR[2]

[1]*Crossover Leadership Partner, 104, Coral Block, Sri Sairam Manor, 8-3-678, Pragathi Nagar, Yousufguda PO, Hyderabad, Telangana – 500045, India, E-mail: debg2006@gmail.com*

[2]*Institute of Public Enterprise, Plot No. 98, Road No. 3, Surya Enclave, Tirumalgiri, Secunderabad, Telangana – 500013, India, E-mail: shulagnasarkar@gmail.com*

ABSTRACT

Digitalization has become inevitable for organizations to survive. It is important to understand the concept of digitalization in the context of people and the workplace. Artificial intelligence (AI) and smart machines are replacing people doing repetitive work. The upside of digitalization and automation is that it could free more people to work on creative, newer, and interesting work. The chapter aims to highlight the attempt of digitalization of the workplace in India. The chapter also elaborates the concepts associated with the AI revolution. India is a developing country, and it will be interesting to understand the perspective of the human resource managers as well as the Generation Y on the scope of AI replacing human interface of organizations in India. The chapter attempts to discuss the factors that determine the scope of AI in replacing human interface in organizations. An attempt has also been made to discuss various cases in successful digitalization implementation across the globe.

1.1 DIGITAL WORKPLACE: AN INTRODUCTION

The workplace is transforming faster than we could imagine! Large scale adoption of technology to enhance the people skills, globalization, culture, and shift in economic balance are defining the new age work styles at all levels. Historically, the adoption of new technology and shift in work culture has always propelled the workplace forward. And it's happening again – even faster! Corporate around the world is discussing about Artificial intelligence (AI) replacing human resources (HR) in various segments. In such a scenario, a concern worth discussing is the agenda of digitalizing the workplace–the means and mechanism. The question that has been traditionally attempted to answer is 'what tasks currently performed by humans will soon be done more cheaply and rapidly by machines?' Yet it is high time we must attempt to venture in answering a new question. What new dimensions people have to develop so that they have better-thinking machines to assist them?

The digital workplace has been defined differently by different organizations and experts. One of the most accepted versions is that the Digital Workplace is meant to be the "virtual equivalent of the physical workplace." However, the rapidly changing definition of workplace adopted by different organizations makes the definition widely varying. Here's an interesting take by Gartner (2018), "The digital workplace is an ongoing, deliberate approach to delivering a more consumer-like computing environment that is better able to facilitate innovative and flexible working practices." However, irrespective of the definitions, the three layers are imperative–people, technology, and management. Given below is a simplified model of an ideal digital workplace (Figure 1.1).

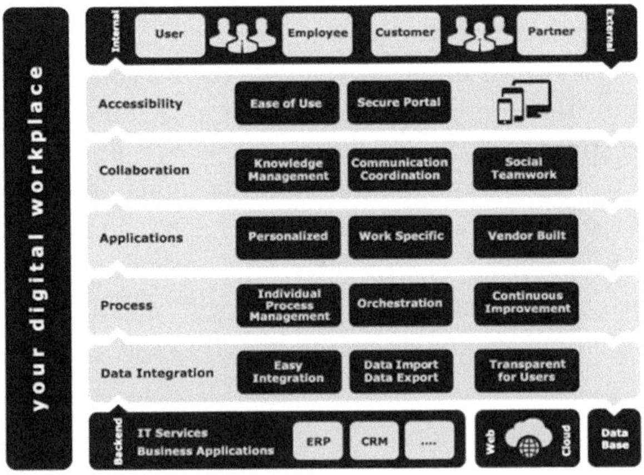

FIGURE 1.1 Ideal digital workplace.
(*Source:* Schichtenmodell Digital Workplace, (2016). Digital workplace. Retrieved from: https://commons.wikimedia.org/wiki/File:Schichtenmodell_Digital_Workplace.png)

1.1.1 ESSENTIALS OF DIGITAL WORKPLACE

- **User Experience:** Digital workplace is an integrated, single source for the user with an enhanced ease of use and better experience, while being more efficient and effective.
- **Communication and Collaboration:** Wider and informal communication options, task/project/organization-specific collaboration through social – Intranet like tools.
- **Multiple Business Applications on Multi-Device Platforms:** Multidevice, multiple business solutions for all user groups providing personalized work-specific applications environment. The applications are vendor-built or individual, providing the same look-and-feel.
- **Business Process and Data Integration:** Technology-generated business-oriented processes using an integrated database, sharing corporate and individual skills and knowledge.
- **Agile Working and Decision Support Systems:** Ability to be productive any time at any place through integrated business applications, database, and transparent user access, providing real-time analytics (Figure 1.2).

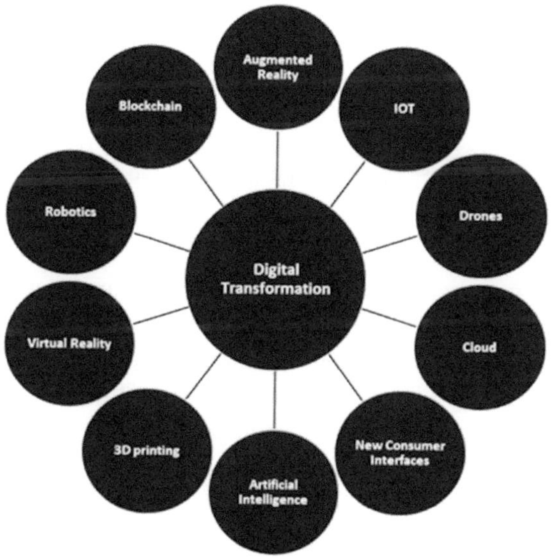

FIGURE 1.2 Essentials of the digitalizing workplace.

A digital workplace to be effective should make physical location independent of work requirements. It's not only to save time to access physical work location, but also will shape where people will work from, how teams will be formed and how people will collaborate to address business situations. According to Klaus Holse (VP – Microsoft), there will be an increasing focus of the modern day's businesses to create an opportunity for their employees to work from anywhere, which can be achieved by integrating people, workplace, and technologies, thereby fuelling productivity and creativity. Information Technology will play a significant role in this direction by empowering the employees to use technology in the way they want to use. This will lead towards competitive advantage and business success. Businesses can achieve this by inculcating a culture of "people first." A digital workplace will fuel adoption of technologies like Chatbots, Robotics, Automation, Internet of Things, Deep Learning, Machine Learning, AI, Augmented Reality, Virtual Reality, etc., which have raised questions about Human Interface at work. Let's explore which industries, jobs, functions will be most affected and how human interface may have to be redefined in the new paradigm.

1.1.2 DIGITAL REVOLUTION AND TRANSFORMATION

The advent of the computer has provided us with a new kind of machine being the computer program. Computer programs are like these more familiar physical machines yet; their behavior is determined by-laws (Bundy, 1988). These programs are different from the physical machines as they are not made of physics. The concept of AI is based on the use of a computer to model and/or replicate intelligent behavior. Companies are working on AI development and analysis of algorithms that learn and/or perform intelligent behavior with minimal human intervention. Minsky (1968) has defined AI as the science of making machines to do things that would require intelligence if done by men. This definition stresses results over methods. Bundy (1988) has defined AI as the building of computer programs which emulate human intelligence, i.e., the striving to create programs that equal or exceed human intelligence without necessarily achieving the intelligence in the same way as humans. The AI programmer works by the pragmatic criterion being computer programming deserves to be treated equally as it acts like a human mind (Bolter, 1984) (Table 1.1).

TABLE 1.1 Three Eras of Automation

Era One: 19th Century	Era Two: 20th Century	Era Three: 21st Century
Machines do routine jobs, which are harmful to humans due to the arduous manual labor involved.	Machines replace the regular work, which is more of routine transactions and involves clerical responsibilities.	Machines do work based on past and present data, using analytics, enabling reliable and faster human choices.

Source: Christensen, Grant, and Govindarajan (2017). HBR's 10 Must-reads 2017: The definitive management ideas of the year from *Harvard Business Review*. Harvard Business School Publishing Corp.

Christensen et al., (2017) elaborated the three eras of automation in 19th, 20th, and 21st century. There have been several studies past three decades attempting to measure efforts and impacts of AI as given in Table 1.1. Autor and Dorn (2013) tracked the effects of automation on labor markets and expressed that journalist and expert commentators have overstated the extent of machine substitution for human labor. He explained that researches do not reflect on the benefits of increased productivity, raises earnings, and augmented demand for skilled labor. He pointed out the challenges of applying machines to any tasks being flexibility, judgment, and common sense. Tasks that cannot be substituted by computerization are generally complemented by it (Autor and Dorn, 2013). Acemoglu and Restrepo (2017) have also studied the impact of robots on industry and occupation-level outcomes. The ultra-rich Wall Street investment bankers and hedge fund titans have also upgraded to automated trading and portfolio management systems. Madia (2018) shares that companies like DSP are also encouraging the same to enhance investment returns within precise decision making. Bolter (1984) expressed that robots will replace human workers at more complicated tasks than usual assembly lines jobs. He also shared that AI and robots can be used for warfare where smart bombs will find their targets with greater accuracy.

Articles from *Business Week* (1986) revealed that the major challenge for business is transforming a new paradigm for manufacturing. Changes in manufacturing since ages have been centered on the implementation of advanced technology. The scope of these technology-driven changes has expanded to more specific issues, such, quality assurance (Harrington, 1987), Inventory control (Klien, 1991), and strategy (Hayes et al., 1988). Work of Dean and Snell (1991) has also explained the need for changes with regards to job design. It is inevitable for companies to make a conscious decision of making a consistent transit to a new paradigm of manufacturing. Evidence from the research of Majchrzak (1988) and Ettlie (1988) suggest that the mismanagement of people is a greater issue in affecting the transition to newer manufacturing techniques and processes rather than technical systems per se. Managing people for implementation of systems driven manufacturing has turned out to be a great challenge. The present HR ecosystem is impacted by the influence of the millennial, tech disruptions a volatile economic environment, and the urgent need for enhanced employee engagement (NASSCOM, 2015). With such a challenging scenario, the role of HR and the way they operate needs to be redefined.

A digital economy could permeate every section of society and transform the way services are carried out by governments, businesses, and consumers. The digital revolution in India is significant as it promises to bring a multi-dimensional metamorphosis in almost all sectors of the society. From digitization in governance to better health care and educational services, cashless economy and digital transactions, transparency in bureaucracy, fair, and quick distribution of welfare schemes all seem achievable with the digital India initiative of the present Government (Singh, 2017). The world has been undergoing several changes, both socially and environmentally, due to which the businesses have remodeled themselves. The incorporation of cloud, mobile, social media, big data, and digital consumer experience into Human Capital Management (HCM technology) is disrupting how organizations manage, enable, and engage their workforce. The disruptive business scenario has forced organizations to make changes to their operatives. The disruptive business scenario is mainly due to the ease of companies to go global. The companies have higher ease as compared to a few years earlier in promoting themselves worldwide with a little effort due to technology advancements. Businesses today are a part of a truly globalized economy.

Companies have realized that it is not always conducible to operate in the home country as other countries might have better support in the operation. Japan has an aging workforce, whereas India has a demographic dividend of the youth population. Thus, if any business requires youth workforce, it would prefer to opt for operations in India rather than Japan. In case the company still operates in Japan, it will attempt for technology-based solutions rather than relying on its aging workforce. Thus, the demographics of a country are a large determinant of acceptance of a system of AI or robotics. Similarly, employability skills and skill diversity of the country also has a large say in the acceptance of a system of AI or robotics. If employability skills and skill diversity is low, it indicates that the workforce is incapable of handling today's job. Companies in such cases are attempting to invest more on AI as an initial cost and reduce the cost on errors, non-performance, quality, and learning and development. Employment cost of a country also impacts the decision and choice of AI. It is evident that lesser the employment cost, lower the acceptance of AI. This is true only in cases where the quality of hire available is good and employable. Demographics of the country, employability skills, employment cost, and skill diversity determines the way and the extent of usage of AI by companies in the country (Figure 1.3).

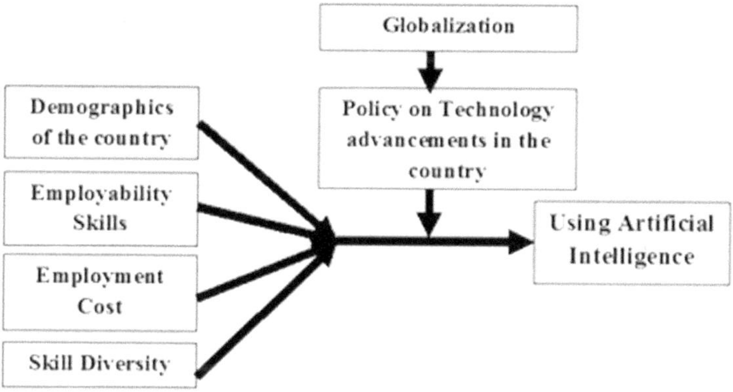

FIGURE 1.3 Factors determining the acceptance of artificial intelligence in a country.

A study by Seth (2018) also considered similar factors like globalization, demographic changes, and adoption of industry 4.0/exponential technologies in predicting the impact of jobs in 2022 in the Indian context. He concludes that technology is expected to be disruptive on IT-BPM and BFSI, automotive, and retail and apparel and leather industry in India. There have been several studies conducted in India by leading consulting firms. NASSCOM's (2015) reports several facts and figures on the growth and acceptance of HR technology in India:

- About 25% of the new Indian Tech-start-ups would deal with Education and HR, and even 23% of those would solely concern HR Technology.

- Demand for specialized skills would be high as more than 40% of the Top 100 IT-BPM firms would start offering SMAC (Social, Mobility, Analytics, and Cloud) solutions.
- 77% of business executives other than CTOs would be the key influencers in technology purchase.
- 85% of the survey respondents opted for tech-enabled HR processes

From the findings, it's evident that HR Technology will assume great proportions in India in the future. As the workforce of the future will be dominated by Millennials and Gen-Ys, constituting about 60% of the total workforce, would look at HR Technology-driven Human Resource Information and Management Systems as the need of the hour. Many HR Technology companies are working in this direction to provide HR services at an affordable cost following the 'pay-as-you-use' or subscription-based business model. PWC (2014) highlights that 31% of HR professionals are worried about their jobs being at risk due to increasing automation. The same report states that 70% of the respondents think that technology cannot be a replacement for the human mind.

1.1.3 CHALLENGES DURING DIGITAL TRANSFORMATION

According to a study conducted in 2017 by Jabil Digital Solutions, USA, ninety percent companies, presently seeking digitalization say they found the transformation to be gradual, time taking and with many constraints. They have still a long way to go. However, the major challenges were listed out as Employee Pushback, Lack of Expertise to Lead Digitization Initiatives, Organizational Structure, Lack of Overall Digitization Strategy and Limited Budget (Jabil, 2018). Employees feel threatened to face a digital transformation. Digital revolution brings in many disruptions in terms of the workplace, nature of work, expertise, and the execution. Human nature is to follow a routine and be comfortable within the operating zone. There's always a threat perception to the very existence of the jobs, for example, in the manufacturing segment. While the process of transformation is unavoidable and the threat perception can't be completely addressed, it's the responsibility of the organization, primarily the HR department to create continuous and comprehensive channels of communication to alleviate such fears. Employees should be able to see a bigger picture and a brighter future, which will motivate them to support the digital transformation strategy of the organization. That calls for a comprehensive strategy.

Talking about strategy, it can't be a piecemeal approach. It needs to be company-wide. Studies show that almost 75% of the companies who have ventured into digital transformation do not have a company-wide strategy. The vision, purpose, goal, and roadmaps should be clearly defined to go through the transformation. Organizations rush in to transform individual business units or product lines even before a company-wide strategy, as high as 38% of companies responded so, which could prove counterproductive. Though roadmaps could be different for different departments, they should be part of a common strategy and goal. So where do you start? Businesses thrive on customers. It's ideal to create interfaces and communicate to the customers first about your thought process and take cues to put a customer-centric digital transformation strategy in place.

Digital transformation is not all about technology transformation; it needs specialized talent to lead. Also, if existing systems and processes are a roadblock, the existing technology partnerships have to be reassessed. It's imperative to create 'digital literacy' before the digital revolution. Companies must be ready with re-skilling of the workforce to meet the challenges. Early investment in people – the readiness is important, and if need be companies need to hire external partners to help them go through the transition. Digital transformation may need to inflict many strategic changes in the organization, which may not be aligned to the conventional wisdom of structuring an organization. It may lead to changing role, changing departments for the employees, and structural changes at a much bigger scale. It is imperative for the HR department to take the employees into confidence that the new structure will support productivity and career growth. The new age organization structure may be fluid as an adaptation of technology, data flow, and customer experience would command such an agile structure. Digital transformation also demands substantial financial resources. The budget has to be meticulously planned, and constraints should be anticipated well in advance. The digital transformation plan could be spread over phases. There is no need for a quick turnaround, and it's wise to take stock of the impact made once in a while through the journey. Overall, the digital transformation strategy should result in innovation through motivation (Table 1.2).

TABLE 1.2 Digital Transformation Challenges – Classified On Company Size

Less Than 100 Employees	100–1,000 Employees
• Lack of expertise to lead digitization initiatives • Employee pushback • No overarching strategy for digitization • Business partners unable to support • Limited budget	• Employee pushback • Organizational structures get in the way • No overarching strategy for digitization • Limited budget • Lack of expertise to lead digitization initiatives
1,000–5,000 Employees	**More Than 5,000 Employees**
• No overarching strategy for digitization • Lack of expertise to lead digitization initiatives • Limited access to the required technical expertise • Employee pushback • Limited budget	• Lack of expertise to lead digitization initiatives • Organizational structure gets in the way • No overreaching strategy for digitization • Limited access to the required technical expertise • Employee pushback

Source: Jabil, (2018). *Digital Transformation Readiness Survey.* Retrieved from: https://www.jabil.com.

1.2 THE HUMAN INTERFACE

1.2.1 DISRUPTIVE HR: THE TECHNOLOGY INTERVENTION

The changing demographics and advent of technology are bringing on marked changes in the way organizations manage their talent. HR leaders and professionals must adapt to these changing times and question the traditional approaches to talent management. They must develop an external view of the latest trends. Thus, HR professionals must adopt the latest technologies and trends existing globally and apply them to the organizational requirements. Only then can organizations accept to build the talents for the future, while meeting present and future needs. The word 'disruptive' has gained popularity in today's world. Luke Williams explains 'disruptive thinking' as a "way of thinking that surprises the market again and again with exciting, unexpected solutions.

A way of thinking that produces an unconventional strategy that leaves competitors muddling to catch up. A way of thinking that turns consumer expectations upside down and takes an industry into its next generation." According to Christensen, products evolve from 'successful' to 'niche' over a period of time. This also sees the usage of the product moving from 'easy level' to 'complicated level' and pushing the pricing up. This creates opportunities for innovative companies to come up with cost-effective and simpler solutions for mass usage. Disruptive thinking is not always to get the best resources on board or spend huge money on training or trying to fuel innovation in the traditional way. Design thinkers seek non-conventional ways of taking the challenges head-on and coming up with unexpected solutions

Similarly, the evolution of HRM systems is taking them beyond technology and expertise, while meeting the aspirations of the new generation, creating an exceptional experience and adding value. Human Resource Leaders have to think out of the box and bring in policies and practices which will be perceived by the management and job market as a value-added before there is a sense of obsolescence about traditional human resource management practices. There is a wave of becoming faster, cheaper, better in an increasingly competitive global scene. HR cannot avoid its influence. Gone are the days that HR was "humanly rich" as a department with all sort of people, mostly dumped by the seniors. The efficiency with regard to optimizing cadre, in clearly identifying the needs and the specific roles associated is of utmost importance. Disruptive HRM is all about essentially challenging the conventional way of doing things. It also involves new initiatives with the use of technology in moving beyond traditional boundaries. 'Disruptive HR' is proactive, whereas handling, disruption in HR is a reactive concept. Disruptive practices are all about initiating change in the existing systems and procedures driving shift, which is as follows (Figure 1.4).

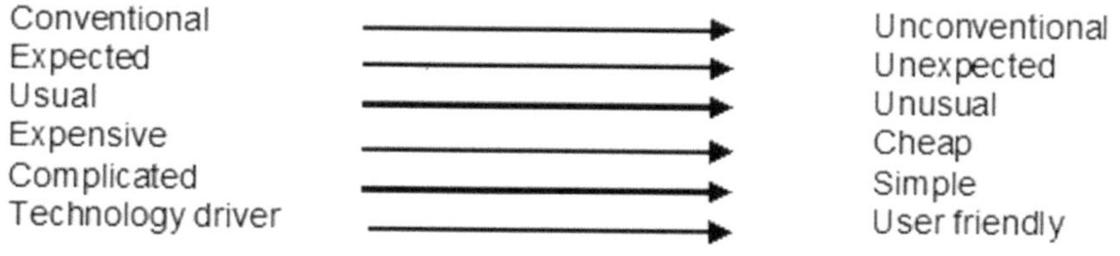

FIGURE 1.4 Driving shift through disruptive practices/initiatives.

Disruptions in HR policies, processes, systems, and technology have become imperative for the organizations to stay competitive and successful, by developing an agile workplace capable of responding to changing needs and demands of business and market seamlessly. The changing workforce demographics and technology innovations are leading to significantly improvised talent management strategies in organizations. HR leaders are faced with the challenge to adapt to the change while ensuring productivity, efficiency, and conducive work culture. However, technology disruption helps to build a 'smarter workforce' in an 'enabled' workplace. Disruptive HR Technology should aim at resource optimization, operational efficiency, and providing solutions aligned to business success. Innovation leading to process improvement through HR technology should be a continuous endeavor of the HR function-driving adoption of the smart, intelligent, team-enabling, people-friendly, and development focused tools and systems. HR Leaders need to accept and lead such disruptions.

1.2.2 SYNCHRONIZING DIGITIZATION AND HR STRATEGY

In 2018, the physical and digital worlds will continue to merge, as the workplace is reshaped by AI, bots, predictive software, and augmented reality. A study by Gartner (2018) explains the key trend in digitalizing HR (Table 1.3).

TABLE 1.3 Trends in Digitalization

Trend	Explanation	Improvement
Blockchain	Recruiters hoping to verify candidates more efficiently and for payroll managers who want to make their organization's global compensation process less costly and timelier.	Improves recruitment and payroll system
AI foundation.	Making systems that learn, adapt, and act autonomously will be a major focus for technology vendors.	Improve decision-making, reinvent work processes and revamp the customer experience

TABLE 1.3 *(Continued)*

Trend	Explanation	Improvement
Intelligent apps and analytics	Virtual customer assistants and bots to improve employee performance, sales, and marketing analysis and security.	Improving the way how things get done, analysis, or to improve a users' experience
Internet of Things (IoT)	"Smart" items such as autonomous vehicles, robots, and drones.	Improving consumer and industrial systems.
Digital twins	Data from multiple digital twins can be aggregated for a composite view across real-world entities.	Improving enterprise decision-making by helping users respond to changes, improving operations, and enhancing performance.
Conversational platforms	Programs leading humans to interact with the digital world.	Improving employee self-service by enabling employees to "talk" to members of your team.

Source: Gartner, (2018). *Gartner Top 10 Strategic Technology Trends for 2018*. Retrieved from: https://www.gartner.com/smarterwithgartner/gartner-top-10-strategic-technology-trends-for-2018/Apple Inc.

Digital transformation is all about formulating data-driven strategy; enabling technology with a focus on process and innovation. The digital transformation is possible only and only if the organization encourages data and analytics for its staff and customer management. Enabling individuals and machines to communicate seamlessly would, of course, make production more cost-efficient. But perhaps more compellingly, digitizing the value chain facilitates innovation and can directly improve the top line (Table 1.4).

1.2.3 HUMAN INTERFACE IN DIFFERENT DEPARTMENTS OF HUMAN INTERACTION IN THE DIGITAL AGE

1.2.3.1 TALENT ACQUISITION

We are seeing increasing interest amongst the new-age companies to address this aspect of human resource interaction through digital and

TABLE 1.4 Digital Transformation Across Value Chains

New Digital Business Models	Digital Engineering	Smart Manufacturing	Digital Supply Chain	Smart Services	Digital Workplace	Digital Sales and Marketing
Augmented digital product player	Digital collaboration in R&D	Production operation and planning and execution	E2E visibility	Smart Asset Management	Efficient order management	Product configurators
Complete Solution /Service provider	Digital modeling mockup and simulation	Quality operations management and analytics	E2E supply and demand planning	Multi-channel customer management	Digital HR	Omnichannel commerce
Data Analytics, Content, and Platform integrator	Digital Research & development pipeline management	Digitalized production (CPC)	Smart warehousing	Predictive maintenance	Internal knowledge sharing	Customer Analytics
Integrated digital ecosystem provider	Integrated digital engineering	Intra plant transportation & material flow	Intercompany transportation & logistics	Augmented reality solutions	E-financing/ controlling	Dynamic pricing
		Predictive maintenance and augmented reality	Sourcing 4.0			Digital Customer relationship
		Plant infrastructure and site management	Supply chain data analytics			E-payments

Source: Sanwalka and Malkani (2018). *Sprinting Towards Digital Transformation.* Mint, Hyderabad edition.

AI-based mediums. The solutions are targeted to address time, cost, and quality of hire, especially a better job fitment through intelligent analytics. This segment may see increasing machine interaction with the job seekers and less human interface. As talent acquisition function is moving from 'job-based' to 'capability, scalability, culture fit-based' assessments, AI is able to provide great insights and helping in job-candidate matching. Organizations are creating fan clubs around their technology, culture, and outlook, so that there is a consistent interest in the organization, creating a natural pipeline of job seekers. Typically social/social like platforms are used for this approach. Organizations are able to tweak the job designs based on the 'capabilities' available to them. For example, 'Textio' is a simple tool helps you to build your job description better, constantly guiding you on the tone of the description, target market, available capabilities, to the extent being able to guide on gender neutrality in your language and suggested improvements.

Entelo, Hiring Solved, Impress are some of the tools which help reducing recruiting time by providing very close matches from the job seekers analyzing what the present employees are doing well and available capabilities in the market. Job seekers are going to interact with AI-powered job assistants like Cyra or JoBo to learn more about the job online and without any human interface. AI-based facial recognition, capable of analyzing emotional intelligence and personality insights like Emotient and gamified psychometric tools are increasingly going to be used for screening the job seekers, which may be a threat to need of human interaction to analyze the culture and job fitment better. However, this kind of applications will depend upon deep learning and susceptible to accuracy, probably calling for human interaction at the time of decision making.

1.2.3.2 *ORGANIZATION DEVELOPMENT*

AI-powered tools can provide insights on job distribution, competency development, diversity, etc. by studying the present job profiles and employees in such roles and comparing between individuals about their capabilities to address them. Online tools can analyze individual behaviors at work advising them on the ways to be more productive as against human observations at the workplace. The department of learning can be a great beneficiary from the learning pattern and learning behavioral analysis of the individuals, making it more personal in nature. It helps in understanding specific developments needs of the individual to be delivering better in the role, based on the better performers' behavior in the same role across the organization.

In the age of agile working, forming teams on the go with the necessary requirements often becomes a challenge for HR and management. AI-based tools are coming to the rescue to analyze and suggest best team combinations, easing out human interference in the process. It's amazing the way AI can help in change management. For example, Humanyze helped Boston Consulting Group to design the new office building based on human behavior tracked over a few months before the movement. They could analyze team behavior, silos, collaboration gaps, etc. by analyzing a huge amount of data generated by the wearable device about what employees are talking about the new office, who they are connecting to, what are they satisfied or dissatisfied about the new office, etc. This otherwise would have been a humongous effort by manually interacting with such a large workforce. AI-powered face recognition software can get down to understanding individual social behaviors by tracking their movements and reactions to different situations and environment.

1.3.2.3 HR OPERATIONS

Chatbots will be very popular in addressing regular queries from employees and providing them with databased responses to queries. At times, smartly analyzing the query and assisting them with policies and procedures. IpSoft's AI-based assistant Amelia is available to the employees anytime during the day, connected through various form factors, including smartphones to address their queries. This drastically reduces the human interface and interdependence. The experience is getting more personalized and entertaining through human-like voice interactions. A lot of repeatable jobs are getting queued using intelligent programs without human interface, like, service certificates, performance records, customized letters, etc., right from generating them to making dispatch ready, omitting any human interference.

1.3.2.4 PERFORMANCE MANAGEMENT

In the digital era, access to knowledge would be abundant through various sources readily accessible to the employees. The performance management is already tending towards a continuous process than a yearly appraisal process. Business agility, productivity, and competitiveness will demand performance management in a more seamless and result-based or achievement-based instantaneous way. It will also demand large data and analytics from internal and external sources, from various stakeholders and interdependent processes. Eventually, an open, transparent, connected system of providing immediate feedback and agile directions to manage performance would replace a conventional approach. At the same time, the interaction will not be limited to face-to-face conversations and would be made available on various digital forms. In order to make better decisions, the HR leaders are required to rightly acquaint themselves with the latest process, technology, data, and analytics. However, the human interface of interpreting data and providing feedback to the employees will continue in the future too (Table 1.5).

TABLE 1.5 Digitalizing the Departments of Human Interaction in the Digital Age

Talent Analytics	Employee Expense Management	Talent Acquisition and Workforce Optimization	Talent Transformation	Managing Performance
Forecast demand, diversity mix	Brach Imprest management	Auto sourcing, Scoring, and ranking	Intelligent leaning path	Competency framework – Building systems
Shape the High potential HERO profile	Employee Flexi-benefits	Resume Parsing	Capability building	Quantification of performance
Forecast engagement trends	Vendor payments	Employee on boarding	Learning management solutions	Maintaining records on performance standards and achievement

TABLE 1.5 *(Continued)*

Talent Analytics	Employee Expense Management	Talent Acquisition and Workforce Optimization	Talent Transformation	Managing Performance
Identifying Re/up/Cross skilling talent	Spend Management Features – Travel request, Reimbursement platform, Settlement dashboard	Workflow automation and workforce allocation	Gamification	Performance dashboards
Identifying Employee Career Growth plans	Expense Report Automation	Internal job matches	Accelerated learning	Feedback systems
Usage of Microsites videos	Rewards and recognition programs		Integrated capability development	

1.2.4 ROLE OF HR IN DIGITALIZATION

Today the role of HR is integral to the business outcome and organizational change aligned to business goals. Hence, HR departments are claiming to provide direction to IT Systems. Businesses need to accept such changes in approach where the employees need to evolve. Today's workforce prefers 'meaningful work' directly contributing towards business success over fat compensation and benefits. The workplace of the future will not be a fixed location and fixed hours, unlike the yesteryears. Subsequently, the future workforce will demand autonomy to decide the place and hours of work. They will like to work from home or co-working spaces within easy reach while deciding the work hours based on their schedules around work-life. We will see a paradigm shift in the definition of the workplace, primarily based on technology, globalization, and work structure. Under these circumstances, the role of HR is becoming more strategic, and they need to understand the nuances of digital era workplace better.

The major functions of HR still remain as talent acquisition, talent development, and talent retention. Managing talent in the age of disruption and continuous change in the market conditions is a prime 'differentiator' towards business growth. The transformation to Digital HR will be primarily through e-learning programs, HR analytics, digital and automated platforms, and processes for employees (Table 1.6).

TABLE 1.6 Digital HR Transformation Using the Four Roles Described Above in the Dave Ulrich Model

HR Role	Ulrich (1997)	Deo (2018)
Strategic partner	Develops and aligns strategies with business.	Digital by aligning the organization's business strategies with the upcoming digital environment.
Change agent	Are the catalysts of change? Given the culture and decision making authorities of the organization, they in still and drive change capabilities.	Learning in the digital era will lean towards 'personalized' learning options, ease of access, and ease of delivering content – like e-learning, mobile learning, video content, etc.
Administrative expert	Administrative Experts continuously work on enhancing HR processes to make them more effective and efficient while addressing business needs, being cost-effective, and managing HR tools and products.	Digital HR will fuel better and quick decision making through analysis of the huge amount of data, providing alternate solutions, being able to calculate ROI on employee initiatives, helping to adopt better people strategy.
Employee champion	Employee champions monitor and implements action plans resulting in better commitment and contribution from the employees establishing unbiased people processes and practices.	Employee Engagement has improved in the organizations who have implemented tools and platforms for online networking, and digital platforms for functions like hiring, on-boarding, learning, rewards, query management, etc.

Exponential growth in technical capabilities, increase in computing power at both enterprise and consumer ends, seamless internet connectivity, other digital form factors will soon see full-time employments with a nine-to-six job at the office, own campus or factory locations-go through transformation. Companies are visualizing increased productivity, cost savings, mobile, and agile connected workforce, and agile business maneuverability in an ever-increasing complex marketplace. Enterprise collaboration with the diverse global workforce will be the hallmark. The employer-employee relationship will go through a paradigm shift, as the employees will assume more power in the relationship by virtue of their ability to work from anywhere and stay connected anytime, through smartphones, tablets, and other mobile devices and collaborate seamlessly with their peers, and gain instant knowledge about the business environment, sometimes faster than the enterprise itself. However, this will also build expectations of being 'connected' round the clock, demanding certain 'digital etiquette.' It may also result in burnout and retention issues. The productivity of the workers presumably will be higher in the new environment. The interaction between the workers and executives will take a new dimension, largely based on the technologies that they will take help from. Companies have to proactively change their policies and systems to accommodate digital culture, thereby ensuring productivity, employee satisfaction, and customer satisfaction. The work components have to be reorganized to give a feel of human-centered organization amongst the deluge of technologies.

The digital workplace will demand technology-savvy workers as the interaction with internal and external stakeholders will increasingly be on devices that face-to-face. Smart collaboration over network and applications like Skype, VoIP, cloud-stored data – will see less expensive travels around geographies and time zones. Agile work methodologies will see more 'on-demand collaboration' between an extensive network of knowledge workers, innovators, and technical experts. It ups the level of agility in the enterprises and fastens execution, which is in favor of both enterprises and workers. When there's a threat that cognitive technologies and smart machines with the push for lesser and more skillful jobs, enterprises should look at integrating smart machines and people in their talent framework. Managers need to be equipped with handling diverse and large workforce and business data. The Millennials will constitute 75% of the workforce by 2030, as predicted by the US Bureau of Labor Statistics, will be connected, mobile, and focused on results and work-life integration. They will be the first generation of 'digital native' in true sense. The digital connectivity alike with their work and friends and families, borderless work options will help the digital workers to be more productive and happy. Digital medium may see bit casual communication with peers, subordinates, and bosses too.

There's a flipside to the digital workplace and human interaction which may possess delicate work-life concerns. How much tech-dependent should we be? How much connectivity is just enough? Will the digital demand cause frequent burn-out? Will it blur the demarcation of personal hours, vacationing, etc.? Digital workplace may raise significant concerns about data security, regulatory compliance, compensation for device usage, and internal governance. This will lead to build a consistent culture of data security and usage, balancing between workers' needs and company's IT framework, government, and web regulations, etc. In the digital workplace, BYOD (Bring Your Own Device) almost becomes a norm but opens up security threats to enterprise data, client data, personal data, etc. The number and variety of digital means will require a substantial and accessible IT workforce managing the workplace 24/ 7.

1.3 AI REVOLUTIONS: THE IMPACT

It's but a known fact that deployment of AI infrastructure, innovation, and research in India has been primarily restricted to the consumer goods sector as of now. To be successful in adopting AI, creating appropriate infrastructure, fueling research and skilling the new generation in AI – Government participation is imperative to make it a large scale success similar to countries like the USA, China, South Korea, etc. It also needs to be in line with broader policy frameworks, investment, and enablement like 'Make in India,' 'Digital India,' 'and Skill India.' These thoughts have propelled the government to adopt an AI policy under National-Artificial Intelligence Mission (N-AIM). It may see a budget allocation of nearly INR 12,000 crores this financial year. With reference to the use – cases of AI in various sectors like Fintech, Manufacturing, Retail, Healthcare, Education, and Security from across the world, it's estimated that the jobs created using AI will, in fact, outnumber the redundant jobs in the process of adapting to AI. The BFSI sector is seeing the maximum AI-enabled jobs being created, up to 47% of total opportunities. However, the bright side is the increasing internet and mobile usage in India, which will fuel large scale adaptation of AI in the near future. The AI's impact can be discussed in the context of two situations prominently being Human–Machine interface and Human–Human interface.

1.3.1 HUMAN-MACHINE INTERFACE

The first and most affected sector in case of Human–Machine Interface would be 'manufacturing sector.' Automation and Robotics may have a significant impact on present era jobs in this sector. However, it's predicted that the extent of job loss in the repetitive and precision activities could be 30% by 2025. The middle-skill labor involved in assembly lines and clerical works will be most affected.

- **Case 1: Foxconn, China (Source: The Verge, 2016)**

The world's largest contract electronics manufacturing took to large scale automation and industrial robots in early 2016, which resulted in the redundancy of almost 60,000 jobs in one of Foxconn's factories in Kunshan alone. The CEO predicted almost 40% job loss over a period of time. Technology innovation, robotics, and AI saw the same number of workforce contributing to 2.5X revenues for the company down two years. India, being a developing nation, needs to balance job creation, employability, and AI-related job dislocation carefully. It's also noted that Information Technology companies in India, typically the new-age ones, are heavily inclined to work in the domain of AI, machine learning and robotics. It's yet to be seen how these companies would be capable of developing required skills and competencies for the next-gen enterprise IT automation.

- **Case 2: Uber, USA (Source: Business Insider, 2016)**

Owning almost 77% of ride-hailing services in the USA, Uber has been test-driving, self-driving cars, and air taxis. Though it's still a bit premature to assume that most of the taxi-hailing will be managed by robotics and AI in future, there could be a significant impact on individuals dependent on their livelihood based on driving skills. India needs to adapt to technology evolution at an early stage, preparing the workforce with necessary up-gradation of skills. As there is a strong global inclination towards AI at the workplace, we need to re-engineer our industry towards a machine-intelligence-driven economy. No doubt, during this transition, we may face challenges of increased job-loss, but eventually, when the employability orients towards the required skills; it will ease out in the long run.

1.3.2 HUMAN-HUMAN INTERFACE

The impact of AI replacing human interfaces could be categorized into Most likely, Moderate, and Least likely scenarios. Few significant factors will determine the uptake of AI like; technical readiness, the opportunity cost of automation, superior results, regulatory acceptance, and 'social acceptance.' In the Indian context, 'social acceptance' plays a major role as we have a deep traditional culture of contact transactions. Industry segments primarily involved in transaction processing and regular and repetitive physical work will be subject to 'most likely' scenarios of automation and AI take over. The industry segments with irregular physical tasks due to changing work environment and managing multiple stakeholders may see a moderate uptake. Whereas jobs involving the exercise of discretionary expertise as decision making, planning, and creative tasks will see the least scope of AI.

Investment and development in AI in Fintech in India will see exponential growth over the next 10 years. It's predicted that India may house some of the major Fintech companies of the world in the next decade. The reason behind the enhanced focus on AI in Fintech is the analysts' predictions of 43% improvement in productivity by replacing human efforts with automation and AI in this sector. India is already witnessing phenomenal uptake of digital transaction processing like Unified Payment Interface (UPI) at 172 million, growing at about 25% per quarter. AI and Machine Learning in Fintech help improving security features reduces risk of fraudulent transactions, reduces processing time for financial transactions requiring multiple scrutiny and entries, eliminates duplicates and human errors, deciphers the huge amount of data producing game-changing insights, helps taking complex business decisions increasing productivity and revenues. Also, AI-powered chatbots help to address regular and complex customer queries.

India also saw significant investment in AI startups in Healthcare in 2017. AI will help in compiling, categorizing, and analyzing the enormous amount of medical records for better accessibility and application of solutions. An efficient AI system can handle repetitive procedures like X-Rays and CT Scans and provide basic medical intelligence around them. An application like Babylon, which is increasingly becoming popular in the UK, uses speech recognition to analyze the patient's illness and suggest simple medication. A Boston Lab has developed a virtual assistant with a human touch which advice on upcoming consultations, medical tests, etc. using Amazon's Alexa. It's already under test usage at Boston Children's Hospital. AI-driven health tracking appliances like Apple, Fitbit, Garmin, etc. can provide predictive health maintenance guidance to users. So it's evident that some of the peripheral healthcare jobs may be handled better with AI and less human interface. However, the critical diagnosis and consultations will still remain with healthcare professionals. The question is, how comfortable are we yet to go through a robotic surgery in India, for example? We have a long way to go.

Similarly, in the retail space, AI is revolutionizing customer behavior analytics, predictive, and personalized marketing efforts, digital marketing, customer service, and sales forecasting, etc. Indian e-commerce giant Flipkart is working on an AI platform aggregating customer reviews helping the prospective customer to take informed decisions. Whereas it's rival, global giant Amazon uses BLINK – the video commerce platform for customers to see the product in use before buying. Another India AI-commerce ventures Capillary Technologies has launched Visitor Metrix with embedded AI, which captures millions of shoppers' behavior in the retail stores and provides real-time dashboards using trend-analysis engines around busy hours, brand positioning and display, efficient allocation of resources, etc. It uses a repository of millions of pictures of stored articles with the ability to distinguish between human and non-human objects and match them during store visits. Advanced AI capabilities like locating inventory and fetching seamlessly based on the customer order and auto-billing etc. will see lesser and lesser human interface in the retail space in the coming days.

1.4 PERSPECTIVES ON ARTIFICIAL INTELLIGENCE (AI) IN INDIA

1.4.1 METHODOLOGY

An attempt was carried out to study the perspective of managers and Generation Y of India on AI replacing human interface in organizations. Both quantitative and qualitative data was collected. A questionnaire was floated online to collect the responses. Eighteen responses were collected from HR managers, of which 61% of respondents belong to

Senior Management and Board level executives. 39% of the respondents were from a middle management position. Response of 94 students was collected who are in the final term and about to join the organization. These students belong to an age group of 21–24 years. These responses include Gen Y from both management and technical background. The Cronbach's Alpha for the questionnaire used was.867. Logistic regression was used to identify the factors affecting the decision of the respondents on the notion that whether AI can replace human interface in Indian organizations.

1.4.2 DATA ANALYSIS

The data above reveals that most of the respondents believe that systems can't rule out human interface in organizations. Though there is a huge variation in the percentage of responses as 89% of HR officials do not agree with the concept of system ruling out human interface in organizations, the Gen Y contrary is more acceptable to the notion. Most managers were of the opinion that AI can be initiated minimally in India from the year 2025 (Figure 1.5 and 1.6).

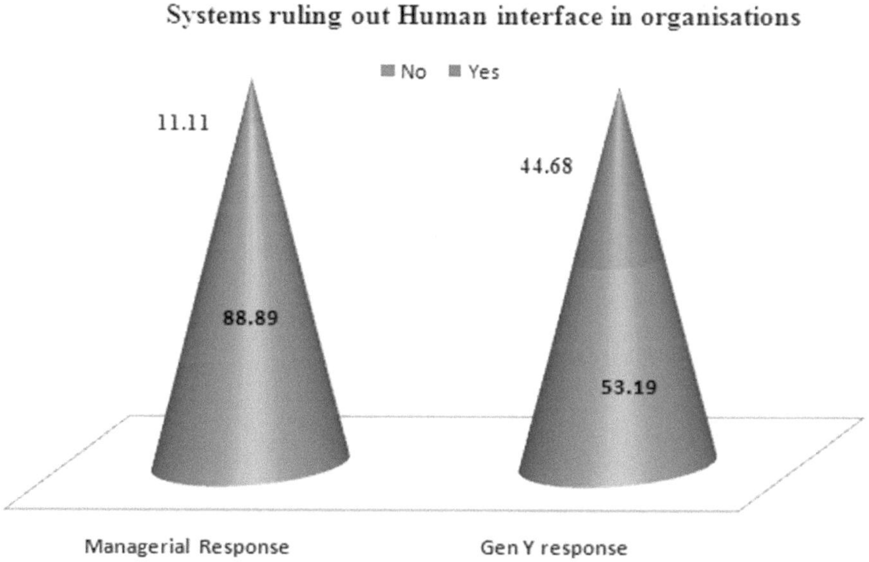

FIGURE 1.5 (See color insert.) Acceptance of AI replacing human interface in India.

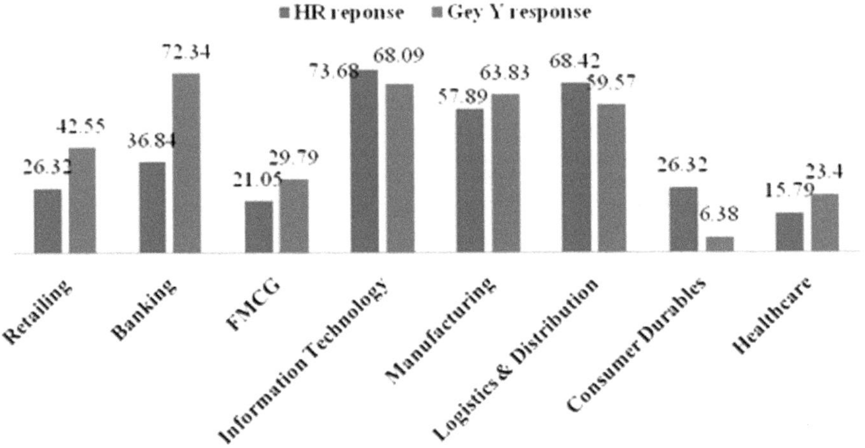

FIGURE 1.6 (See color insert.) Possibility of AI replacing human interface in various sectors.

There is not much variation with the responses with regards to the possibility of AI replacing human interface in various sectors except in banking and consumer durables. Most of them believe that the possibility of AI-related operation is possible in sectors like Information Technology, Manufacturing, and Logistics and distribution (Figure 1.7).

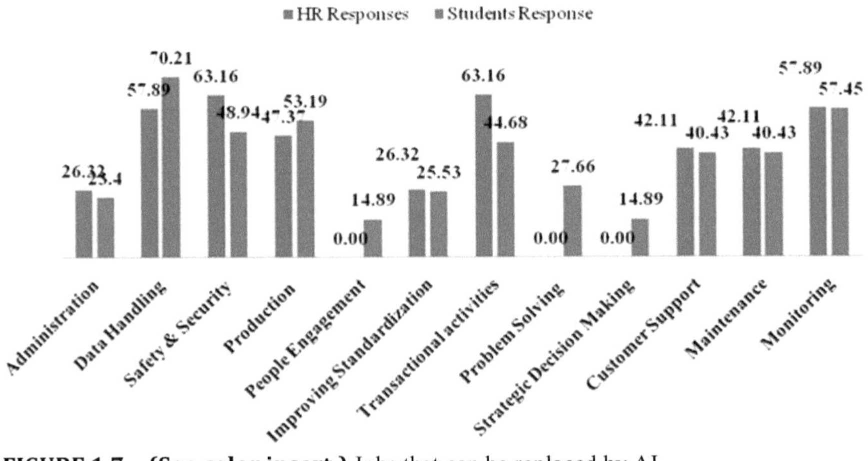

FIGURE 1.7 (See color insert.) Jobs that can be replaced by AI.

The above graph reflects that there is almost an equal conviction on jobs that can be replaced by AI. Almost both the HR officials and Gen Y opinionated that activities like Monitoring, Data Handling, Administration, Safety, and Security, Improving standardization can be well-taken care by using AI. Though the no HR officials expressed that People Engagement, Problem Solving, and Strategic Decision Making can be handled by AI, yet 15% of Gen Y believes People Engagement, 28% for Problem-solving and 15% for Strategic Decision Making can be taken care by administrating AI (Figure 1.8).

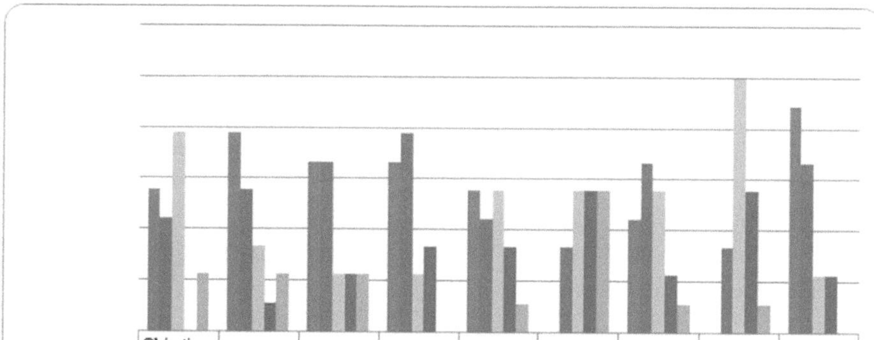

	Objective observation / Evaluation	Mentoring	Flexibility	People Orientation	Business logic	Focus on productivity	Intutive Capabilities	Control	Team Responsibility
Not Much	27.78	38.89	33.33	33.33	27.78	0.00	22.22	0.00	44.44
Little	22.22	27.78	33.33	38.89	22.22	16.67	33.33	16.67	33.33
Somewhat	38.89	16.67	11.11	11.11	27.78	27.78	27.78	50.00	11.11
Much	0.00	5.56	11.11	16.67	16.67	27.78	11.11	27.78	11.11
A great deal	11.11	11.11	11.11	0.00	5.56	27.78	5.56	5.56	0.00

FIGURE 1.8 (See color insert.) Capabilities of robotics -HR viewpoint.

The HR managers are of the opinion that production, control, and objective evaluation can be handled easily by the robotics; whereas most of the officials believed that robotics is very little capable of handling jobs of mentoring, people-orientation, flexibility, team responsibility. Capabilities of working on Business logic and intuitive capabilities are possible only to some extent (Figure 1.9).

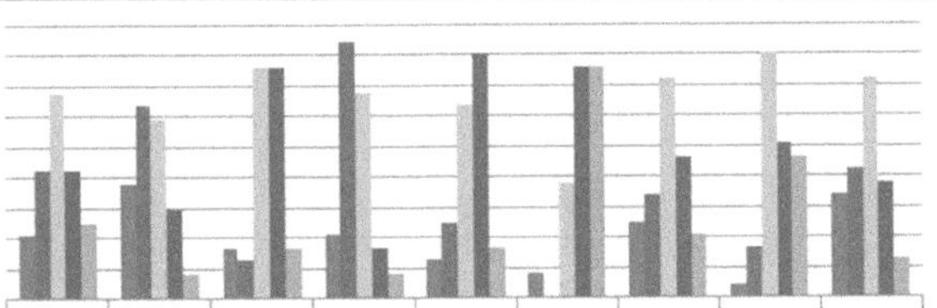

	Objective Evaluation	Mentoring	Flexibility	People Orientation	Business Logistics	Focus on Productivity	Intutive Capability	Control	Team Responsibility
Not Much	10.64	19.15	8.51	10.64	6.38	4.26	12.77	2.13	17.02
Little	21.28	31.91	6.38	42.55	12.77	0.00	17.02	8.51	21.28
Somewhat	34.04	29.79	38.30	34.04	31.91	19.15	36.17	40.43	36.17
Much	21.28	14.89	38.30	8.51	40.43	38.30	23.40	25.53	19.15
A Great Deal	12.77	4.26	8.51	4.26	8.51	38.30	10.64	23.40	6.38

FIGURE 1.9 (See color insert.) Capabilities of robotics -gen Y viewpoint.

On the other hand, generation Y is of the opinion that production, control, and objective evaluation, business logic, and flexibility can be handled easily by the robotics. Whereas, most of the others believed that robotics is very little capable of managing mentoring, people-orientation, team responsibility. Capabilities of working on Business logic and intuitive capabilities are possible only to some extent. Mr. Susanta Mishra, CEO, Nicefit Consulting Services Pvt. Ltd. says, "AI will replace human effort in all tasks that are human error-prone." Mr. Harish TR, Director Exemplary People, says "Long time to come but being a People touch intensive Country I don't see the connect missing ever." AI can work a large number of responsibilities, yet some behavioral attributes cannot be managed. An attempt was made to segregate the activities that can be managed by AI in organizations (Table 1.7).

TABLE 1.7 Areas of Work That Can Be Replaced By AI

Can Replace	Cannot Replace
Individual accountability	Team responsibility
Control	Autonomy
Intuitive capabilities	Behavioral differences
Focus on Productivity	Focus on HR development
Technical logic	Business Logic
Task Orientation	People orientation
Efficiency	Flexibility
Managing employees	Mentoring employees
Objective observation/evaluation	Subjective observation/evaluation

AI-powered workplace – The future is NOW! Though the general perception is, the advent of AI will take away jobs in some of the HR functions; it's time to analyze how human-oriented AI may actually benefit the companies and individuals to be more accessible, efficient, productive, and predictive. For example; Retention of employees has been a great concern for most of the organizations across due to 'inability to predict employee disengagement effectively' and 'high replacement cost to reach the same productivity level within a reasonable time.' Using AI-driven data analytics and machine learning, even considering the complex human dimensions, we can achieve a highly accurate predictive intelligence without really making human interfaces redundant. On the other hand, retention today is a direct outcome of 'employee experience' which can be enhanced by AI-driven intuitive intelligence and interfaces powered by cloud computing and form factors.

AI at the workplace may not be able to replace human intelligence and the ability to study different circumstances. Rather it will supplement better, smarter, faster decision making. By being able to analyze massive amount of data of repetitive and unique behavior, combined with acquired knowledge, emotional intelligence, and creativity – we shall be able to take more data-driven decisions than guesses and intuitions. Similarly, the selection of the right resources for the right roles will become easier by applying deep analytics and algorithms. This will help to understand behavioral, skill growth, situational response, logical thinking, and decision-making capabilities rather depending on education and experience records. To a large extent, these tools will help avoiding wrong hires and increasing the efficiency of the hiring function, not necessarily replacing the interfaces. The traditional talent development has always involved system supported analytics of training need and periodic training interventions, which essentially addresses 'groups of individuals.' However, there's a growing need for understanding business situations, roles, progression, skill requirements, behavior, and engagement quotients to identify right training at the right time to 'individual employees' to help their career path and progression, thereby improving business performance significantly. AI, coupled with deep analytics and cloud computing, training-as-a-service model can help HR in skill development, improving engagement, and resulting in sustainable revenue growth.

For gauging employee satisfaction index, the present practices of one-time/annual feedback and analysis will become obsolete with the advent of AI at the workplace. HR Department will be able to measure employee loyalty and satisfaction more accurately by capturing and analyzing metadata from emails, feedback platforms, work-hour patterns, communication patterns, participation in engagement events, networking patterns, mood representations, leave, and vacation patterns, workplace aspirations, etc. This helps to build trust, mentor for career progression and retention, facilitates predictive counseling towards successful career mobility. The effectiveness analysis, measures, communication-related to many of these processes need human intelligence and interface to convert into appropriate action planning and execution. This doesn't render human interface redundant, but throws new challenges to be 'AI-powered digital workplace' – ready with enhanced skills to use and interpret AI and machine learning in a highly disruptive business environment to sustain growth, optimize performance and create value for internal and external customers.

1.5 SUCCESS STORIES IN THE DIGITAL WORKPLACE

An attempt has been made to identify various success stories in the digital workplace. The case area compiled based on information shared in Deloitte Insights (2017), Global Human Capital Trends:

- **RehabCare**, a healthcare major operating in over 1200 hospital facilities across 43 states in the USA, had a major challenge in managing patients' database. As the nature of services required the employees to record minute-by-minute patients' detail around diagnostics, treatment, and appointments during rehab and post-acute-care, it needed an easy-to-use, always accessible, process-driven applications on gadgets like iPhone and iPad, etc. Apple Inc., in their customer success stories, quoted that, mobile devices equipped with such applications helped the healthcare staff to capture all relevant clinical information quickly and helped the staff to access patient information anytime, anywhere. This improved staff efficiency manifold and helped in bringing down operational cost and clumsy paperwork significantly, while increasing employee satisfaction and retention in due course, thereby resulting in huge savings for the organization.

- **Standard Chartered Bank**, one of the premier international bank and financial services organizations, with about 1800 branches in six continents, was faced with a typical challenge of reinforcing customer confidence during the 2009 global financial crisis. And clearly 'use of technology' in serving the customers better and creating long-lasting relationships was one of the key focus areas. Their suite of mobile banking and lifestyle applications "Straight2Bank" was unleashed to retail and corporate customers to manage their complete banking portfolio better and more efficiently, anywhere, anytime. They went on building several in-house applications meant for personalized banking service, seamless, and secure financial transactions, approvals, etc. leading to a better interaction between the customers and banking staff. Some of the processes were streamlined, including back-end approvals, which could be done online, therefore increasing customer loyalty and satisfaction.
- **Alaska Airlines**, with a fleet size of 117 aircraft plying between 91 cities in the USA, and with close to 10,000 employees, was grappling with the challenge of pilots having to carry a huge number of hard copies of the flight manual, every time they flew. Also, it was difficult to include any change in the manual, as, and when it was made. As a solution, all the pilots were equipped with tablets holding all flight manuals, updates, and additional information in one place. This gave a cheaper and lighter alternative to the airlines, particularly -where flying to Alaska was weight-sensitive.
- **Cisco**, a multinational corporation, with interests in information technology, networking, electronics, and communication technologies, has over 70,000 employees worldwide. As a large global organization, Cisco had realized the importance of collaboration, connected, and integrated workforce and global mobility. Cisco launched communication and collaboration platforms, enabling, and empowering their employees to connect across the world. The applications are user-friendly and customizable to personal preferences and business priorities. Cisco reportedly saved over USD 1 billion by rolling out these collaboration suites, which also resulted in seamless, connected, empowered workforce.
- **IBM**, a global multi-national Information Technology and Services Company, with around 400,000 employees across the globe, has

taken steps towards digitalization of HR, by way of introducing AI in performance management. When an internal brainstorming session and hackathon resulted in a thought process of adopting a new performance management system "Checkpoint," it drastically changed the feedback process, increasing engagement, alignment, and goal management. The application is available on mobile devices for anytime access and instant feedback. IBM's new digital learning platform can recommend role and experience-based learning, employees can publish important content, and it can integrate external sources, including internet-based learning. IBM's Cognitive Human Interface Personality (CHIP) is a smart chatbot capable of answering over 200 commonly asked questions, also learning new questions in the process. This is becoming more and more popular with the employees, and the dependency on human call centers have significantly gone down.

1.6 FUTURE RESEARCH DIRECTIONS

The chapter shares elementary research on the perspective of the human resource managers as well as the generation Y on the scope of AI replacing human interface of organizations in India. A worldwide perspective on the impact of AI at the workplace would be interesting. Magnitude of the company operations would also be an important element of deciding the digitization of the workplace and its impact on employees, being a vital area of research. It is assumed the penetration of digitalization in developed, developing, and least developed countries would be different. Studies can be undertaken to understand the penetration of digitalization of HR in various countries.

1.7 CONCLUSIONS

The future workplace will see the Human Resource Department to be an integral part of the business and working in tandem with the other departments. Digitalization of businesses will make it seamless to interact between different departments. HR departments are increasingly taking the ownership of planning for the IT systems; especially the kind of S-a-a-S based People Management Applications. This is helping the HR departments to play a significant role in influencing business outcomes and driving organizational change. The increase of IT systems in HR and digitalization of HR is helping organizations to save cost and time, hence leading to better profitability.

The advent of large-scale automation has its fair share of concern amongst the workforce. However, it's only going to increase in the future. Instead of resisting the trend, the workforce needs to upskill itself to handle the challenges of moving up the value chain. It is good to automate yet better to identify where existing employees can perform better using automation rather than focusing on replacing people with machines. If we do not identify enough tasks for our existing employees, it might lead to psychological ills of joblessness resulting into economic recession leading to youth unemployment to individual crises of identity. Instead of seeing work as a zero-sum game with machines taking an ever-greater share, we should focus on the growing possibilities for employment. We could reframe the threat of automation as an opportunity for augmentation.

Augmentation is a concept encouraging identification of what humans do today and figuring out how the work could be deepened rather than diminished by greater use of machines. Technology alone cannot be substituted with a human touch. There has to be the right usage of the technology. Technology is ever-rowing. No technology stays as the USP of an organization for many years. So it cannot be the only tool for substitution of employees rather should lead to complementary effect to the organizational performance. Some of the regular, repeatable, system-driven work may move to AI or Automation. That will be a small portion of HR responsibility. Most of the HR functions will remain human-dependent. It's majorly because machines will not be able to understand variations of human emotions and as a culture, Indians would like to 'interact' with HR resources rather than solely depending on automation, etc.

India is a developing country and has very high social issues of employability and unemployment. The quantitative and qualitative data reveals that the acceptance of AI replacing human interface in organizations is not acceptable in the Indian scenario. Though Gen Y is open to the idea yet, implementation in real-time would be a great challenge. Human interface is important as human ensures that the computer is doing a good job and makes it better. The world is focusing on STEM (Science, Technology, Environment, and Management) based learning and robotics cannot replace the STEM. It is only expected to complement the STEM in all phases of operation. It will be interesting to understand the changes required in the education system in developing countries to implement advanced forms of AI at the workplace; as still developing countries are lagging behind in preparation of future workforce.

KEYWORDS

- **digital HRM**
- **digital work culture**
- **digital workplace**
- **digitalization**
- **HR transformation**
- **human interface**
- **technology**

REFERENCES

Acemoglu & Restrepo, (2017). *Robots and Jobs: Evidence from US Labor Markets.* Retrieved from: https://economics.mit.edu (Accessed on 28 August 2019).

Autor, D. H., & Dorn, D., (2013). The growth of low-skill service jobs and the polarization of the U.S. labor market. *American Economic Review, 103*(5), 1553–1597.

Bolter, (1984). Artificial Intelligence Daedalus. *Anticipations, 113*(3), 1–18.

Bundy, A., (1988). Artificial intelligence: Art or science? *RSA Journal, 136*(5384), 557–569.

Business Insider, (2016). *We Rode in Uber's First Self-Driving Car-Here's What it Was Like*, Retrieved from: https://www.businessinsider.in (Accessed on 15, August 2019).

Business Week, (1986). *High Tech to the Rescue*, 100–108.

Christensen, C. M., Grant, A. M., & Govindarajan, V., (2017). *HBR's 10 Must Reads 2017: The Definitive Management Ideas of the Year from Harvard Business Review*. Harvard Business School Publishing Corp.

Dean, J. W. Jr., & Snell, S. A., (1991). Integrated manufacturing and job design: The moderating effect of organizational inertia. *Academy of Management Journal, 34*, 776–804.

Deloitte Insights, (2017). *Global Human Capital Trends*. Retrieved from: https://www2.deloitte.com (Accessed on 28 August 2019).

Deo, T., (2018). The four roles of HR and digitalization transformation. Retrieved from: https://www.capgemini.com (Accessed on 28 August 2019).

Ettlie, J. E., (1988). *Taking Charge of Manufacturing*. San Francisco: Jossey-Bass.

Gartner, (2018). *Gartner Top 10 Strategic Technology Trends for 2018*. Retrieved from: https://www.gartner.com/smarterwithgartner/gartner-top-10-strategic-technology-trends-for-2018/Apple Inc (Accessed on 28 August 2019).

Harrington, H. J., (1987). *Quality and Reliability*. New York: M. Dekker; Milwaukee: ASQC Quality Press.

Hayes, R. H., Wheelwright, S. C., & Clark, K. B., (1988). *Dynamic Manufacturing*. The Free Press, a Division of Macmillan, Inc., New York.

Jabil, (2018). *Digital Transformation Readiness Survey*. Retrieved from: https://www.jabil.com (Accessed on 28 August 2019).

Klein, J. P., (1991). A reexamination of autonomy in light of new manufacturing practices. *Human Relations, 44*(1), 21–38.

Madia, C., (2018). *Going Ahead, We Want to Go More Digital; AI to Help Fund Managers: Chairman of DSP BlackRock Investment Managers*. Financial Express.

Majchrzak, A., (1988). *The Human Side of Factory Automation*. San Francisco: Jossey-Bass.

Minsky, M., (1968). *Semantic Information Processing*. Cambridge: MIT Press.

NASSCOM, (2015). *HR in Digital Age*. Retrieved from: http://www.nasscom.in (Accessed on 28 August 2019).

PWC, (2014). *6 Indian-Workplace-of-2022*. Retrieved from: https://www.pwc.in/assets/pdfs/publications/2014/indian-workplace-of-2022.pdf (Accessed on 28 August 2019).

Seth, (2018). *A 5-Year Outlook on the Future of Jobs in India*. People Matters, Retrieved from: www.peoplematters.in (Accessed on 28 August 2019).

Singh, G., (2017). *Digital Revolution in India*. Retrieved from: http://pib.nic.in (Accessed on 28 August 2019).

The Verge, (2016). *iPhone Manufacturer Foxconn Plans to Replace Almost Every Human Worker with Robots*.

Ulrich, D., (1997). *HR Champions: The Next Agenda for Adding Value and Delivering Results*. Boston: Harvard Business School Press.

CHAPTER 2

Helping Entrepreneurs and Small Businesses Make the Digital Transformation

SUGANDHA SONI CHOPRA[1,2]

[1]*Rama University, Management, EWS-4111, AWAS Vikas-3, Panki Road, Kalyanpur, Kanpur, U.P. – 208017*

[2]*676-A, Near Canara Bank, Gandhi Nagar, Jammu – 180004, India, E-mail: sugandhasoni101984@gmail.com*

ABSTRACT

The main focus of this chapter, draw attention to that in what way entrepreneurs of Small and Medium Enterprises (SMEs) with inadequate potentials and restricted assets operated digital transformation in their companies. We think carefully about qualitative research on digital transformation, including the movement of E-Commerce experience by 7 SMEs on the Alibaba digital program. We relating to obtaining a process in this chapter that targets to mark out and make clear in what way SMEs entrepreneurs, withhold up from the digital platform good turn supplier, work digital transformation by managerial cognition replacement, managerial social money growth, business group building, and institutional capacity building. This chapter enlarged our understanding of both digital entrepreneurship and digital transformation. It also shows the new capability to gain a correct into in what way digital platform good turn suppliers can help SMEs aspect and strive to gain.

2.1 INTRODUCTION

With the changing scenario of the state, today, many dimensions in the society are also changing, and this is actually leading to many evolving practices when the situation entails to choosing of a career. It draws on previous days that people hand to hand only one notion of finishing their education and focusing a proper 9 to 5 job with a fixed salary increase each year. But now the youth or rather the mindset of the society is really changing, with the finest of the startup companies which are truly economic impact in the society. Nuti (2017) says, "there was happiness for me to see my small businesses to grow up and meeting imaginative entrepreneurs." He has given some examples:

1. A boutique store manager, can immediately see target customers and giving the priority and offer them digital loyalty discounts at the point of sale.
2. Seeing the neediness of the customer, a restaurant manager can instantly see – what's selling – and what's not – to make real-time changes, from the shop, at the market, or at home.

The digital transformation, its all right information hidden by it, can actually be a smooth, highly-beneficial process. Again he explained that his company understands the move to a real economy better than other company. With the boon of the internet, we undergo our business to present a new, Omnichannel world, where people selling and purchasing by online and phone. Today, we could help our customers travel over this world. Helping small businesses make this process is decisive. According to the census bureau of the US, small businesses account for 60% of all new jobs. This is very impressive; companies with fewer than twenty workers account for the selective perception of completely the U.S. companies. Although quite a relevant point yet during the discussing U.S. recovery body had more view endpoints like:

- The new budding companies would be creating the majority of net new jobs enhancing the economy.
- The local people would be more involved in such a small-scale business, as the residents of the areas and neighborhoods.

- The quality of living in a society is thoroughly enhanced and uplifted by this new age small startup companies.
- It is also believed that the global economic status would have much stronger and deeper roots within the ground with the spread and growth of the small-scale business houses, with proper strategies been applied.

2.2 BACKGROUND

As with entrepreneurship, entrepreneur too has been having multiple ways by various researchers, varying in the aspects or attributes under study. Entrepreneurship is a multiplex process of selection and adaptation that happens over a prolonged ahead of time, of days or years, as against the perception of a sequence of unrelated activities. Montanye (2006) considers it as a fact price, product innovation, and risk-based to what place entrepreneurial reimbursements are tied to apprehensions and benefits. Entrepreneurship-related activities arise from a separate's perception about the alive of market opportunities, capacity, and economic viability of those business opportunities. Uddin and Bose (2012) describe economic and monetary union benefits to entrepreneurship through contributions to innovative market and industry diversifications, development of novel technology, the creation employment, and an overall surge in productivity enhanced the standard of life, decrease in poverty and better social positioning.

Digital transformations go beyond process redesign and technology implementation and involve structuring of new business operations to facilitate and leverage the core competence of an organization through digital technology (Susarla et al., 2012). Social technologies like social media marketing are transforming the structure of social relationships in both the consumer and the enterprise space offering new ways for collaboration. Scholars emphasize the needy digital ages in which the IT strategy of a company is not only considered functional-level strategy but integrated with the business strategy. The major technological trends related transformation in the contemporary literature described social, moving, analytical, and cloud computing. Digital technologies have become a source of disruption infant industries as they are changing business models and processes, and even affecting the rules of business.

2.3 MAIN FOCUS OF THE CHAPTER

2.3.1 STATEMENT OF PROBLEM, OBJECTIVES, AND METHODOLOGY

With the new innovations and the fast growth of technology, it is surely believed that the new age entrepreneurship and digitalization would surely continue to bring much better transformations in the scenario of business and new industries. The impact from the tenant for years has been more on consumer-oriented industries such as it is the music industry, the consumer products, the publishing industry, retailing, financial companies, and largely the manufacturing industry. So, it estimated that it would be quite a benefit-oriented clause to adopt the organizational changes which are more perspective oriented and follow the technology trends. The objectives are as follows:

- The main thing in the present research is to analyze the present status of how digital natives are helpful for entrepreneurs and small businesses (SMEs).
- Determining the drive towards enhancing technology upgrading through digitalization transformations lean manufacturing companies starting from small to medium.
- Analyzing the better opportunities focused on proper digitalization for complete be-all and end-all-sized manufacturing companies in the coming few years.
- The study would help in facing challenges been encountered in the procedure if hi-tech digitalization of the small-scale industries.

This is the new age form of research technology which is now mostly adopted researched is found on secondary data collection, which is attached to the needed and syllabus study materials. Various books, journals, research papers dealing with the concerned topic and relevant websites have been considered order taker study effective one, and believed that it would help in a deductive nature in future (Chopra, and Yadav, 2018).

2.3.2 NEW WAYS OF ENTREPRENEURSHIP AND DIGITALIZATION

2.3.2.1 TECHNOLOGY ENTREPRENEURSHIP

Technology entrepreneurship is an accepted academia. Therefore, those call provision and advancement, which recent developments have completed in the condition of digital currencies (Zhao and Collier, 2016). People would like to management development of discussion between entrepreneurs, policymakers, and researchers on the effect of digitization on entrepreneurship, and set a future analysis aims and possible achievements, according to the recent definition of terms.

2.3.2.2 NEW FORM OF TECHNOLOGY ENTREPRENEURSHIP

This is not easy to decide whether people are still an inner part of the original invoices bondage because of not being present of a characterization of technology entrepreneurship. It is a fusion of entrepreneurship and technology-based revolution (Giones and Brem, 2017). Similarly, this entrepreneurship aims to explore chances connected to advances in science and engineering.

2.3.2.3 DIGITAL TRANSFORMATION FOR ENTREPRENEURS

The digital age and the plethora of digital channels obtainable have generated so many chances for entrepreneurs and Small and Medium Enterprises (SMEs). They have more ability than ever to attain business victory, should they determine to combine technology into their business representation. Digitalization is not something that you execute once and not remember about. This is so energetic that an organization must look at the business plan steady. The plan must be proactive and comprise paces on what way to reinvent its digital extend to come up with the predict technological advances (ICAEW, 2016). All necessary objects that an entrepreneur should basis on when thinking about transformation are:

- To increase the consumer incident by taking command, as much as feasible, of their consumer's interactions with the company through their OmniChannel experience.

- To judgment, look at your business and your consumer's digital trip so that you assess whether you are conveying a frictionless experience.
- To make sure all your requests, whether a website, marketing, or E-Commerce platform, is modernizing and connects your consumer's requests. This will make sure that any future modernizes to digital badges can be done swiftly and efficiently.
- To investing people who have digital abilities as usual up-ability yourself.
- To have a digital-first attitude at the top down. Transformation must be imported from the CEO of the business right down to each staff member.
- To uninterruptedly path and monitor digital development through metrics and KPIs.

2.3.2.4 DIGITAL BUSINESS

Present time, digital badges mainly are handed down in a context of concerned with business optimization, integration/convergence, disturbance, and digital transformation. This is an incalculable amount that the communal business, marketing orientation and – as people can never fear the human relations that risks being ignored the further occurrence of something of the latest technologies and the digitalization – humanization. A key bit of it all is facts – put at work, which needs a concerned with Information Management approach – and carrying value to generate more value, throughout the ecosystem.

2.3.3 DIGITAL STRATEGIES THOSE TRANSFORM

The capability to re-imagine the business digitally/is the firm decision in big part without problems digital plan helped by leaders who encourage a culture able to become different and originate the new. These kinds of business consist earlier technology change and development; there is something unusual to digital changing is that leaders can take the risk to become a cultural audit as digitally proceeded companies need latest levels of competitive advantages. Employees willing to job for businesses that are extremely performed in digital development; this is equally important for all age classes. Company leaders must have born this in the brain in sequence to attract and to keep the best talent (Kane et al., 2015).

- **Digital Strategy Provides Power to Digital Maturity:** At the early parts of which we say digital maturity, only 15% of respondents from organizations – a workplace where digital has altered procedures, business models and talent engagement – say that their workplace has without the problem and logical digital plan among the digital maturing, above 80%.
- **The Strengths of a Digital Transformation Strategy:** Digital strategy is developed in transforming business. This is a mature organization.
- **Digital Organizations Who Matured Build Ability to Realize the Strategy:** According to overall findings, at the previous parts of digital maturity the capability to mental concepts that how digital badges can influence the business is an adroitness not having the quality in so many commercial businesses.
- **Employees Impress By Digital Leaders:** Employees willing to job for digitally authorized organizations, companies through age classes from 22 to 60. Employees are looking for the highest quality digital chances/and businesses which they have to do continuously for retaining the digital work and entice them.
- **The Possibility of Loss Becomes a Cultural Norm:** The digitally maturing workplace is more relax-able to take possibilities than less digitally mature employees. To make their workplaces less risk, business leaders have to accept non-success as something that must exist for success. Business leaders necessary also address the probability that workers may be just as the possibility of loss-averse as their managers and will require help to become brave.
- **The Digital Aims are Led From the Uppermost Point:** Organizations, who are maturing, are nearly two times as like as less digitally mature business institutions to have a one-person or unit leading the attempt. Employees of digital maturing workplaces are full of assurance for their leaders regarding digital fluency. Digital movement, however, doesn't want mastery of the technologies. Instead, it needs the capability to fulfill the importance of digital technology for the time ahead of the organization.

2.3.4 MODES OF DIGITAL TRANSFORMATION

2.3.4.1 CREATING A STRATEGY THAT TRANSFORMS

This may be the best ways to come near to turn the long-established strategy development method on its head/when developing a progressive digital strategy. Mennecke et al., (2007) would preferably begin by considering commerce and business and then effort backward. There is the probability of a new solution which is creating by new ability, and required solution stimulates demand for advanced abilities. For the sample, Mennecke et al., (2007) mark to the seasoning and flavor constructor McCormick & Company. McCormick re-finds Flavor Print, an algorithm defining the flavors of the company as an aim of fifty data points, after given the priority of personalizing and digital hi-tech capability to supply it.

2.3.4.2 THE CULTURE OF DIGITAL BUSINESS TRANSFORMATION

Digital revolution is conducive by a culture, is an authentication mark of mature companies. This kind of organizations has a strong natural bias to behave in an appropriate way to motivate risk-taking, encourage alteration, and growth collaborative work environments. Charki et al., (2018) said, "Culture needs to support collaboration and creativity," which targets the outcomes related to the action of an undertaking social web at a cosmetics company in Europe.

2.3.4.3 LEADING THE DIGITAL TRANSFORMATION

Respondents more than half, from digitally developed organizations, say that one person or company led the digital program at their companies. In 2018, Charki et al., say that "It is very important for Managers to go above and say – Digital is excellent and does it" and "Managers required to do it themselves and perform boldly themselves." Leading by lesson is a bit of the performance. There are various human beings who engage by Twitter and other social platforms, but then also stand in the presence of 100 people and be genuine. All-day long, people come back and forth at the interval almost and physical platforms. Employees of digitally developed workplaces are confident of their leaders' they have the capacity to play that digital game. Their leaders have adequate skills to guide the digital strategy, say more than seventy-five percent of defendants from these companies. Only 90% add that their chiefs understand digital development and technologies.

2.3.5 WHY DIGITAL TRANSFORMATION IS THE GREAT ACCELERATOR FOR SMES?

The key vision from the address is that there is no "correct way" for digital revolution – the finest businesses had a broader strategy in area, but the target was on strategic shape-shifts; the small emphasis of small change rather than a big, group-wide overhaul. For SMEs, the internet has been of an extent level; non-selective in its influence/and conduct forth a gesture of the chance for all companies. It means even the little of businesses can take part in a world phase, growing more mature, providing that propel the borderline and bringing forward latest change. While golden opportunity plentiful, the shift has come an arrangement of provocations as businesses discover themselves with a broader, global ruthless landscape. This small change has also shifted suppositions, with a requirement to construct an experience that not only is online/but lives and respire digital.

2.4 SOLUTIONS AND RECOMMENDATIONS

2.4.1 OVERCOMING THE CHALLENGES OF DIGITAL TRANSFORMATION FOR SMES

Digital transformation is generating unprecedented golden opportunities for development and innovation. For decades, broad companies have carried the ruthless benefit in terms of extent, transportation, and the capability to reply fastly to changing market state. At present, the work surfaces are turned off. Improvement in technology is attending small scale companies with remarkable chances to obtain a foothold in markets previously controlled by large workplaces. Cloud computing is comfortably reachable to small-scale companies, and its pliability and absence ahead investment make it an ideal fit for businesses travel over new ground. It also separates require for recurrent, costly software improves/and has inbuilt security.

Move-ability makes it less essential for businesses to layout on office rentals, with platforms and cooperation tools accessed from smartphones, desktops, or tablets. And corporeal mail hi-technologies are now being right-sized, so small-scale businesses have entrance to superior, quicker, more fertile technology non-segregated with brilliant digital platforms (Miller, 2016).

2.4.2 START-UP – INNOVATION INSPIRING BUSINESS TRANSFORMATION

According to World startup report, the world had almost 140, 000 startups businesses, of which more than 15% were found outside America, at the ending of the previous year. McKinsey Institute Research foretells that by 2025, more than 45% of the Fortune Global 500 will be founded in energetic, appearing markets contrast to 17% in 2000. In 2025, it approximates 15,000 companies will have $1 – B plus in yearly revenues contrast to 8,000 in 2010, and 40% of recently developed broad companies will be located in the appearing world. A software company – "The Business Performance Innovation (BPI) Network" with Tech Mahindra, specializes in consulting, business re-engineering solutions, digital transformation and with TIE (The Indus Entrepreneurs), a non-financial gain global entrepreneurs network of 14,000 contributors to the other side of 17 countries, this is a thought of leading enterprise groups.

2.4.3 ADVANTAGE AND BUSINESS BENEFITS OF DIGITAL TRANSFORMATION

Digital transformation hand over the sorts of benefits certainly that businesses long-established benchmark. When thoughts about your own business, though remember that digital transformation is about alteration and benefits may take in new areas, such as:

- Increased customer betrothal and make better customer conversion;
- Improved customer, supplier, and employee cooperation;
- Make superior grade control processes;
- Flexible in a work;

- Personalization of products;
- Co-production of products and services.

2.4.4 CONCEPTUAL MODEL, FINDINGS, AND STRATEGIES

Recently the approach of the clustering of services is really helping in focusing on different issues from a holistic and better viewpoint, for the planning committees the production planning, decimation, and profits sharing with the proper use of knowledge, are the basic attitude behind the idea of clusters learning. So, it is much more in store by giving much priority of planning knowledge rather than focusing on issues, and this is seen as a standard grade rate towards the success of these companies all-around the globe. The focal point is on the utilization of the advanced equipment as a successful way of medium starting from the lower levels of technology which includes the traditional industries and handicrafts too (Alikhani et al., 2013). There are some findings and strategies shown below:

- RDS (Rural Development Strategy) development of rural entrepreneurship; "first to attract more business, secondly to retain."
- To generate a rural development strategy, this would include both the hierarchical approach towards a structured settlement and also the integrated rural development of sustainable development in the area.
- It would help in the employment strategies of the rural people giving them proper income and accelerating their prospects.
- It would also help in the appropriate financial and communal services through the interface of rural areas.
- An immense growth in the tourism industry.
- Planning, designing, and implementation of diverse schemes for the provision of public services in rural areas.
- The cluster approach to the development, i.e., developing the industrial clusters, technology clusters, and rural knowledge clusters.
- Rerating knowledge-based society with all amenities, strengthening the infrastructures in the rustic areas.
- Evolution of new age-enhanced technologies in rustic areas.

- Beholding the traditional technologies in rural areas with advanced and new technologies.
- More focus on participatory management in rustic areas.
- It would be beneficial if the components of a constructive culture can be applied to the evolution of the entrepreneurship in rustic areas through various educational programs and advocacy.

2.5 FUTURE RESEARCH DIRECTIONS

The limitations of this research follow from the nature of the study. Other than the general limitations of a qualitative, semi-structured interview-based scholarly paper, within a less developed research field, as well as the respondents' biases towards their understanding, helping entrepreneurship, interpretation of digitalization and organizational changes related to digital transformation, most of the arguments in this chapter are in need of further grounding in relevant theory. The understanding of digital modification strategies and the organizational change processes involved is also contingent on the researcher's limited, restricted, and prior knowledge and bias towards the literature on the subject. The qualitative nature of the study also separates the feature of the generalization of the consequences. In future research on digital transformation, it would be fascinating to further traverse the cultural features of digital modification in businesses, not just the more 'technical' or 'process-related' characteristics of the phenomenon, which was mainly done in this research.

2.6 CONCLUSION

The above study talks about the new age and modern technical analysis and its effect on the small and medium-sized institutions a fact which is the face of business in this fast pacing world. It would help in more explorations of transforming challenges to victories and help these institutions and the employees with understandable and skills for betterment. In this chapter, I have been cautious to mention to institutions as digitally developed rather than already mature. The digital modification of business is the latest phenomenon, and no institute has still arrived at the conclusion state nor clearly stated describe it. But the outlines are becoming understandable, as are the applications that precede institutions forward.

KEYWORDS

- **digital entrepreneurship**
- **digital transformation**
- **entrepreneurs**
- **small and medium enterprises (SMEs)**
- **start-up innovation**

REFERENCES

Alikhani, M., Khodayari, J., Dehnavi, M., & Verijkazemi, J., (2013). *A Conceptual Model of Entrepreneurship Development and Employment of Rural New Approach* (Vol. 1, No. 5).

Charki, M. H., Boukef, N., & Harrison, S., (2018). *The Successful Implementation of an Enterprise Social Media Platform is a Technical and Cultural Task.*

Chopra, S. S., & Yadav, A., (2018). *Women Innovation and Entrepreneurship in Jammu* (Vol. IX, Special Issue, e-ISSN: 0976–7967, p-ISSN: 2319–2322).

Giones, F., & Brem, A., (2017). Digital technology entrepreneurship: A definition and research agenda. *Technology Innovation Management Review, 7*(5), 44–51. http://timreview.ca/article/1076 (Accessed on 28 August 2019).

ICAEW, (2016). *Digital Transformation-the Next Steps: A Business Guide to Digital Change Management.*

Kane, G. C., Palmer, D., Phillips, A. N., Kiron, D., & Buckley, N., (2015). *Strategy Not, Technology, Drives Digital Transformation.*

Mennecke, B., Roche, E. M., Bray, D. A., Konsynski, B., Lester, J., Rowe, M., & Townsend, A. M., (2007). *Second Life and Other Virtual Worlds: A Roadmap for Research.* 28th International Conference on Information Systems (ICIS). Available at SSRN: https://ssrn.com/abstract=1021441 (Accessed on 28 August 2019).

Miller, S., (2016). *Why Digital Transformation is the Great Accelerator for SMEs.*

Montanye, J. A., (2006). Entrepreneurship. *The Independent Review, 10*(4), 547–569.

Nuti, B., (2017). *Helping Entrepreneurs and Small Businesses Make the Digital Transformation.*

Susarla, A., Oh, J. H., & Tan, Y., (2012). Social networks and the diffusion of user-generated content: Evidence from YouTube. *Information Systems Research, 23*(1), 23.

Uddin, M. R., & Bose, T. K., (2012). *Determinants of Entrepreneurial Intention of Business Students in Bangladesh* (Vol. 7, No. 24).

Zhao, F., & Collier, A., (2016). *Digital Entrepreneurship: Research and Practice.*

CHAPTER 3

Digital Transformation Using ICT Tools for Blended Teaching and Learning

S. KARKUZHALI[1] and S. SENTHILKUMAR[2]

[1]*Kalasalingam Academy of Research and Education, Computer Science and Engineering, Anand Nagar, Krishnankoil, Tamilnadu – 626126, India, E-mail: vijikarkuzhali@gmail.com*

[2]*Ayya Nadar Janaki Ammal College, Chemistry, Srivilliputtur Main Road, Sivakasi, Tamilnadu – 626124, India, E-mail: senthilkumar@anjaconline.org*

ABSTRACT

Combining the advantages of online learning and traditional classroom learning environments have been leading to a new learning environment, often referred to as "Blended Learning (BL)." BL integrates face-to-face interaction with technologically mediated interactions among students, teachers, and learning resources. The role of teachers is essential in the growing educational environments that are being formed with the help of Information Communication and Technology (ICT). ICTs allow students to take control of innovative learning, solve on practice, and identify the framework that allows them to cope with new learning methodology extensively. The combination of the use of MOODLE, spoken tutorials, flipped classroom (FC) and website creation for content management system should foster the student's active participation so that students become involved in their learning through an impulsive process to acquire a high level of competencies. ICT for teaching is an enabling technology with a possibility for increased learning outcome in education.

3.1 INTRODUCTION

Digital transformation is the change associated with the application of digital technology in all aspects of human society, usually transformation of business and organizational activities, processes, competencies, and models to fully influence the modifications and possibilities of a mix of digital technologies and their accelerating impact across society in a strategic and prioritized way. Digital transformation also includes the integration of digital technology into all areas of a business, that will confront as workers, and organizational leaders adjust to adapt to unfamiliar technologies. This chapter has been focused to concentrate the strategies to be adapted to convert the conventional physical classroom to online classrooms. Towards this coal, one should be familiar with teaching-learning in a physical classroom (Learner, Content, and Teacher), and provide the corresponding elements, interactions, and features (Learner, Content, and Facilitator). The transformation can be possible when the facilitator provides Learning Dialogs (LeD), Learning by Doing (LbD), and Learning eXperience Interaction (LxI). In this chapter, the digital transformation in education technology is discussed. The learning objectives of the chapter are as follows. After reading this chapter, the learner will be able to create online assignments and quizzes, develop spoken tutorials, develop flipped classroom (FC) using visual presentation, and create a personal webpage.

3.2 BACKGROUND

There are several reported results which argue that there is a need for new theories in this age of digital innovation and digital transformation. Ueda and Nakamura (2017) analyze data for the learning management system used in the year 2015 in Japan and get feedback in real-time. Ifinedo et al., (2017) designed to explore the effects or roles of usability factors (i.e., perceived ease of use, perceived usefulness, and satisfaction) and external support (i.e., teacher, and peer support) on undergraduates' use outcomes of MOODLE in a blended learning (BL) environment. Doobashi (2017) describes a data integration method for MOODLE course logs and pivot table functions to analyze the behavior of students' material page views in face-to-face BL using MOODLE course materials. Munoz et al., (2017) is present about the state of plug-in, which provides useful information to teachers and use social network analysis to make decisions to improve and promote participatory education.

Multimodal competence plays an important role in the valuable teaching-learning process. This study explores English-Medium Instructor's (EMI) choice and combination of modes when instantiating a university interactive lecture pair work activity (Morell, 2018). Hung and Chen (2018) used collective intelligence available in the online discussion forum to help instructors create meaningful Interactive Learning Activities (ILAs) to be embedded in video lectures for improving learners' comprehension and retention. Szpunar et al., (2014) represented video-recorded lecture is a central feature of most online learning platforms. Memory test and video recorded lectures help to interpolate actual with predicted performance. El-Sayed et al., (2013) teaching human anatomy and physiology courses to student nurses using video-based lectures are related to higher exam results, which represents the objective approach to measure actual usage of the video-based instructional materials.

Wang et al., (2018) employed a quasi-experimental approach by comparing pre- and post-class results of students who underwent three different learning methods, namely: Model-based Flipped Classroom (MFC) supported by Modern Teaching Technology; FC; and model-based classroom. Sergis et al., (2018) aims to address this identified issue and investigate, in an exploratory manner, the impact of Flipped Classroom Model (FCM)-enhanced BL environments not only on students' cognitive learning outcomes, but more importantly on their internal dispositions (level of satisfaction) and 'needs' for competence, autonomy, and relatedness. Castañeda et al., (2018) identified the level of satisfaction by learners is high and they benefited from the experience in different ways: reinforcement of their knowledge and capacity to perform innovative research in One Health (e.g., using digital epidemiology), visits, and meetings with experts in Global Health (e.g., Institute of Global Health and World Health Organization in Geneva, Public Health Institute and Swiss Tropical in Basel), and emerging research collaborations, etc. Munir et al., (2018) discusses the reasons for flipping a classroom in an Engineering course and for including cooperative learning. The results showed that FC helped to develop and improve students' learning and critical analysis skills.

Sojayappan et al., 2018 has employed in the study consisted of a FC model with online group investigation; a website for FCs; and a learning management plan. The data collection instruments comprised a self-assessment form of team learning ability and a team learning behavior observation form. Becker and Birdi (2018) has used the technique of classroom flipping, by which the sequencing and nature of learning and teaching are configured in novel ways, has started to make an impact on the teaching of economics. Mohammed and Lamia (2018) implement FC as an element of the Internet of Things (IoT) into the learning process of mathematical logic course. Intelligent tutoring systems help to improve the problems faced by students in an out classroom activity in FC. Wang (2017) promote engagement and achievement in FC has been a critical research issue. Gilboy et al. (2015) illustrate how to implement the FC for two undergraduate nutrition courses. The template designed can use before and after class activity using all levels of Bloom's taxonomy. Mohamed et al., (2015) identified the factors that motivate lecturers to use online teaching tools, such as Web 2.0 tools. There are a variety of online teaching tools available to the lecturers. The popular tools are online whiteboard, WordPress, YouTube, Facebook, Edmodo, and others.

The instruments used are classroom observation and interview sessions, which are used to measure four motivation variables: self-confidence, usefulness, enjoyment, and ease of use, based on the ARCS Motivation model. Lin and Jou (2012) aim to apply well-known web applications to propose a web application supported the learning environment for supporting classroom teaching and learning activities that can seamlessly enable instructors and students to meld with their daily use of Web 2.0 tools. Hinings et al., (2018) identified digital organization forms, institutional infrastructure, building blocks for digital transformation.

Driven by the exponential growth in computing power, digital technologies have successfully transformed most sectors of the economy – ranging from finance to entertainment – with one prominent exception: healthcare (Walsh and Rumsfeld, 2017). The digital transformation in health care technology is the most emerging technology. The active and passive roles in health care technology established in the American College of Cardiology (ACC). Choosing the latter, the ACC has launched an Innovation Strategy, advocating that digital health technologies must be effective and safe, improve the efficiency of care, improve patient-clinician interaction, and be supported by payment models. Al-Ruithe et al., (2018) highlights the cloud computing roles to achieve the digital transformation in Saudi Arabia, and it seeks to empirically examine the cloud computing concerns in the public sectors organizations of Saudi Arabia. An online questionnaire was launched aiming at the existing cloud computing adoption and its concerns in Saudi Arabia. Security, privacy, and loss of governance are still the main obstacles for adopting cloud-computing technology, although some organization lack of knowledge in this area.

Goerzig and Bauernhansl (2018) analyzed the content of the digital transformation. Main challenges are the high complexity of the approaches and the infeasibility for the fast development of new solutions. That is why we presented a new lightweight, agile approach. Industrie 4.0 and digital transformation in the production area are relatively new concepts that are still being addressed on a feasibility projects level. ICT developers and business consultants are dominating the narrative regarding how companies' journey should be while undergoing such a crucial transformation (Issa et al., 2018). Leipzig et al., (2018) addresses the model is based on a continuous improvement cycle, and also includes triggers for innovative and digital thinking within the enterprise. The model was successfully validated in the German service sector.

The "Mittelstand 4.0-Digital Production and Work Processes" initiative by the Federal Ministry for Economic Affairs and Energy in Germany supports small and medium-sized enterprises to become digitized, to network and to start using industry 4.0 applications. The "Mittelstand 4.0 Competence Center Chemnitz" is part of this initiative. It provides information, practical trainings, test environments, and application projects for the small and medium-sized companies in the region. The center's goals, structures, and service offers are described by Muller and Hopf (2017). The rapid proliferation of technology in modern society is changing the way people behave and is dramatically increasing the traceability of individuals. In particular, new portable and miniaturized technologies put powerful capabilities in everyone's pocket, and capture large quantities of detailed information in a digital format (Casey et al., 2018). Ferreira and Pantidi (2018) aim to develop community resilience, social regeneration, and transformation and their considerations around digital participation, and technology adoption. This early stage in their technology adoption process proved to be an opportunity for careful deliberations over the perceived characteristics of the digital technologies, and over the trade-offs and impacts of its use and nonuse. Haggarty (2017) described digital transformation and security in patient care.

Li (2018) describes the digital transformation in industries and business models. Based on the experiences in the case studies, it can be concluded that ICT tools are often introduced rather late during the planning process. And when introduced, the tool is used to evaluate the performance with regard to very specific aspects (Oregi et al., 2015). Duta and Martinez-Rivera (2014) consider that students recognize the need and importance of ICT as a tool for collaborative learning and as see there are great similarities between them about training methods, their application is mainly aimed at seminars on the use of multimedia equipment teaching-learning-assessment by some teachers and students, thought-provoking that there is an area that teachers are not fluent enough in Romania and therefore, the request is in line with this.

Gámiz-Sanchez (2017) elaborated the pedagogical tool to make the active learning and teaching activities, the research experience performed at the Faculty of Educational Sciences of the University of Granada promoted with the European Higher Education Area (EHEA). Tvenge and Martinsen (2015) identify some possible factors that enable or inhibit teachers' use of ICT in their teaching practice. The results are based on seven semi-structured interviews conducted amongst teachers at manufacturing engineering education, the analytical framework derived from the main research study; an international education monitoring project. Sanchez and Aleman (2011) report on the results obtained from a teachers' opinion survey on the use of ICT tools to support of attendance-based teaching. In order to carry out this study, it was necessary to design a questionnaire to collect data among all in-service teachers with access to the university virtual campus. Ramirez et al., (2017) elaborates the brief review on tools used in ICT for effective teaching and learning. The systematic review is based on the methodology of Kitchenham. The starting point of the review are the research questions, then keywords, selection of the databases, definition of the inclusion and exclusion criteria, the definition of the search chains, search process and selection of papers, the analyzes of the paper and the results of the systematic review to answer the questions posed.

Rohatgi et al., (2016) generated new knowledge about the relations between ICT use, ICT self-efficacy, and CIL achievement. ICT self-efficiency plays a particular role for the use of ICT and CIL achievement. Ruiz-Calleja et al., (2011) proposes to collect data from third party sources, align it to a vocabulary understandable by educators, and finally publish it to be consumed by educational applications. A case study with practitioners has been carried out to evaluate whether the information about ICT tools provided by this dataset is understandable and useful for educators.

Vega-Hernandez et al., (2018) consider ICT are a very important help tool in their learning process, and they value positively it. Nevertheless, there are significant differences in the areas of knowledge. Science students distinguished by use educational portals, Health Science university students by search engines in the network, manipulate some web browsers and consider ICT such as an important element in their academic training, essential in today's society.

The ICT tools provide information immediately, are easily accessed, and are user-friendly for all. However, it is important to understand which tools the community would prefer to use and does everyone have access to these tools and will the community trust the information (Wright et al., 2018). Zylka et al., (2015) addresses this issue by integrating them into the concept of ICT engagement. Its theoretically deduced dimensions of social exposure to ICT, self-concept related to the use of ICT, and ICT-related interest were analyzed in an explorative study assessing 445 students aged between 14 and 17 years in the German federal state of Baden-Wuerttemberg.

Online and BL methods are growing exponentially in use in higher education and the systematic evaluation of these methods will aid understanding of how best to teach clinical supervision. BL provides added pedagogical value when compared to online learning in terms of teaching undergraduate nurses clinical supervision skills (McCutcheon et al., 2018). Tosun (2015) aimed at investigating the effects of BL strategy in teaching vocabulary and the students' perceptions of BL approach in learning vocabulary. Forty samples from two different schools are involved in this study. Ngan et al., (2015) study provide an example of adopting BL in anatomy teaching to facilitate active learning among students. The active access and favorable attitude towards courseware features in addressing the existing learning barrier, although no inference could be drawn on the influence of the courseware on learning due to the limited sample size. Prasad et al., (2018) develop a testing mechanism to measure the extent to which international students have built up digital capital. We use the Unified Theory of Acceptance and Use of Technology (UTAUT) as a framework for this investigation, built around social influence (behavioral intention) performance and effort expectancy (attitude), and facilitating conditions (ease and usefulness), using a critical theoretical approach.

Boelens et al., (2018) planned BL to address student differences developed from the data. Firstly, disregard: instructors considered no extra support within the BL arrangements to coordinate students' needs. Secondly, adaptation: instructors believed that increased support in the existing BL arrangements was sufficient to match students' needs. Thirdly, transformation: instructors thought that BL arrangements ought to be outlined in a totally different way, and be custom-made to the characteristics of the students. Clausen et al., (2018) developed practice-oriented, interdisciplinary, interactive, and multimedia-prepared online case studies that students can work through in their own time and combining online case studies with face-to-face classes in a BL concept, designed by the teachers. A blended teaching-learning model which has online video lectures develops learner's knowledge as well as skills (Coyne et al., 2018). Jokinen and Mikkonen (2013) consider teachers' experiences of teaching undergraduate nursing students in a BL environment. The basic idea of the study program was to support students to reflect on theory and practice and provide with access to expert and professional knowledge in real-life problem-solving and decision making. Kharb and Samanta (2016) aimed to address the blended teaching-learning to facilitate both faculty and student by showing facilities available in the blended teaching-learning module (Marchalot, 2017).

3.3 ICT TOOLS FOR BLENDED TEACHING AND LEARNING

Digital literacy is an absolutely essential component towards preparedness for the workplace to efficiently and effectively use the technology, making an effective search, understanding the search results which leading to the selection of best result out of it. The understanding of digital literacy can able to use the technology everywhere to support the students. It can also be able to help prepare the teacher to be more productive and efficient, effective in the workplace. In this chapter, the following tools have been used for the development of online content towards the digital transformation in blended teaching and learning: MOODLE (Modular Object-Oriented Dynamic Learning Environment) for design and development of Lesson Activity, WordPress for webpage development, and Microsoft PowerPoint for visual presentation and Screencast-o-Matic for screen recording.

3.3.1 *DESIGN AND DEVELOPMENT OF LESSON ACTIVITY USING MOODLE*

MOODLE is a learning platform or Course Management System. It is a freely available open-source software package designed to help educators to create effective online courses with video lectures, assignment, and quizzes. As open-source software, it is really easy to download, modify, and redistribute the MOODLE. The MOODLE can be used in all platforms like UNIX, Linux, Windows, Mac OS X, Netware, PHP, and web developers. The data can be stored in a single database, such as MySQL and PostgreSQL are best supported, but it can also be used with Oracle, Access, Interbase, ODBC, and others. The content developed by the teacher in MOODLE can be easily shared with their students for effective teaching and learning.

The salient features of the MOODLE are user-friendly customization, MOODLE mobile app which can be used in a handheld device, live engagement, assessments, content repository, accessibility, analytics, tool variety, and course management (for both teacher and administrator). The MOODLE has in-built conversation functionality. Some of the add-ins for the majority of the digital learning tools such as Blackboard Collaborate and Ultra, Big Blue Button, Zoom, and WebEx. In order to use these learning tools, additional licensing may be required. The assessment of a student can be effectively carried out using MOODLE through the automation and customization. The MOODLE's teacher quiz interface is really a challenging task; very often, teachers can take valuable time to learn it. The MOODLE have an excellent content repository which allows teachers to upload files in a particular course through various file warehouse including Dropbox, Flickr, Google Drive, and conventional file uploading from computer hard disk. The MOODLE can be easily accessible and usable for all users (students) irrespective their ability. It is teachers (instructors) responsibility to ensure that their learning materials such as pdf, video, and PowerPoint files which are accessible easily by the students.

The extensive analysis of the performance of the students can be done using MOODLE. The plenty of reports and logs available in the MOODLE can be allowed to track the progress and success of the students. These kinds of analysis can be done for an individual course. In order to use it effectively, a teacher should have a proper training. The MOODLE has a number of tools which can be turned on or off at the system level, like publisher integration. The customization of the course setup can be managed by the course management system. The teachers can use MOODLE as a file repository. For an organization, several courses can be integrated through the course management system. Because MOODLE has a huge number of customization options.

3.3.1.1 ADVANTAGES OF MOODLE

As an organizational tool, it gives great advantages for teachers to maintain different courses. The multiple courses have been conducted in an organization for their employees and students. The end-user can navigate and manipulate the MOODLE's structure. The facilitator and student can use a different type of formats in MOODLE. MOODLE provides a platform for the instructor to share all their materials, assignment, and quizzes in online as well as the offline classroom. MOODLE allows for communication to between instructor and student as well as student and another student. MOODLE allow the instructor and student to communicate via chat, blog, discussion forums in the course page. This ensures the MOODLE is an effective tool for the teaching-learning process.

3.3.1.2 DISADVANTAGES OF MOODLE

A disadvantage of MOODLE is that it is a technological tool which can be a dispute for some students. If someone has some practical problem in accessing technology, MOODLE may be a problem for the student. If the student does not have access to the internet or some form of technology to utilize MOODLE, then this becomes a disadvantage for the student and also for the instructor. Another disadvantage to MOODLE is that some student, more specifically, adult learners are resistant and hesitant to using any form of technology for learning. Even though MOODLE allows for a great place to organize and store learning content which will be provided to the students, some adult learners are very resistant to utilize technology for learning, and they will not use it. If MOODLE is used for online and distance learning, these students will not be keen on taking the course unless it is required. These students prefer face-to-face instruction and interaction, which MOODLE does not provide. I think if the face-to-face interaction is available in MOODLE, some student would be open to this technology for learning.

3.3.2 SPOKEN TUTORIALS USING SCREENCAST-O-MATIC

The URL for creating the spoken tutorial using Screencast-o-Matic is https://screencast-o-matic.com. It does not require to sign up or login to use it that is it can be used without having an account in it. It provides user-friendly, fast, and free screen recording. By click on the "start recording," an external pop-up will come up. Then, by accepting or launching or anything that is affirmative that just click on launching the recorder. A separate transparent window will appear entitled "Record," containing screen, webcam, and both. The option "screen" can be used to create the spoken tutorial through the screen recording. The option "webcam" can be used to make spoken tutorial by focusing the person appearing in front of the camera where the content on the screen cannot be seen. The option "both" can be used to record the spoken tutorial with the help of screen as well as webcam. Further, in the same window, the maximum time that the recording can be done is seen, for free version maximum of 15 min can be recorded. However, several spoken tutorials can be recorded, e.g., tutorial-1, tutorial-2, tutorial-3, etc. The "size" of the screen can also be varied, such as 480p, 720p, and full screen, where the dashed line enlarges its size upon increasing the size. The full-screen mode will record everything shown on the screen. The final option present in the window is "narration," which will detect the microphone connected to the computer. Then, upon clicking the record button, it gives the countdown to start talking and describing the content of the spoken tutorial.

3.3.2.1 FEATURES OF SCREENCAST-O-MATIC

- **Fundamental Mechanics:** The software is easy to download and use, the options available in the tool is easy to understand and use by an end-user. Since you don't even have to install it on your computer, it makes things even easier for the user.
- **Publish in YouTube:** We can publish the video recorded in software to you tube directly. It's very effective and efficient for the end-user to upload the video, and learner can also access the video easily.

- **Able to Record Screen and Webcam Simultaneously:** With Screen-O-Matic software, you can also record screen and webcam simultaneously. This means you will not have to use multiple software to combine your recordings.
- **Supplementary Features in Paid Version:** Screencast-O-Matic PRO is the paid version; we can add more features in the video like onscreen drawing. In PRO version we can add zooming option, can add animation, edit audio, and video features.
- **Cost-Effective for PRO Version:** For using additional features, we need to pay 15 dollars per year. It will enhance your work experience without taking a toll on you in terms of money.

3.3.2.2 ADVANTAGES OF SCREENCAST-O-MATIC

- It is free to Internet users for the basic features.
- The website is user-friendly and simple to create a screencast.
- It requires little to no training for new users.
- The website includes screencasts that demonstrate how to use the features of Screencast-o-Matic and includes screencasts on solutions to troubleshooting.
- Screencast-o-Matic captures any program or website that is on the computer screen.
- Users can record video from a webcam that is included in the screencast.
- The user can select a specific area of the screen to be recorded, thus focusing the learner's attention to the part of the screen necessary for the screencast.
- The website has the ability to publish the screencast in a variety of ways.
- Educators can present lessons in an engaging way to students, use as a form of assessment, or provide feedback via a screencast.

3.3.2.3 Disadvantages of Screencast-O-Matic

- Users cannot edit the screencast on the website or record more than 15 minutes without purchasing the Pro edition with additional features for 15 dollars per year.

- To run the software we need to install Java applet.
- The screen and video recording is not much interactive for students. There are no features to include quizzes, polls, discussion forum, and animations.

3.3.3 FLIPPED CLASSROOM (FC) USING VISUAL PRESENTATION

The FC is the new strategy used in the teaching-learning process to make the process simple and effective. There are two types of activities in-class and out-class activity. The in-class activity is the activity planned during class hours like quiz, brainstorming, think pair share, team solo, and tutorials. The out-class activity is sharing a link to the video uploaded in youtube which is prepared by Screencast-O-Matic. FC s also redefine in-class activities like homework problem with more examples, assignments in maths, virtual laboratory exercises, analyzing a document, discussion, project demo or presentation, peer learning, poster presentation, or concept map development.

3.3.3.1 FEATURES OF FLIPPED CLASSROOM (FC)

- **Provide New Interactive Exposure to Students Before Class:** The mechanism used for learning is varying for textbook reading to interactive video lectures.
- **Provide an Incentive for Students to Prepare for Class:** The assignment can vary; the examples above-used tasks that ranged from online quizzes to worksheets to short writing assignments.
- **Provide a Mechanism to Assess Student Understanding:** The pre-class assignments ensures the students understanding and also increases the interest of the students to attend the class.
- **Provide in-Class Activities That improves on Higher Order Thinking and Cognitive Activities:** The students gained basic knowledge about the subject in the interactive video from out-class activity and deep learning from activities done in in-class activity.

3.3.3.2 ADVANTAGES OF FLIPPED CLASSROOM (FC)

- **Self-Paced Learning:** When students prepare their class, they can work whenever they want and take whatever time it takes to finish (as long as it's before the deadline).
- **Dive Deeper into Subject:** By in class and out class activities, the subject knowledge gets deeper, and interest towards the subject get stronger. You can offer more learning material to students who are looking for a challenge.
- **Better Prepared (You Know the Results and Statistics):** Students can follow courses where teachers put on homework students have to prepare. Teachers are able to track the progress of students and view their results.
- **Lectures can be Reused:** Once we prepared the materials, we can able to reuse the materials prepared for next year as well as we can improve in next year.
- **Transparency for Parents:** Parents can able to view the videos and help their child in improving their knowledge about the subject.

3.3.3.3 DISADVANTAGE OF FLIPPED CLASSROOM (FC)

- No access to the internet in some remote areas leads to issues in accessing data.
- The familiarization of these new concepts to teacher and student may take time to understand.
- Students need to complete the task assigned for them at home, so the students who all have a lack of motivation will suffer.
- The level of understanding of slow learners and fast learners are not addressed in FC.

3.3.4 WEBSITE CREATION USING WORDPRESS

The content management system used to store all contents like animation, video, image, and text in the webpage. It helps in editing, publishing, and modifying the content of the website. WordPress was initially developed by Matt Mullenweg and Mike Little on May 2003. In October 2009, it will become open source. The Simple and Multifunctional website can be created using WordPress.

3.3.4.1 FEATURES OF WORDPRESS

- **User Management:** It allows managing the user information such as changing the role of the users to administrator, editor, author, contributor, or subscriber; delete or create the user; and change the user information and password. The main role of the user manager is Authentication.
- **Theme System:** It allows modifying the site view and functionality. It includes images, stylesheet, template files, and custom pages.
- **Extend With Plug-In:** Several plugs-in are available which provides custom functions and features according to the users need.
- **Search Engine Optimization (SEO):** It provides several SEO tools which make on-site SEO simple.
- **Multilingual:** It permits translating the whole content into the language favored by the user.
- **Importers:** It allows importing data in the form of posts.

3.3.4.2 ADVANTAGES OF WORDPRESS

- Can add media files like video and audio easily.
- WordPress is the free content management system.
- Customization is changing; updating the contents is very quick and efficient.
- Administrator can view the daily activities and number of visitors.
- Can create different roles like administrator, editor, and contributor.

3.3.4.3 DISADVANTAGES OF WORDPRESS

- Sometimes software needs to be updated to keep the WordPress up-to-date with the mobile devices, and current browsers. Whenever we are updating the WordPress, the contents of the web page should get back up.
- Modifying the media files after uploading is somewhat difficult.

3.3.5 INCLUSION OF CREATIVE COMMON LICENSE

Whenever we are creating a resource and sharing the resource, need to mention Creative Commons copyright license, this is the simple and standard copyright permission to create, use, reuse, distribute, and redistribute.

3.4 FUTURE RESEARCH DIRECTIONS

Future studies will have new opportunities to study the latest technological innovations in online learning, such as virtual reality. The analysis should be conducted to make the best selection of online tools and activities for each particular group of students. BL is the most effective teaching and learning methodology than in-class learning. Further research is needed on the most effective ways of incorporating video-assisted learning materials into a BL model and how to complete an unbiased evaluation of the resource.

3.5 CONCLUSION

This study aimed to investigate the teachers' experiences in planning and teaching in a course based on an ICT tools and BL approach. The integration of face-to-face learning, online learning, and learning in and about work provides opportunities to enhance learning in blended-learning programs. This is because such programs support student motivation through the presence of "real-life" and their relevance to the students' own places of work. Students and faculty much benefited from BL by innovative methodology and resource sharing platform. BL is about ways of providing students with the best possible learning and teaching experiences, as well as supporting teachers in their role as facilitators. A blend of traditional methods (i.e., face-to-face classroom teaching), with technology-based instruction, can help in overcoming time constraints of traditional classroom teaching and enhance learning. Students and teachers express a positive attitude towards active use of ICT as a support for their teaching and learning experiences. Although some students were initially a bit reluctant to learn new tools, the ease and continuity of use across several years led the students ultimately to accept the tools in their personal learning environments. Another way of working that we are beginning to try with the team of participating teachers is the use of FCs as a strategy to foster the student's active work. In this method, the student prepares some of the more theoretical contents (readings, videos, etc.) outside the classroom and uses the class to perform activities that put the competencies acquired into practice. The classroom becomes a collaborative workspace in which the student can use all of the resources at his/her disposition (laptops, tablets, cell phones, etc.) to do the work assigned.

KEYWORDS

- **blended learning**
- **blended teaching**
- **digital transformation**
- **flipped classroom**
- **ICT**
- **MOODLE**
- **visual presentation**
- **webpage development**
- **WordPress**

REFERENCES

Al-Ruithe, M., Benkhelifa, E., & Hameed, K., (2018). Key Issues for embracing the cloud computing to adopt a digital transformation: A study of Saudi Public Sector. *Procedia Computer Science, 130*, 1037–1043.

Becker, R., & Birdi, A. (2018). Flipping the classroom: Old ideas, new technologies. *International Review of Economics Education, 29*, 1–5.

Boelens, R., Voet, M., & De Wever, B., (2018). The design of blended learning in response to student diversity in higher education: Instructors' views and use of differentiated instruction in blended learning. *Computers & Education, 120*, 197–212.

Casey, E., Ribaux, O., & Roux, C. (2018). Digital transformations and the viability of forensic science laboratories: Crisis-opportunity through decentralisation. F*orensic science international*, 289, e24–e25.

Castañeda, R. R., Garrison, A., Haeberli, P., Crump, L., Zinsstag, J., Ravel, A., & Bolon, I., (2018). First global flipped classroom in one health: From MOOCs to research on real world challenges. *One Health, 5*, 37–39.

Clausen, P. H., Stelzer, S., Nijhof, A., Krücken, J., Von Samson., & Himmelstjerna, G., (2018). Established and novel approaches for teaching and learning of veterinary parasitology in Berlin. *Veterinary Parasitology, 252*, 58–61.

Coyne, E., Rands, H., Frommolt, V., Kain, V., Plugge, M., & Mitchell, M., (2018). Investigation of blended learning video resources to teach health students clinical skills: An integrative review. *Nurse Education Today, 63*, 101–107.

Dobashi, K., (2017). Automatic data integration from MOODLE course logs to pivot tables for time series cross section analysis. *Procedia Computer Science, 112*, 1835–1844.

Duţă, N., & Martinez-Rivera, O., (2015). Between theory and practice: The importance of ICT in higher education as a tool for collaborative learning. *Procedia-Social and Behavioral Sciences, 180*, 1466–1473.

El-Sayed, R. E. H., El-Hoseiny, S., & El-Sayed, E., (2013). Video-based lectures: An emerging paradigm for teaching human anatomy and physiology to student nurses. *Alexandria Journal of Medicine, 49*(1), 215–222.

Ferreira, J., & Pantidi, N., (2018). Going digital: Integrating digital technologies in local community initiatives. *In Digital Participation Through Social Living Labs* (pp. 55–76).

Gámiz-Sanchez, V. M., (2017). ICT-based active methodologies. *Procedia-Social and Behavioral Sciences, 237*, 606–612.

Gilboy, M. B., Heinerichs, S., & Pazzaglia, G., (2015). Enhancing student engagement using the flipped classroom. *Journal of Nutrition Education and Behavior, 47*(1), 109–114.

Goerzig, D., & Bauernhansl, T., (2018). Enterprise architectures for the digital transformation in small and medium-sized enterprises. *Procedia CIRP, 67*, 540–545.

Haggerty, E., (2017). Healthcare and digital transformation. *Network Security, 2017*(8), 7–11.

Hinings, B., Gegenhuber, T., & Greenwood, R., (2018). Digital innovation and transformation: An institutional perspective. *Information and Organization, 28*(1), 52–61.

Hung, I. C., & Chen, N. S., (2018). Embodied interactive video lectures for improving learning comprehension and retention. *Computers & Education, 117*, 116–131.

Ifinedo, P., Pyke, J., & Anwar, A., (2018). Business undergraduates' perceived use outcomes of MOODLE in a blended learning environment: The roles of usability factors and external support. *Telematics and Informatics, 35*(1), 93–102.

Issa, A., Hatiboglu, B., Bildstein, A., & Bauernhansl, T., (2018). Industrie 4.0 roadmap: Framework for digital transformation based on the concepts of capability maturity and alignment. *Procedia CIRP, 72*, 973–978.

Jokinen, P., & Mikkonen, I., (2013). Teachers' experiences of teaching in a blended learning environment. *Nurse Education in Practice, 13*(6), 524–528.

Kharb, P., & Samanta, P. P., (2016). Blended learning approach for teaching and learning anatomy: Students' and teachers' perspective. *Journal of the Anatomical Society of India, 65*(1), 43–47.

Li, F., (2018). *The Digital Transformation of Business Models in the Creative Industries: A Holistic Framework and Emerging Trends*. Technovation.

Lin, Y. T., & Jou, M., (2012). A web application supported learning environment for enhancing classroom teaching and learning experiences. *Procedia-Social and Behavioral Sciences, 64*, 1–11.

Marchalot, A., Dureuil, B., Veber, B., Fellahi, J. L., Hanouz, J. L., Dupont, H., & Compère, V. (2018). Effectiveness of a blended learning course and flipped classroom in first year anaesthesia training. *Anaesthesia Critical Care & Pain Medicine*, *37*(5), 411–415.

McCutcheon, K., O'Halloran, P., & Lohan, M., (2018). Online learning versus blended learning of clinical supervisee skills with pre-registration nursing students: A randomized controlled trial. *International Journal of Nursing Studies, 82*, 30–39.

Mohamad, S. N. M., Salleh, M. A. M., & Salam, S., (2015). Factors affecting lecturers motivation in using online teaching tools. *Procedia-Social and Behavioral Sciences, 195*, 1778–1784.

Mohamed, H., & Lamia, M., (2018). Implementing flipped classroom that used an intelligent tutoring system into learning process. *Computers & Education, 124*, 62–76.

Morell, T. (2018). Multimodal competence and effective interactive lecturing. *System, 77*, 70–79.

Muller, E., & Hopf, H., (2017). Competence center for the digital transformation in small and medium-sized enterprises. *Procedia Manufacturing, 11*, 1495–1500.

Munir, M. T., Baroutian, S., Young, B. R., & Carter, S., (2018). Flipped classroom with cooperative learning as a cornerstone. *Education for Chemical Engineers, 23*, 25–33.

Muñoz, A., Delgado, R., Rubio, E., Grilo, C., & Basto-Fernandes, V., (2017). Forum participation plug-in for MOODLE: Development and discussion. *Procedia Computer Science, 121*, 982–989.

Ngan, O. M. Y., Tang, T. L. H., Chan, A. K. Y., Chen, D. M., & Tang, M. K., (2018). Blended learning in anatomy teaching for non-medical students: An innovative approach to the health professions education. *Health Professions Education, 4*(2), 149–158.

Oregi, X., Roth, E., Alsema, E., Van Ginkel, M., & Struik, D., (2015). Use of ICT tools for integration of energy in urban planning projects. *Energy Procedia, 83*, 157–166.

Prasad, P. W. C., Maag, A., Redestowicz, M., & Hoe, L. S., (2018). Unfamiliar technology: Reaction of international students to blended learning. *Computers & Education, 122*, 92–103.

Ramirez, G. M., Collazos, C. A., & Moreira, F. (2018). All-Learning: The state of the art of the models and the methodologies educational with ICT. *Telematics and Informatics, 35*(4), 944–953.

Rohatgi, A., Scherer, R., & Hatlevik, O. E., (2016). The role of ICT self-efficacy for students' ICT use and their achievement in a computer and information literacy test. *Computers & Education, 102*, 103–116.

Ruiz-Calleja, A., Vega-Gorgojo, G., Asensio-Pérez, J. I., Bote-Lorenzo, M. L., Gómez-Sanchez, E., & Alario-Hoyos, C., (2012). A linked data approach for the discovery of educational ICT tools in the web of data. *Computers & Education, 59*(3), 952–962.

Sanchez, J. J. C., & Aleman, E. C., (2011). Teachers' opinion survey on the use of ICT tools to support attendance-based teaching. *Computers & Education, 56*(3), 911–915.

Sergis, S., Sampson, D. G., & Pelliccione, L., (2018). Investigating the impact of flipped classroom on students' learning experiences: A self-determination theory approach. *Computers in Human Behavior, 78*, 368–378.

Sojayapan, C., & Khlaisang, J. (2018). The effect of a flipped classroom with online group investigation on students' team learning ability. *Kasetsart Journal of Social Sciences.* (Article in press)

Szpunar, K. K., Jing, H. G., & Schacter, D. L., (2014). Overcoming overconfidence in learning from video-recorded lectures: Implications of interpolated testing for online education. *Journal of Applied Research in Memory and Cognition, 3*(3), 161–164.

Tosun, S., (2015). The effects of blended learning on EFL students' vocabulary enhancement. *Procedia-Social and Behavioral Sciences, 199*, 641–647.

Tvenge, N., & Martinsen, K., (2016). Selection of ICT-tools for manufacturing education. *Procedia CIRP, 41*, 1096–1100.

Ueda, H., & Nakamura, M., (2017). Data analysis for evaluation on course design and improvement of "cyber ethics" MOODLE online courses. *Procedia Computer Science, 112*, 2345–2353.

Vega-Hernandez, M. C., Patino-Alonso, M. C., & Galindo-Villardón, M. P., (2018). Multivariate characterization of university students using the ICT for learning. *Computers & Education, 121*, 124–130.

Walsh, M. N., & Rumsfeld, J. S. (2017). Leading the digital transformation of healthcare: the ACC innovation strategy, 2719–2722.

Wang, F. H., (2017). An exploration of online behavior engagement and achievement in flipped classroom supported by learning management system. *Computers & Education, 114*, 79–91.

Wang, J., Jou, M., Lv, Y., & Huang, C. C., (2018). An investigation on teaching performances of model-based flipping classroom for physics supported by modern teaching technologies. *Computers in Human Behavior, 84*, 36–48.

Wright, D., Hammond, N., Thomas, G., MacLeod, B., & Abbott, L. K., (2018). The provision of pest and disease information using Information communication tools (ICT); an Australian example. *Crop Protection, 103*, 20–29.

Zylka, J., Christoph, G., Kroehne, U., Hartig, J., & Goldhammer, F., (2015). Moving beyond cognitive elements of ICT literacy: First evidence on the structure of ICT engagement. *Computers in Human Behavior, 53*, 149–160.

CHAPTER 4

A Critical Analysis of E-HRM as a Strategic Capability Builder

B. C. SINHA

Rakshpal Bahadur College of Engineering and Technology, Management, Knowledge Park, Part III, Greater Noida, UP–201310, India, E-mail: sinhachandra99@gmail.com

ABSTRACT

Electronic-Human Resource Management (E-HRM) as a true business partner, must be able to increase organization's strategic capability by employing a shared mindset and accountability to key performance indicators by standardizing its service delivery, reducing the administrative burden of HR professionals, improving employer brand image, creating ready to change workforce, increasing competence of employee, and supporting knowledge management (KM). This research paper, in its endeavor, examines the role of E-HRM as a tool in increasing the competitive edge of Indian organizations, the response has been sought after from employees of eight organizations using the structured questionnaire as an instrument. The study reveals that all attributes of financial contribution are not above the test value, and there exists a difference in level for some attributes of financial contribution in context to private vis-a-vis public and manufacturing vis-a-vis services.

4.1 INTRODUCTION

E-HRM is always on the radar and examined whether it can succeed to change the outlook of HR function from reactive to proactive, cost center to the bottom-line contributor, facilitator to a business partner, transactional to transformational or strategic. E-HRM facilitates in better serving and management of every institution's most important strategic capability builder, innovator, thinker, server of client-employee. Digitization of HR work gets free of redundant activities, presents more timely and precise HR information and perhaps most vital, mechanizes the time killing, fault inclined human resource paper story. It leaves human resource professionals with more time to focus on strategic tasks and manage in the best possible way the company's most important resource – its people.

E-HRM is supposed to shift the focus of HR function from employee welfare to strategic contributor by reorienting human capital, social capital, organizational capital, intellectual capital and be a facilitator of knowledge management (KM), and at the end provider of competitive edge to the organization. E-HRM attributes, tools, mechanism, the system can be a big game-changer and pathfinder in providing the strategic capability to the HR function. As per Verma (2010), E-HRM enables better management of every enterprise's most important competitive edge: the thinking, idea-generating, and customer-serving human resource. E-HRM eliminates redundant activities, provides more accurate and timely personnel information and – perhaps most important – automates the time consuming, error-prone human resource paper trail. It leaves human resource professionals with more time to focus on strategic tasks and manage in best form the company's most important resource-its people.

Technological change may be a key driver for HR change, giving the foundation to bolster HR's developing key focus. Especially, network and internet technologies have already given employees direct contact with each other, to HR, and to business information with such ease and intelligence that every worker can contribute more directly to business results. It is seen that there is a radical change in human resource management function and there is a transition from a secretarial, maintenance, utility department to the core of strategic orientation and the application of technical know-how is one of the main ingredients of this massive change. Budding technical knowhow has resulted in new scope in the direction of human resource by means of formation of virtual staff, dropping levels of hierarchy and rewriting the position of human resource as a strategic associate in the age of information system. Establishments associating human resource function along with the other functions like marketing, finance, operations, information technology of company have the prospective to add useful value to the strategic assimilation of human resource function with establishment's overall strategy. E-HRM technology is a harbinger of KM and creates a positive organizational culture and high-performance systems. KM is supposed to be biggest wealth creator, and it may be explicit or implicit inside the establishment or outside the establishment, if it is utilized in the proper way can deliver gold both for the organization and individual employee. Information and Communication Technology (ICT) is supposed to be a facilitator of knowledge exchange and has been largely automated through the latest internet and web technology. Now a day's lot of organization has initiated to clinch human resource automation and computerization as it has a widespread benefit to the organization. E-HRM not only automates the HR function but provides a lot of leverage to the business by being effective and efficient.

E-HRM is the, in reality, a strategic plan to outshine the business in the computerization era, by interlinking HR function along with the organizational strategy of the establishment and create congruence between business strategy and human resource policies which jointly enhance the attainment of bottom line, competitiveness, and performance. Strategic HRM helps the organization to achieve strategic fit with its market environment. E-HRM facilitates HR function in liberating itself from daily operational, monotonous requirements, and making it more in tune with the decision making, understanding of the employees, preparing the change-ready workforce. E-HRM is a big contributor and facilitator in employer brand building. As a strategic facilitator, E-HRM standardizes the service delivery, reduces administrative burden, and improves the competence of the employee. Alignment of HR policies/practices with the business strategy of the organization is one of the major characteristics of E-HRM. E-HRM attributes, tools, mechanism, the system is a big game changer and pathfinder in providing the strategic capability to HRM function if implemented properly.

4.2 BACKGROUND

1. E-HRM as Strategic Enabler: As per Sanayei and Mirzaei (2008), "technology-enabled model of HRM is frequently started as an cost-cutting program, but it quickly progressed into a foremost medium of competitive edge, and in the meantime human resource function of few organizations utilize some virtual HRM utilities, others try hard to reengineer, computerize, and incorporate almost all of their HRM activities." Walker (1992) states human resource strategies are departmental strategies like marketing, operations, or information technology strategies; however, these are unlike in the logic that they are entangled with all other strategies in the establishment. E-HRM operates as an instrument that assists human resource management in attaining strategic edge.

2. **Administrative Burden Reduction:** As per Sacht (2003), it is clear that HR must find a way to relieve itself of administrative burdens (without abdicating their administrative role) in order to maximize its contribution to business strategy. As per Wright and Dyer (2000), the HR function has had to play a more strategic role in the organization. The only way to achieve this is to relieve much of the burden of transactional human resource activities in order to free up time so that HR can concentrate on traditional and transformational HR activities. This is done either by outsourcing some of the human resource function, but what would be more relevant to this study, is to utilize information technology in the form of E-HRM. New systems have been developed that allow management and employees to manage much of their own human resource activities, such as leave application and approval, personal data changes, etc. Thus, E-HRM systems aid in freeing up time for the HR function so that there can be a greater focus on high-value strategic initiatives.

3. **Change Management:** According to Gloet and Berrell (2003), "the development of ICTs has radically changed our social and economic lives, and has had a profound effect on the way organizations are managed." According to Hitt and Brynjolfsson (1997), ICT enables firms to introduce organizational changes in the areas of re-engineering, decentralization, flexible work arrangements, outsourcing, lean production, teamwork, customer relations and it also allows firms to produce with shortened product cycles, and greater flexibility to fulfill shifting buyer preferences and in turn, these organizational changes are basic for realizing the complete benefits of ICT.

4. **Knowledge Management (KM):** As per Biesalski (2003), to begin with, the growing consideration of companies on the factor knowledge is primarily driven by the advancement of information technology. Information systems like E-HRM arrangements that network information, empower companies to induce a steady concept for their KM. According to Newman (1999), the availability of new Information and Communications Technology (ICT), particularly the World Wide Web, has been

instrumental in catalyzing the KM movement. KM may be a collection of processes that administer to create, to disseminate, and to utilize knowledge in an organization.

5. Employer Brand Image: As per the committee Rijn Van, HRM could play a significant role in strengthening the position of the public sector in the labor market. However, this required the renewal of the HR system and the use of IT. As per CedarCrestone (2013) HR Systems Survey report, with HR technology, especially business intelligence, organizations are able to retain more talent resulting in competitive advantage.

Bhatnagar (2007) sees E-HRM as providing a powerful brand identity in the external recruitment area, which would not be possible with traditional approaches. E-HRM can help in a positive way by providing the ability to access, collect, and disseminate information, giving individuals greater access to information about job opportunities, benefits, and setting expectations about an employer. As per Bondarouk et al., (2012), "Promoting an organizations employment brand often occurs via different media sources, of which corporate web-sites and social networking sites recently gained in popularity." As per Alleyne et al., (2007), HR intranet had an influence on satisfaction with the overall HR function. These internal perceptions of the organization and the HR function are also be shaped by the use of E-HRM through improved communications, better internal job application processes, and access to policies. External applicants are also likely to form impressions of the organization as they pass through the recruitment process.

6. Standardization of Practices: Lepak and Snell (1998) stated that IT could be of support when the HR function is to be integrated as different parts of the HR function are provided by different parties (managers, employees, HR professionals or even other organizations). As per Scott (2008), in case one would isolate the HR work into two wide components, namely non-transactional and transactional activities, then it is simple to conceive the transactional components being e-enabled. Components being e-enabled can be implemented within subsidiaries as a standard tool in performing HR functions. Bjorkman and Lervik (2007) put forward three dimensions of ascending levels of transfer success-implementation, internalization, and integration of diffused HR practices. According to Marler (2009), organizational goals for E-HRM investments include cost reduction through streamlining HRM operations.

4.3 MAIN FOCUS OF THE CHAPTER

The literature review reveals very little empirical study exclusively on E-HRM contributing strategically. The limited research that has been undertaken is mostly based on western countries not from emerging countries like India, which is altogether different from western countries, so there is a possibility of a different result. The present research in its endeavor identifies research gap and formulates objectives:

- To assess the present level of E-HRM attributes of strategic capability in Indian organizations.
- To analyze the difference in the level of E-HRM attributes of strategic capability based on public vis-a-vis private organizations.
- To analyze the difference in the level of E-HRM attributes of strategic capability based on manufacturing vis-a-vis services organizations.
- To examine whether there exists a significant difference in the mean value of strategic attributes among different select Indian organizations.
- To examine whether there exists a significant association between strategic attributes and strategic capability.
- To examine E-HRM strategic attributes having a significant impact on strategic capability.

1. **Research Hypothesis:** Research hypothesis has been classified as per the six major research objectives. For convenience and better understanding, these research hypotheses have been mentioned and tested with a separate heading.
2. **Research Model:** Research model developed on literature survey envisage that HRM function delivered in electronic form supported by web-based, wireless, multimedia, software, and hardware tools and instruments generates mechanism and process resulting in E-HRM attributes or power of HR function, leading to increase in strategic capability as final output. From the literature review, a total of seven E-HRM attribute has been identified, and the model has been portrayed, taking an increase in strategic capability as a dependent variable (Figure 4.1).

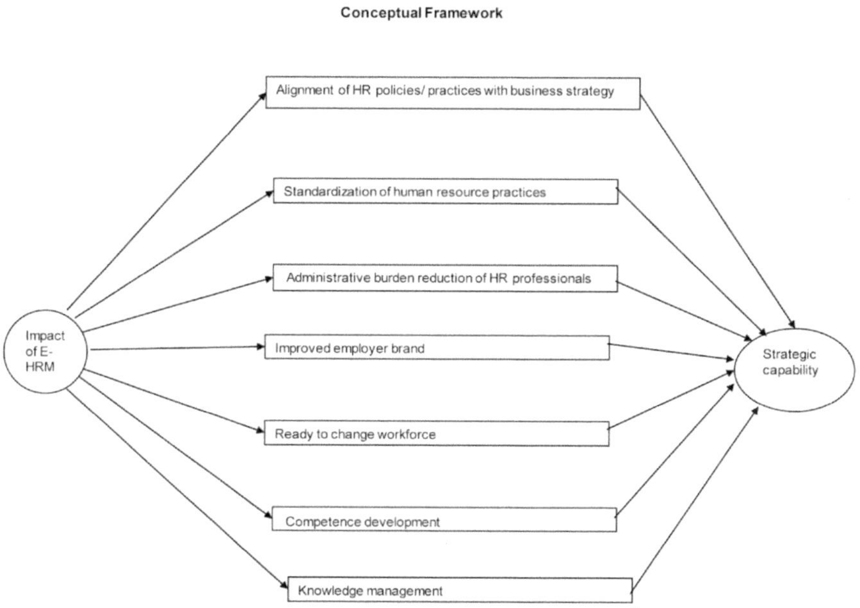

FIGURE 4.1 E-HRM strategic capability model (portrayed).

3. **Sample Unit and Sample Size:** The researcher has chosen8 Indian organizations as test organizations, and it comprises of private and public organizations both from manufacturing/mining division and services division in equal numbers (Table 4.1).

TABLE 4.1 Nature of Business

Sample Organization	Manufacturing/Mining	Services
Public	1. Power generator	1. Bank
	2. Mining	2. Insurance
Private	1. Automobile manufacturer	1. Bank
	2. Tech-manufacturing company	2. Software developer

These blue-chip organizations are having sound HR practices and market leader in that particular segment. Since primary data is being used for this study, so the identity of the organizations has been not disclosed. Target respondents are managers of these organizations.

4. **Data Collection:** Structured questionnaire with a 5-point response scale (not at all, very little, to some extent, to a great extent, very much) has been used as a research instrument. The statement of the questionnaire is as follows:

- To what extent E-HRM standardizes its service delivery to different departments, business units in your organization?
- To what extent E-HRM reduces the administrative burden of an HR professional in your organization?
- To what extent E-HRM improves the employer brand of your organization?
- To what extent E-HRM creates ready to change workforce?
- To what extent E-HRM increases the competence of employee in your organization?
- To what extent E-HRM supports KM in your organization?
- To what extent E-HRM aligns HR policies/practices with the business strategy of your organization?
- To what degree, E-HRM increases strategic capability (competitive edge) of your organization?

Guessing 33% (one out every three) as good response rate, total numbers of 485 questionnaires were sent for the survey. Out of 110 questionnaires (surveys) send online, 22 respondents filled the questionnaire and submitted, i.e., a response rate of 20%. Out of 375 questionnaires (surveys) provided to respondents in hard copy format, 156 responses were collected, i.e., a response rate of 41.6%. Total 176 responses out of a total number of surveys of 485, i.e., overall response rates of 36.3%. Out of 176 responses, 162 responses were found valid. Response having discrepancy and incomplete, not fit for further processing were discarded. Out of 162 valid responses, for compatibility and convenience in data analysis and hypothesis testing, 20 responses from each organization (Total: 20*8 = 160) were entered for final statistical analysis and interpretation.

4.4 SOLUTIONS AND RECOMMENDATIONS

1. **Strategic Attributes Value:** To test the hypothesis C1 (There is no significant difference in present mean value of strategic attributes of

E-HRM and test value.), one sample t-test has been used applying a hypothesized test value of 3.5, assuming that a mean value of 3.5 will offer support to the null hypothesis C1 (Table 4.2).

TABLE 4.2 One-Sample T-Test of Strategic Attributes

Attributes	Test Value = 3.5 Degree of Freedom =159		
	Mean	T	Sig. (2-Tailed)
Standardization of HR practices	3.6375	1.639	.103
Administrative burden reduction	3.8562	4.311	.000*
Employer brand (improves)	3.7250	2.803	.006*
Ready to change workforce	3.6188	1.489	.138
Competence of employee (increases)	3.6812	2.252	.026*
Knowledge management (supports)	3.7125	2.690	.008*
Alignment of HR practices with business strategy	3.8875	4.741	.000*
Strategic capability (increases)	4.0250	7.523	.000*

[Significance level with*- Null hypothesis rejected or else accepted (DF 159, sig 5%)].

Administrative burden reduction, Employer brand (improves), Competence of employee (increases), KM (supports), Alignment of HR practices with business strategy, Strategic capability (increases) significance level is below assumed significance level of 0.05 indicates that there is a significant difference between the test value and the observed mean value hence null hypothesis is rejected. A significance level of the strategic attribute, Standardization of HR practices, and Ready to change workforce is above the significance level of 0.05, hence null hypothesis is accepted. Over all null hypotheses, C1 is partially rejected.

2. **Strategic Capability (Public vis-a-vis Private):** To test the hypothesis C2 (There is no significant difference in the mean value of strategic attributes of E-HRM between private and public organizations), paired sample t-test has been used as a statistical tool (Table 4.3).

TABLE 4.3 Paired Sample T-Test of Strategic Attributes (Private vis-a-vis Public)

Pair	Attributes	Mean Private	Mean Public	Mean Difference	T	Sign.
1	Standardization of HR practices	3.912	3.387	.5250	2.952	.004*
2	Administrative burden reduction	3.925	3.812	.1125	.725	.470
3	Employer brand (improves)	3.750	3.687	.0625	.419	.676
4	Ready to change workforce	3.637	3.600	.0375	.253	.801
5	Competence of employee (increases)	3.737	3.637	.1000	.611	.543
6	Knowledge management (supports)	3.712	3.700	.0125	.078	.938
7	Alignment of HR practices with business strategy	4.175	3.600	.5750	3.956	.000*
8	Strategic capability (increases)	4.287	3.762	.5250	3.749	.000*

[Significance level with * – Null hypothesis rejected or else accepted (DF 79, sig 5%)].

For null hypothesis-C2 for strategic attributes standardization of HR practices, alignment of HR practices with business strategy is rejected as the significance level is below the assumed significance level of 0.05. So for all these strategic attributes is not same for private and public organizations. Null hypothesis-C2 for the strategic attribute, Administrative burden reduction, Employer brand (improves), Ready to change the workforce, Competence of employee (increases), KM (supports) is accepted as the significance level is above the assumed significance level of 0.05 hence, and these attributes are same for private and public organizations. Overall null hypothesis C2 is partially rejected.

3. Strategic Capability (Manufacturing vis-a-vis Service): To test the hypothesis C3 (There is no significant difference in the mean value of strategic attributes of E-HRM between manufacturing/mining and services organizations), paired sample t-test has been used as a statistical tool (Table 4.4).

TABLE 4.4 Paired Sample T-Test of Strategic Attributes (Manufacturing/Mining vis-a-vis Service)

Pair	Attributes	Mean Mfg/Min	Mean Service	Mean Diff.	T	Sign
1	Standardization of HR practices	3.512	3.787	–2750	–1.697	.094
2	Administrative burden reduction	3.812	3.925	–1125	–773	.442
3	Employer brand (improves)	3.500	3.937	–4375	–2.806	.006*
4	Ready to change workforce	3.462	3.775	–3125	–2.062	.043*
5	Competence of employee (increases)	3.587	3.787	–2000	–1.321	.190
6	Knowledge management (supports)	3.450	3.962	–5125	–3.494	.001*
7	Alignment of HR practices with business strategy	3.837	3.937	–1000	–638	.525
8	Strategic capability (increases)	3.912	4.137	–2250	–1.738	.086

[Significance level with*- Null hypothesis rejected or else accepted (DF 79, sig 5%)].

For null hypothesis C3 for strategic attributes, Employer brand (improves), Ready to change the workforce, KM (supports) and Strategic capability (increases) is rejected as the significance level is below the assumed significance level of 0.05, so all these strategic attributes is not same for manufacturing/mining and services sector. Null hypothesis C3 for strategic attributes, Standardization of HR practices, Competence of employee (increases), Alignment of HR practices with business strategy, Strategic capability (increases) is accepted as significance level is above the assumed significance level of 0.05, hence these attributes are not significantly different for manufacturing/mining and services sector. Overall null hypothesis C3 is partially rejected.

4. Strategic Capability (Organization vis-a-vis Organization): To test the hypothesis C4 (There is no significant difference in the mean value of strategic attributes among different select Indian organizations), one way ANOVA has been used as a statistical tool (Table 4.5).

TABLE 4.5 One Way ANOVA of Strategic Attributes (Organization vis-a-vis Organization)

ATT	Mean (Organization Wise)								Mean	F	Sig.
	CIL	HCL	ICICI	LIC	MB	NTPC	SBI	TM			
SZ	2.5	4.15	3.6	4	3.8	3.65	3.3	4.1	3.63	6.43	.000*
AB	2.9	3.95	3.65	4	4	4.2	4.15	4.1	3.87	3.65	.001*
EB	2.5	3.55	4.2	4	3.7	4.25	4.05	3.55	3.72	8.19	.000*
RC	3	3.6	3.8	4.15	3.7	3.7	3.55	3.45	3.62	2.20	.037*
CT	2.8	3.75	3.8	3.95	3.7	4.15	3.6	3.7	3.68	3.33	.003*
KM	2.9	4.2	3.9	4.1	3.7	4.15	3.7	3.05	3.71	5.98	.000*
Al	2.8	4.05	4.35	3.9	4.1	4.25	3.45	4.2	3.89	6.20	.000*
SC	3.1	4.5	4.15	4.1	4.2	4.05	3.8	4.3	4.02	5.54	.000*

[Significance level with* – Null hypothesis rejected or else accepted (DF 7/152, sig 5%)].

[ATT – Attribute, SZ – Standardization of HR practices, AB – Administrative burden reduction, EB – Employer brand (improves), RC – Ready to change the workforce, CT – Competence of the employee (increases), KM – Knowledge management (supports), Al – Alignment of HR practices with business strategy, SC – Strategic capability (increases)]

For significance level of all strategic attributes of E-HRM Standardization of HR practices, Administrative burden reduction, Employer brand (improves), Ready to change workforce, Competence of employee (increases), KM (supports), Alignment of HR practices with business strategy, Strategic capability (increases) is below the assumed significance level of 0.05 hence null hypothesis is rejected and it can be said that these strategic attributes are different in Indian organizations hence null hypothesis (H0) is C4 rejected.

5. **Strategic Attributes Association:** To test the hypothesis C5 (There exist no association between strategic attributes and strategic capability), correlations analysis has been used as a statistical tool (Table 4.6).

TABLE 4.6 Correlations Coefficient of Strategic Attributes

ATT	SZ	AB	EB	RC	CT	K M	Al	SC
SZ	1							
AB	0.776 ** (00.00)	1						
EB	0.596** (00.00)	0.736** (00.00)	1					
RC	0.569** (00.00)	0.593** (00.00)	683** (00.00)	1				
CT	0.632** (00.00)	0.680** (00.00)	0.767** (00.00)	0.683** (00.00)	1			
KM	0.690** (00.00)	0.659** (00.00)	0.684** (00.00)	0.627** (00.00)	0.763** (00.00)	1		
Al	0.691" (00.00)	0.682** (00.00)	0.731** (00.00)	0.604** (00.00)	0.695** (00.00)	0.748** (00.00)	1	
SC	0.742** (00.00)	0.741** (00.00)	0.709** (00.00)	0.590** (00.00)	0.751** (00.00)	0.764** (00.00)	0.802** (00.00)	1

[** Statistically significant at 0.05 level (2-tailed)]

There exists a significant relationship between all the strategic attributes and strategic capability at 0.05 significant levels; hence, null hypothesis C5 is rejected.

6. Strategic Attributes Impact: To test the hypothesis C6 (Strategic attributes of E-HRM have no significant impact on the strategic capability of Indian organizations), regression has been used as a statistical tool (Table 4.7).

TABLE 4.7 Regression Coefficient (Dependent Variable: Strategic Capability)

Attributes	Unstandardized Coefficients		T	Sig.
	B	Beta		
Constant	0.808		5.255	0.000*
Standardization of HR practices	0.142	0.171	2.456	0.015*
Administrative burden reduction	0.131	0.151	2.040	0.043*
Employer brand (improves)	0.018	0.020	0.271	0.787

TABLE 4.7 *(Continued)*

Attributes	Unstandardized Coefficients		T	Sig.
	B	Beta		
Ready to change workforce	–0.048	–0.055	–0.944	0.346
Competence of employee (increases)	0.181	0.209	2.854	0.005*
Knowledge management (supports)	0.136	0.154	2.161	0.032*
Alignment of HR practices with business strategy	0.290	0.339	4.899	0.000*

[Significance level with*-Null hypothesis rejected or else accepted, D.F. 159, sig 5%]

For attributes Employer brand (improves), Ready to change workforce is not statistically significant and doesn't fit in the module. As per statistics, a tentative module of attributes having a significant impact on strategic capability comprises of 0.808 (Constant) +0.142 (Standardization of HR practices) +0.131 (Administrative burden reduction) +0.181 (Competence of employee) +0.136 (KM) +0.290 (Alignment of HR practices with business strategy). Constant is statistically significant as its value is 0.00 below 0.05. It can be said that five attributes have a significant impact on strategic capability; hence, null hypothesis C6 is partially rejected.

4.5 CONCLUSION

Result shows that present mean value of strategic attributes "Administrative burden reduction," "Employer brand," "Competence management," "KM," "Alignment of HR practices," "Strategic capability" is significantly high but mean value of strategic attributes, "Standardization of HR practices," and "Ready to change workforce" is not significantly different to mean test value, hence it becomes evident that E-HRM had not been able to standardize HR practices and even not capable to prepare ready to change workforce in Indian context. As these two attribute are important, so Indian organizations should make changes in their approach, system, and hardware and software so these two could be inculcated and could provide maximum benefit to the Indian organizations. It is obvious from the study that level of E-HRM strategic attributes is not significantly different in public and private organizations except for the attributes "Standardization of HR practices," "Alignment of HR practices with business strategy" and "Strategic capability." Mean values with significant difference indicate private organizations are ahead of public organization for strategic attributes "Standardization of HR practices," "Alignment of HR practices with business strategy" and "Strategic capability."

Similarly, the level of E-HRM strategic attributes is not significantly different in manufacturing/mining and services organizations except for the attributes "Employer brand," "Ready to change the workforce," "KM" and "Strategic capability." Mean values with a significant difference are high in the service sector than manufacturing/mining sector. Research statistics show that there exit organization wise differences in mean values of strategic attributes, so it can be said that the presence of strategic attributes are not uniform in Indian organizations. It is evident from research statistics that there exists an association between all the strategic attributes and strategic capability. Result indicates five attributes or independent variables "Standardization of HR practices," "Administrative burden reduction," "Competence of employee," "KM," "Alignment of HR practices with business strategy" fits in regression model and shows a significant cause and effect relationship but independent variables like "Employer brand," "Ready to change workforce, had been omitted. "Employer brand," "Ready to change workforce," were not good predictors for dependent variable strategic capability. Out of seven parameters, five have a significant impact on strategic capability.

The HR function cannot become strategic just by the digitization of HR function, but it has to be reengineered and aligned with strategies of the organization, then skillful application of information technology can pay dividends. HR Professionals has to take note of this aspect of the HR function. The hallmark of E-HRM decentralization, harmonization, and standardization should be taken into account while developing E-HRM architecture. Culture and language can be an impediment for a multinational company. E-HRM is seen as an important source of strategic capability building. E-HRM is supposed to shift the focus of HR function from employee welfare to strategic contributor by reorienting human capital, social capital, organizational capital, intellectual capital and be a facilitator of KM, and at the end provider of competitive edge to the organization. Now the responsibility lies with the implementer of E-HRM to make it capable of providing a strategic edge; otherwise, it will become just a management fad. E-HRM tools, mechanism, system, the process should be so designed that it should be compatible with the hardware and software of the organization. There must be a change in the approach of the end-users, and E-HRM must be treated as a competence of the organization.

KEYWORDS

- **bottom line**
- **E-HRM**
- **electronic**
- **regression**
- **strategic capability**
- **T-test**
- **transactional**
- **transformation**

REFERENCES

Alleyne, C., Kakabadse, A., & Kakabadse, N., (2007). Using the HR intranet: An exploratory analysis of its impact on managerial satisfaction with the HR function. *Personnel Review, 36*(2), 295–310.

Bhatnagar, J., (2007). Looking from the organizational learning lens at technology enabled HR in Indian organizations. *International Journal of Human Resource Development and Management, 7*(1), 37–52.

Bieasalski, E., (2003). *Knowledge Management and E-HRM.* Retrieved from: https://km.aifb.kit.edu/ws/LLWA/fgwm/Resources/FGWM03_08_Ernst_Biesalski.pdf (Accessed on 28 August 2019).

Björkman, I., & Lervik, J. E., (2007). Transferring HR practices within multinational corporations. *Human Resource Management Journal, 17*(4), 320–335.

Bondarouk, T., Ruël, H., &Weekhout, W., (2012). *Employer Branding and its Effect on Organizational Attractiveness via the World Wide Web: Results of Quantitative and Qualitative Studies Combined.* Paper presented at the 4th International E-HRM Conference, Innovation, Creativity and E-HRM, Nottingham Trent University, UK. Retrieved from https://research.utwente.nl/en/publications/employer-branding-and-its-effect-on-organizational-attractiveness (accessed on 25 august 2018)

CedarCrestone., (2013). *CedarCrestone 2013–2014 HR Systems Survey: HR Technologies, Service Delivery Approaches and Metrics* (16th Annual Edition). Retrieved from:

http://www.cedarcrestone.com/media/whitepapers/CedarCrestone_2013-HRSS-HRTech-100713.pdf (Accessed on 28 August 2019).
Gloet, M., & Berrell, M., (2003). The dual paradigm nature of knowledge management: Implications for achieving quality outcomes in human resource management. *Journal of Knowledge Management, 7*(1), 78–89.
Hitt, L., & Brynjolfsson, E., (1997). Information technology and internal firm organization: An exploratory analysis. *Journal of Management Information Systems, 14*, 77–99.
Lepak, D., & Snell, S., (1998). Virtual HR: Strategic human resource management in the 21st century. *Human Resource Management Review, 8*(3), 215–234.
Marler, J. H., (2009). Making human resource strategic by going to the net: Reality or myth. *The International Journal of Human Resource Management, 20*(3), 515–527.
Newman, K. W. C., (1999). *A Framework for Characterizing Knowledge Management Methods, Practices, and Technologies*. The Introduction to Knowledge Management, George Washington University Course EMGT 298.T1. Retrieved from https://www.semanticscholar.org/paper/A-Framework-for-Characterizing-Knowledge-Management-Newman-Conrad/00a65b1c1c4d3930ee315d43301cfc1ffe33ad1c?citationIntent=methodology#citing-papers (accessed on 25 august 2018).
Sacht, J., (2003). *E-HR Strategy: An Electronic Human Resource Strategy is Attainable by Small and Medium Sized Business*. Retrieved from: http://www.workinfo.com/newsletter/newsletters/vol2no9june2003print.htm (Accessed on 28 August 2019).
Sanayei, A., & Mirzaei, A., (2008). Designing a model for evaluating the effectiveness of E-HRM (case study: Iranian organizations). *International Journal of Information Science and Technology, 6*(2), 79–98.
Scott, R., (2008). *HR Technology Needs to be Thought of as Strategic*. Retrieved from: http://robertscott.wordpress.com/2008/10/09/hr-technology-needs-to-be-thought-ofas-strategic (Accessed on 28 August 2019).
Verma, S., & Gopal, R., (2010). *The Implications of Implementing Electronic Human Resource Management (E-HRM) Systems in Companies* (pp. 168–169). PHD thesis submitted to Dr. D. Y. Patil University, Department of Business Management, Mumbai.
Walker, J. W., (1992). *Human Resource Strategy* (p. 378). New York: McGraw Hill, ISBN 0 07 067846 4.
Wright, P., & Dyer, L., (2000). *People in E-Business: New Challenges, New Solutions*. Working paper- Center for Advanced Human Resource Studies, Cornell University, Retrieved from: http://digitalcommons.ilr.cornell.edu/cgi/viewcontent.cgi?article=1088&context=cahrswp (Accessed on 28 August 2019).

CHAPTER 5

3Cs of International Market Penetration Through IMC: Culture, Change, and Connect

JYOTI KUKREJA

Jagannath International Management School, Management, MOR, Pocket-105, New Delhi – 110019, India,
E-mail: jyoti.kukreja@jagannath.org

ABSTRACT

When change usually occurs, things fall apart but the culture to connect for any marketing strategist is pertinent. It is an untold story that marketers not only create needs but to create trends. In most cases, they make customers more conscious and keenly aware of unfelt or unrecognized needs. The fact that many new products in the market become successful is due to the fact that their marketers undertake to fabricate and at times, mold facts. These are a few of the promising phenomenon for being a successful and progressive marketer. This piece of writing also includes a case let and a case study on the marketing strategies for successful global industry pie. Technology is not a choice but more of a necessity nowadays. Technology-led marketing or marketing technology had been one of the few costly questions to ignore for the marketers. Technology: makes or breaks marketers.

5.1 INTRODUCTION

Many of the existing products or brands tend to amiss to notice the unmet needs and thus are the deficit in creating a need or demand for the new product. Embracing change and innovations is what all organizations aspire for, but unfortunately, only a few are prepared for. An excellence in communication can help to overcome the said hindrances. Two innovative examples deserve to be mentioned here. Indians are known to have not a sweet tooth but sweet teeth. The country had been ranked during the 1990s among the top ten diabetic countries in the world. It was then that the sugar-free tablets were introduced as a replacement of sugar which was very carefully targeted at diabetic patients. Both diabetic patients and health-enthusiasts bought to experiment with the effect. Second illustrative example is that of deodorants and perfumes. The middle-class section of the society was just burgeoning during this time of the 1990s when millennial were growing up and who had been trained to be into discipline. During this time, the women emancipation movement in the economy had commenced. Fashion was the new feature as more television screens were reaching the households and not only good dress, but fragrances too came on the shopping list of family budgets through Odonil and talcum powders and others. Soaps switched from being a cleansing agent to beauty and fragrance bars. Gender segregation was introduced in the economy through P&G and HUL. Different kinds of toothpaste for children too came up. By and large, this bifurcation initiated the art of segmenting and targeting customers.

Marketing campaigns that become successful and memorable are usually the ones which echo to overcome dissonance with the existing products and brands in the market. Market-oriented organizations design and deploy rigorous activities under consumer behavior and thoroughly research consumer psychology. Not only this, they studied product attributes and weighed them objectively against their own offering. The marketing mix of the firm is consistently worked out to squeeze profits with immediate effects. An extended range of products to price to advertisements to payment modes all needs to be well planned for many of the organizations begin with identifying needs, distinguishing between met and unmet needs, immediate, and deferred needs. Further, these spend their thoughts on criteria such as relevant and irrelevant needs, recurring, and nonrecurring needs. An in-depth study is then undertaken depending on the brand cycle and product life cycle, if those needs could be translated as 'wants.' Could these wants be made premium? If yes, then these wants have further potential to be considered and converted as 'demands.'

Many times, the needs are generated, they don't exist, which are also subjected to the extent of being superficial if it is overly stressed or easily accessed by the source from which the needs emanate. The business environment is dynamic and strategizing to successfully manage change is more than important. All marketers in all industries alike are diced a split-challenge to either change market or tweaking marketing strategies. Fetching business on a global level means that the organization is ambitious for growth which follows a complex amalgamation of processes converging into one focal point of market share. Global marketing requires excellence in leadership, communication, and strategic skills. Working across the global contours requires ambidextrous hands-on operating in the most Volatile, Uncertain, Complex, and Ambiguous (VUCA) environment. The business world is dynamic and fluctuates at the speed of thought. Preferences and priorities differ from buyers to investors.

Marketing is juxtaposed next to the concept of celebration, and it is glamour that makes people and events memorable. The big fat Indian weddings are swelling in their budgets year on year that entices global marketers to consider it an opportunity. Although nurture of culture, language, weather, technology all impact the mindset of citizens of a particular region yet impulses, value system, zones of tolerance, perceptions, ego, and attitudes make a difference.

The lines and zones of control can make or break a business, especially in today's time where technology has made everyone powerful and random. Be it expensive vacations or fat-free goods, and in addition to technology, words make an indelible impact. Further, it is a language that etches into the memory and grinds the logic. Interests and preferences are framed with a choice of language. When nestle came to Asia, it communicated its brand strength by deploying the local language. The ready-to-cook Maggi became an instant delight to not only pallets of children but the then, burgeoning segment of working women. This format of instant cooking received an overwhelming response since it made lives easier and meals tastier. Advertising is that pinch of the indispensable tool of marketing that gives maximum results. Maggi broke the ice by being digital on television screens talking in the native language. The advertisement 'Mummy mummy bhook lagi, Maggi chaiye mujhe abhi' (Mom, I am hungry, and I want you to serve me Maggi instantly). The choice of dipping the jingle in the regional language of the country: Hindi had been a strategic approach of Nestle. To comprehensively discuss the fast-food category, even PepsiCo and coke introduced their slim versions in the form of diet Pepsi and diet coke.

For a country that is struggling to be one with the western culture, to accept, adapt, and adopt is simultaneously aspiring for its native culture to thrive, bloom, and prosper. In Asia, especially India is known for its secularism. 'Unity in diversity' is no longer a myth. It's not about being in difference but agreement to differences that have blossomed the concept of multiculturalism.

5.2 BACKGROUND

Philip J. Kitchen, in one of his studies in 2010, studied the concept of Integrated Marketing Communication (IMC) and cited the benefits and barriers to its implications. IMC is a lucrative concept which stimulates synergy that makes an organization flourish financially. The impediments range from the holistic perspective into collaborating efforts of all the powerful tools of advertising, distribution channels while keeping in mind target customers. Further, he identified issues such as vague positioning, lack of shrewd vision, and grudges towards acceptance to the stimulus of change as those barriers which hamper change (Kitchen and Burgmann, 2010). Marketing communication develops a brand by bridging the gap between an organization and its target audience, thus inducing information procession activities of stakeholder (Blythe, 2006; Ouwersloot and Duncan, 2008). Interaction enhances the communication process, which is achieved by the influence of customers (Steward and Pvalou, 2002). Nowadays, the internet proves to be a great interactive medium which is globally accepted as a medium for communication (Janoschka, 2004). The growth of internet users is 7% year-on-year, and in 2018 it is 4.021 billion (Kemp, 2018).

Promotion mix for the organizations has gone phenomenally digital, and communication in the form of online marketing has grown by leaps and bound in the form of email, blogs, and videos for usage and selling (Adegoke, 2004). The messages are structured and crafted in a way that educates consumers and or potential buyers to enable them to either appreciate or make an informed decision (Halloway and Robinson, 1998). Tian and Borges (2011) describe cross-cultural marketing as communication between marketers and consumers who differ in education, regions, religions, language, values, social background and norms, and lifestyle. Oosthuizen (2004) identifies three of the most crucial aspects that act as enablers to the otherwise complex process of IMC: hitting the bull's eye to needs and wants of target buyers, dialect or the parlance among the potential subscribers to the products or service to be made accessible, identifying grounds of empathy, emotions, knowledge, and sign language. Emotions are no one's property, and experiences are personal and are common to human beings irrespective of their continent origins; however, emotional response patterns and attachments differ in the context of cultural-setting (Aaker and Maheswaran, 1997; Tian and Borges, 2011; Orth et al., 2007).

5.3 MAIN FOCUS OF THE CHAPTER

Management is a concept that unifies varied threads and dovetails into one whether it's Cadbury 'Celebrations' on Raksha Bandhan, the Indian festival where sister ties a string of thread on the wrist of her brother requesting defense from all enemies and a promise of support in times of hardships or the festival of lights and sweets called Diwali where Goddess of Wealth, Lakshmi is worshipped. These are few of the annual events that India celebrates, and there are various event campaigns that are organized and managed by the brands. Mother's day is one such celebration that started with preparatory schools. These events with emotional tangent were a huge hit as it was based on the theme of women empowerment and secondly supporting the 'Save Girl Child' campaign. India had been a parochial minded patriarchal society and female-feticide had been much prevalent malpractice. Marketing is not set then to change buying behavior but also the psychology of people. Thus the tripod of taking up the challenge of changing the culture of people is crucial. Archie's changed the way people exchanged birthday greetings. It had been successful in capturing emotions and all occasions of inventing a greeting card for all relations, whether of saying sorry, thanksgiving, anniversary, etc. Words have a soothing power and leave an indelible mark, but this cohort is limited to teenage, and when they escalate across age, they hardly keep up with this habit of cards. Thus, Archie's stretched its portfolio from a birthday greeting to birthday gifts which changed the thinking style and greeting etiquette of the educated layer of the society.

Seth Godin remarked, 'Marketing is a contest for attention.' This is because in today's world, where there are ample options weighing features against each other is a time-consuming process. It's not enough to seek attention but arrest the focus. The question that lurches is: gaining and enriching attention and focus easy, fruitful, and impactful? This calls for a transformation. For every new product, there is always a willy-nilly reaction. An impulse of 'yes, go for it' to transcendence to that decision of 'I want it for a lifetime' goes for marketing and after-sale services. With that one transaction becoming a successful and immediate trigger to the transformation of making that one-time need-driven buyer into a repeated lifetime customer calls for the strength of marketing.

1. **Experiential Technology:** A marked shift was marked when during 1970s Upton entered the Indian market into the television category and the government of the nation came up with a channel known as Doordarshan. The serials to watch were no less than a family gets together occasion. All residents of the block used to view it together. Five decades later, 3D screens and home theatres have been introduced into the market. Watching a movie in 3D and now 7D makes the viewer experience technology in a different format.
2. **Brand Addiction:** Getting habitual to a product is great, but even greater is the addiction to brand. It's like a gift to any marketer. If the customer buys nothing but only your brand, it is a matter of huge brand success. This further entails that a customer is not only addicted to the product but the brand the reason for which could be more than many. Taste, price, attributes, or loyalty points: the ranking may vary person to person and from one product category to another. To decipher what forms as the criteria for selection is difficult to scrutinize.
3. **Reactive or Proactive:** Successful marketing lies not in being reactive but proactive to the changing needs of the customers and competitors strategies.
4. **Ideapreneurs:** The evolution in telecom industry emerged into a new platform from landlines to wireless connections, from notes taking to SMS, Emails, and SnapChat. This digital model has changed the way youth feels and decides. The product doesn't change, but the idea that wraps around it does; therefore, value delivered changes. The Indian market has evolved from 'aur kya milega' (what more will I get) to 'kitna aur milega' (how much more will I get). Inspiring and experience-centric example to be cited here is of Spacex. Struggling for vacation venues got

a lot easier with technology where reservations had been facilitated by the virtual world and digital payment modes. Booking rooms in hotels and seats in theatres had been quite tedious a job in summer holidays for parents. In the recent past, there had been a shift from roadways, railways to airways and then water holidays in a yacht. Though this trend of exotic vacations had taken a full swing and there had been phenomenal changes in the aspirations of passengers, the out of the box thinking is still not out of fashion. Vacation designers thought of delivering a stark contrast to an improvement from amusement parks to amusing, exciting, and innovative, thrilling experience of vacations in space. Experiential marketing is one of the most abstruse but most profitable forms. A single ticket of Spacex costs INR 25 lac which is a whopping amount for any upper-middle class, but this bizarre sort of vision has created space for itself not only in the hotspots but in the wish list of many people to aspire for at least once in a lifetime experience.

5. **Social Wealth:** Humanities and emotional drives go hand in hand. Twitter and Facebook became instant hits the digitally active generation. 2G has moved to 4G in 2016 introduced by Reliance Jio after the economic breakdown due to demonetization. Not only is the identity of the individual is built and recognized through these platforms, but the reality check with recruitment decisions are influenced and filtered through LinkedIn, Monster.com, and such websites. These websites weave a story for brand experience in the form of blog writing that covers movie to a book review to employer and employee posts. Just as any developing economy that is heavily reliant on family ties and social rituals, marketers who can capture the business eye on these, succeed. The number of contacts in the phone book and social networking sites actually claim for contacts, associations, and wealth or better said net-worth of an individual.

6. **Fashion Wealth:** Consumerism is the next market phenomenon that has hit the Indian market. Accessories and cosmetics is no more a minority market, but several courses such as dress designing and hair-cuts and nail art and nail spa have become beauty referral zones to look in shape and in fashion. The art to put different pieces together is crucial to be in vogue. Customer endorsement had been the new theme picking up with dove being pioneers into the concept of using the product and clicking a selfie and uploading on the website made the marginalized section of the society a boost in their self-image and further implementation of the concept of women empowerment.

7. Lifestyle Marketing: Fitness centers and spa are now a matter of keeping health and life in style. Brands such as Patanjali and Dabur are known to keep health in check. This pendulum started to oscillate with Kellogg's and Saffola other such brands. Cornflakes and oats took over the breakfast routine of the middle age generation, and soon it penetrated to the ambitious people who valued health with wealth. It has been observed that health has major implications for people. Without health, more high life spans will be like a cake without creams. Thus to have life, health marketing became a norm. The aids into backache and other ailments such as steam pressure, foot spa all are into marketing life in lifestyle.

8. Snowball Marketing: Marketers understood here that Asians especially Indians who are known to have extended kith and kin family formats could be explored to market once for the product, but the association could be long-tenure fetching given the strength and width of relationships.

9. Viral Marketing/Word of Mouth Marketing: MNCs in India became a much sought after brand not only because of the rich hygienic processes while manufacturing, but with consistency in quality delivery. Unlike many of the domestic brands that were continual and monopolistic in the market structure, these foreign brands delivered quality in no time.

10. Internal Marketing: Many-a-times employees are nurtured in organizations with certain terms and conditions of serving the departments and customers. Based on the premise that employees are as important as customers, with not only their training and development but also their development and growth plans mapped across months, insurance for them and their family members had been adopted as a key practice by HR Department. The first marketing of organizational values, psychology, and culture is first done to employees and then customers.

11. Diffusion Process: Any new product remains in the spotlight but is diffused to be discussed among customers through the marketing strategies. This process varies whether the product is perishable or not, and if it is a high involvement category. The various adopter categories range from early-bird registrations to laggards. Thus, diffusion happens quickly if the communication is credible and creatively packaged. Personal influence also exercises a significant proportion to the mind and heart share zones. The product had been bought with a particular sphere of influence and the personal experience and thought-process molds perception of the brand.

12. Reference Group Influence: Technology is an interception over financial and human preparation for adaptation. Creativity, strategizing, and media all count for marktech. The challenge of overriding mental blocks being navigated as cultural impediments and making people scientifically credible is more than difficult. It is that technology influences a purchase decision when a relative or a contact had been subscribing to a particular brand and had built a desirable memory.

To earn goodwill in the market, we need to practice the principle of being consistently quality conscious. This often mentioned. But the thumbnail to this principle is not only attracting customers but credits to credibility. It is known that marketing is more about how you make a sale than how you make a product. It is more than significant to strike a difference in every product, whether it is a catalog of durable commodities or menu card of non-durable items. To fetch a sale, you must tell a tale. It is difficult to earmark a difference when the category is as compound as FMCG in which every item has more or less same features and fulfills broadly the same purpose. The relevance of communication skills in this category or for that matter, any category cannot be discounted. Marketing operates on the premise of the need to build change in the mindset of people and their buying-decision-making processes. What influences the decision and what not, which factor positively influences and what causes an aversion, what is considered to be urgent and important and what is that criteria that are overlooked and compromised are the factor sieved out through successful market research.

A universal truth that cannot be ruled out is that organizations, whether with their scale of operations in India or abroad, suffer from language barriers. However, translation can be a rescue to the problem in the short term, but the comfort of a talk and building a repertoire is slightly missing. In a world of brands where everything seems promised aiming to give a feeling of being secure, relationship-based marketing had been one of the significant shifts in the domain. Brands had been able to navigate and penetrate aligning with the schema of culture-based marketing. There had been another movement marked from a push strategy to pull strategy, meaning thereby, if products are sold as commodities, the indispensability of it is not marked but if it is sold as a solution after identifying the unmet needs gap or as a problem foreseen in future, the connect with the product is incontinent. Many of the sellers who have not been able to build a market share under the umbrella company picked upon the habit of making people buy not only as a stop-gap arrangement but as a habit. Selling is both a science and an art now underlying with a new philosophy of not to make a sale but relation.

5.3.1 CASE STUDY: FROOTI GETS FIZZY, PEPSI GETS PEPPY

GST made a very loud and bottom rocking impact on all industries and business alike but most notable was in the F&B industry. Both Frooti and Pepsi came with the changes in their product manufacturing strategies one after the other. There had been two-fold reasons for this change from the perspective of sugar drink getting fizzy: an increase in tax on sugar made a decision to continue with the traditional format of the product a little sour for Frooti and hence, a change was passed by the board of brand management. The secondary reason that supported this change was that India had become the second-largest diabetic country in the world just next after China. From a healthcare perspective to a pocket-friendly product from the eyes of manufacturers, the decision lay in favor in the laps of Frooti.

Pepsi too, on the same stream, thought of picking up a route on the road less traveled and implemented its thought of making the product a little peppy for its Indian consumers. Being an artificially flavored sugar drink, Pepsi too wants to ensconce itself in the segment of health-friendly product category. Moreover, Indian palettes are fonder of spices so, this idea could be a better move. It seems that these decisions have not been taken in the silos but from a very practical and grounded realities viewpoint. Parle Agro, this Asian giant is known worldwide for its massive production in the food and beverages sector. The manufacturing, packaging, and promotion of the mosaic of the product portfolio including of biscuits and recently acquired beverages brand Appy Fizz is known to do well in the market. The company operates under three business verticals:

- Beverages (fruit drinks, nectars, sparkling drinks, and carbonated soft drink);
- Packaged drinking water and foods; and
- PET packages, bottles, and containers.

Each of these verticals operates as independent entities. Under the penetration strategy, they have also set up an exclusive division for beverages in Returnable Glass Bottles (RGB) format. As part of the infrastructure, the company has 76 manufacturing facilities in India and overseas. For catering more than 6 lakhs outlet in India only, the company has 3500 distributers. Business Statement of the company is "We are in the business of refreshing India with our products, refreshing the market with new categories and refreshing ourselves through innovation." Vision of the company is "To be the leaders in business. We will stand apart from the competition by being the first in the market to innovate." The organization is reigning for its massive market share on the basis of rich experience and expertise of its staff, the commitment to maintain and escalate quality, commitment to customers, and by being competitive and innovative.

India did not see PepsiCo undergoing such exercise of brand establishment as when it entered the Indian market much of its quality-credibility had already been recognized. The two brands are not exactly alike, but similar as Frooti focuses on kids, Pepsi targets college students. 'Youngistaan' had been the tag line with Ranbir Kapoor, Bollywood actor, signed in as the brand ambassador. After much criticism that Frooti had been high in its sugar quantum and Pepsi need to change its composition from artificial flavors to serving the spicy-fond Indian palettes both the companies manifested and dived into the right stride with GST (Goods and Services Tax) that made an impact on both sugar and soda and spices. A change was inevitable, but it was much deployed for advantage (Parle Agro, 2017).

5.3.2 CASE STUDY: VICKS DOES IT IN A COOL WAY!

Vicks, a known producer in the industry of pharmaceuticals, is a global aspirant. It is popular for the cure for cold and cough. Known for Nimulid tablet for fever, Cetrizet for vitamin D, Vicks inhaler and a few more, the brand had been hiring senior and experienced applicants as it was new in the year 1970s when it was eager to ensconce a name among people. The idea was to ensure trust among people.

1. **Introduction to the Case:** Vicks as a brand is working on the concept of social inclusion. Transgender who are still marginalized in our otherwise robust democratic political structure are denied the human rights in general. These suffer a lot of rejection, ridicule, and mockery from society. They are forced to live in inhuman conditions aren't provided

with any separate schools, colleges, and vacancies. What if God, denied them, the right to be biological parents; what if their speech and looks are different from the normal-looking heterosexuals: are these causes enough to make them feel and treat them inhuman?

2. **About the Product:** Vicks VapoRub ointment is a mentholated topical ointment. Vicks VapoRub is demonstrated for usage on the throat, back, and chest for cough suppression due to the common cold or on joints and muscles for minor pain and aches. The product is also seen as a mosquito repellant than just being an ache-killer. More recent uses of the product have been found that extend beyond what is just stated. It is being used as an ointment that facilitates fat loss, and that lightens stretch marks.

3. **Case in Point:** A recent advertisement of this brand has proposed a revolutionary idea. The ad very subtly dovetails two mutually-exclusive concepts of child adoption and transgender-social-inclusion. When the advertisement director, Mr. Neeraj Gheywan, BCCL, was quizzed upon the brewing up of this superb mindful idea, he mentioned an instant assent to the idea of the advertisement by ad agency, Publicis Singapore. The concept had been highly thought-provoking and more than contemporary. This ostracized section of the society which is least respected in the society, has no special schools and colleges for them. Reservations are made on the basis of caste and a regular tussle between men and women but none to wink a thought upon their fate and fortune. The idea of a transgendered person taking care of a child has been amazingly portrayed. It is about the girl traveling in a bus to her school and reminiscing the time spent with her mother. The script (including the voice-over) was in English, and the screenplay with the agency had been rewritten by BCCL. Further, the Hindi voiceover had been recorded.

There has been a successful attempt in breaking traditional brand imagery. Fortunately, the target audience is ever-evolving. Moms, worldwide, have more awareness about global and local issues and are more vocal about them. In this post-modern world, the same principles of brand science may not hold good. The brand had always been portraying successfully its image of 'Touch of Care' by mother who is taking care of a child fallen sick due to change of weather and is a victim of lack of immunity and habits such as eating with dirty hands. It is essential that brands change as per times, widen their scope, and show inclusivity. The double-edged sword of purpose-driven branding is required. In case of this campaign, was genuinely felt that it was essential to acknowledge that the emotion was bigger than the necessity to have the brand presence itself. In pure advertising terms, it was a deliberate plan to keep the brand secondary and keep the play of emotions between Gauri and Gayatri (the daughter) at the forefront. If there had been some overt brand-elements, it would have robbed the film of its integrity. From a functional brand point of view, the idea was to forge an even bigger emotional attachment, which was a conscious strategy.

It is the most appropriate time to challenge the stereotypes in advertising. The conventional mode of child-bearing and child-rearing has paved its way to pick up on the unconventional mode of child adoption and emotional fulfillment. The point is 'What is really a dysfunctional family'? Aren't we all dysfunctional at some level? Why is the image of a heteronormative smiling-through-the-dinner table family, the hallmark of ideal advertising? Why can't we question our understanding of what comprises an ideal family? When such powerhouse brands take such bold steps, it starts a dialogue. Motherhood is universal. Abnormalities make the wish list more acute and strong. This advertisement highlights themes of women empowerment and female child adoption. The strength of the protagonist to be so unusual and different from society is highlighted here (Vicks, 2018).

5.3.3 CASE STUDY: HOW CAN WELSPUN INDIA DO WELL?

Exports in any developing nation are supported and appreciated and in most of the cases awarded. One of the renowned home textile manufacturers, Welspun India which is certified by the Egyptian cotton goldfield certification: Cotton Egypt Association (CEA), after Canada-based Cambridge towels and Mundo textile of Portugal which is known for its quality excellence worldwide. Five months ahead, on August 20, the B. K. Goenka-controlled company ceded the merit only to be mocked and ruined. The organization faced tumultuous pressure against its international competitors so much so that the allegations caused imminent brand management issues. The battle worsened with legal issues brewing up from all corners. The compromise on integrity had been deteriorating quality of Egyptian cotton towels and bedsheets. Target corporations, the US-based second-largest discount retailers, that was a significant contributor to 10% or so to say $90 million of the total revenue of Welspun stroked-off the chord with this black sheep Indian firm, alleged to mislabel 750,000 sheets and pillowcases. Target, acting more ethical, on the other hand, went a mile ahead to call back the products sold under the 'Fieldcrest' label and offered refunds to customers.

In a random testing from retail shelves in New York, the CEA, too, found falsely labeled goods. Indeed, DNA-based authentication revealed that 90% of labeled products did not have a wisp of the premium fiber. Subsequently, other leading customers of Welspun including Walmart, JC Penny and Bed, Bath, and beyond, have initiated independent investigations in their stocks. As soon as the news spread, panicked investors started selling off the company's shares, eroding marketing capitalization of Welspun by Rs. 5700 crores in a matter of 10 days. The price of share that was known to escalate ten times in the past two years slashed by 55% to Rs. 46.50. The news was not very new for $3 billion Welspun Group to be popularized for such unfair trade practices. Further in 2010, Indian market regulator of Stock markets, SEBI prohibited the promoters of the four group companies for their alleged role in stock manipulation and insider trading. In January 2015, Welspun India flashed in the news for violating takeover norms and conceded to pay a whopping settlement fee of Rs. 2.48 lacs to SEBI. Further, to make the matter worse, the group had also worked up a debt of approximately Rs. 10000 crores. To add fuel to the flaming fire, an aggressive plan of diversification to steel and power sectors was chalked out. Later, it was decided to sell a stake to increase assets base and step out of debt bait.

Amidst this, Target corporations initiated to consider an alliance with Rs. 3700 crore Trident group, a Punjab based home textiles manufacturer that was also supplying terrycloth in forms of towels to the US Company. While Egyptian cotton constitutes a percent only out of global cotton production that built a steep demand-supply divide. There had been a prevalent malpractice among the industry players internationally to adulterate low-quality fiber with Egyptian or organic cotton to bridge the demand pool. It was always precautionary to curate quality and most significantly, reputation.

Celebration of silver jubilee for the brand across the border of the entrepreneurial sojourn of Mr. Goenka, who had acquired a reputation of being a turncoat entrepreneur. It is not facile to propagate a business, but our connoisseur protagonist had a fair appetite for risk and availed incentives offered by the government. When most of us brood at career choices,

Mr. Goenka took over the reins of a family-owned loss-generating movie hall into a commercial complex and fast food corner. The path of success continued as he eyed to set up a textile unit and reported higher profits than the major Indian textile houses such as Bombay Dyeing and Raymond. In the late 1990s, he committed himself to manufacturing pipes too. Unfortunately, manufacturing plants in the UK, USA, Mexico, Argentina, and Portugal turned out to be a boomerang upon Mr. Goenka which culminated him to shut down the Mexico and Argentina plants, while Portugal unit found its way back to its previous owners. Further, the organization was drained out by Rs. 180 crores. The textile business splurged a high swing in 2011 after Goenka consolidated his core businesses and restructured the non-core units. Though an aggressive attempt had been made under the compulsive instincts of entrepreneurship, acquired VikramIspat, an arm of iron business of Aditya Birla, for Rs. 1085 crores and pumped another Rs. 300 crores as the preemptive cost for establishment expenses. Finally succumbing to innumerable trials to secure environmental endorsements, the unit was sold to JSW steel for Rs. 3035 crores in the third quarter of 2013.

Target corporations, in its recent press release, announced its decision to terminate its 10-year contractual relation with Welspun. The announcement was followed by a very meticulous research. Welspun hired E&Y to inspect the matter and operational issues. A validation from the third party would be extremely crucial to save its magnanimous reputation. Its fortune is decided to be set in a few weeks from now. High levels of anxiety and stress abound in the entire industry with the linen of Welspun being washed in public. Though last five years witnessed the growth of 183% for the firm but presently, it is not only the face of this company but the entire nation that is in question (Welspun India, 2014).

5.4 SOLUTIONS AND RECOMMENDATIONS

5.4.1 RESEARCH METHODOLOGY

This study is based on the premise that a market would not take long to connect with an organization that markets it well. While marketing is all about generating demands, logistics is a means to meet those demands and research in this domain helps to unveil the various patterns of navigation tools across the market and the marketing criteria. Central point of the study is to explore the extent of difference in the linguistic criteria for adoption of marketing strategies from one country to another and subsequently from one continent to another continent. The research methodology adopted for the chapter is a base of the empirical survey. The respondents were both marketers and customers. This chapter comprises an analysis of five organizations in the Food & Beverage industry, namely: Mc Donald's, Burger King, PepsiCo, Coke, and Cadbury. The findings indicate why these brands did become an instant hit and what made them strike right at first sight with the market.

The survey had been inclusive of respondents whose demographic profile is grouped in the age bracket of 15–35 years. The reasons for taking this age profile had been more than many. Consumers under this age bracket are more agile and responsive to change. It is this cohort is more vulnerable and gets easily influenced by the marketing activities. Secondly, these are the opinion and thought-leaders for the next generation to be marketed. Their experiences and guidelines would be found to be more conducive for the upcoming world. Thirdly, it is the youth that aims to talk freely to all alike. Discussions are the beds of exchange for ideas and feelings unlike the senior cohort of the society who indulge in conversations which are just satiating for a while but hardly trigger any new thought or ideas. The respondents also included youngsters in the occupational characteristics as students and working professionals where their income ranged from INR 5–10 lakhs. Both the genders had been explored, but there were more responses from the females than men. As the age bracket had been of people who had been gadget-friendly, an attempt had been made to have insightful discussions on what do they perceive of a certain brand and why? The sample had been profiled from the geographical location of Delhi and NCR, which is the capital of India. The study was carried out for four months of a year, i.e., from October 2017 to January 2018.

5.4.2 RESEARCH FINDINGS

Focuses in depth interviews were conducted, and analysis was drawn. The marketers of Mc Donald reported that youth is a substantial market for them not only among customers but employees too. They are more agile to take and deliver orders whilst the self-service format, which was easy to upset the then-existing schema of serving. There had been a transition from it being a fast food retail outlet to being a family restaurant. Not only food and taste buds, but there had been a change in the way we commute. Both Ola and Uber have changed the way employees travel to work. The system of waiting for local taxi reservation has led the way to instant booking and plastic money payment which was most successful during the period of demonetization. Dining out has been made a convenient option than the expensive format of having dinner.

Several questions were asked to moms as to the frequency of demands posited by their children, and the number of times they are allowed for dining out had been found to be 67.29%. Many of the women being working or due to elderly-care responsibility find them at bay cooking healthy or tasty meals. Secondly, they themselves pick up on the unhealthy options due to taste and ready-to-eat snacking. Friends during the youth and connectivity though various apps and food have made eating with friends more than with family a popular option. Domino's and pizza hut have found that even the mature crowd is getting fond of organizing these items in their menu cards. Youth has one such characteristic of flaunting and influenced by the brand tags, which gets further manifested by prestige issues and connotes being inclusive in the changing social trends.

5.4.3 RECOMMENDATIONS

The health-conscious dimension of minds is so crucial into framing choices. People had been so insecure and worried about their health concerns as much of the diseases are vulnerable to them due to poor weather conditions such as pollution and climate change. Marathons and other such health-friendly parks have been widely acclaimed. The not so popular oats in the recent past had gained such popularity within a matter of visual merchandising, and attractive packaging that the youngsters would appreciate more if their favorite brands such as McDonalds and Domino's would serve oats burger and oats crust pizza respectively. In addition, a cherry to the cake would be if these fast-food-retail outlets would serve fresh juice instead of air-carbonated drinks. Another innocuous innovation these Generation Y would want is that of sprouts salad being served on the metro-located snacks bars as healthy-snacking is preferred over tasty-meals.

The difference is among the techniques and focus. Marketing is all about making products and services likable and popular. Influence marketing deserves to be mentioned here. To move from a spiral effect of making an unknown product familiar to convince to influence to persuade and making a one-time purchase to make it a family name and indulging in a repeated purchase is more than a difficult process. A careful study of consumer behavior proves useful. Consumerism and hedonism are two of the prime characteristics of Asian markets. Customer pyramid that segregates buyers into the categories of lead, gold, and platinum generates a different set of customer-lifecycle management software under the CRM ambits and reward of customer loyalty. This shelving is based on the creditability and credibility of the buyers along with their tenure of subscription to the products and services. It is recommended that for your business to be royal, you must make your customers loyal. The pareto principle asserts that 80% of the business is extracted from 20% of customers, and the vice versa also holds true be it any industry or company.

5.5 FUTURE RESEARCH DIRECTIONS

Any study in this direction can be more extensive of being either specific to any one of the industries referred in the chapter or can include a few more industries to make the findings more comprehensive. Not only the industries but a broader database of respondents can enrich findings for future research studies. An enhancement in terms of different geographic locations would also prove to be more advantages and strengthen findings.

5.6 CONCLUSION

The 3Cs of challenging the culture and changing it through the principles of communication with an expertise in becoming extensively and thoroughly competitive is crucial to survive in the product life cycle and enriching the market share. 'Changing your product can make your existing market happy, but if you can't change your product, you necessarily need to change your market.'

KEYWORDS

- **communication**
- **digital marketing**
- **lifestyle marketing**
- **marketing**
- **technology**

REFERENCES

Aaker, J. L., & Maheswaran, D., (1997). The effect of cultural orientation on persuasion. *The Journal of Consumer Research, 24*(3), 315–328.

Adegoke, Y., (2004). Web still fastest growing channel in marketing mix. *New Media Age, 22*, 12–19.

Blythe J (2006). Principles and Practice of Marketing, Jennifer Pegg Cambridge. *Madrid*, pp. 6–7.

Halloway J. C. and Robinson C. (1998), 'Marketing for Tourism', Addison: Wesley Longmann, pp. 72–74.

Janoschka, A., (2004). Web Advertising: New Forms of communication on the internet. *Pragmatics & Beyond New Series*, 131.

Kemp, S., (2018). *Digital in 2018: World's Internet Users Pass the 4 Billion Mark.* Retrieved from: https://wearesocial.com/blog/2018/01/global-digital-report-2018 (Accessed on 28 August 2019).

Kitchen J. Philip and BurgmannInga, '*Integrated Marketing Communication*', Wiley International Encyclopedia of Marketing, edited by Jagdish N. Sheth and Naresh K. Malhotra. Copyright © 2010 John Wiley & Sons Ltd, pp. 1–23.

Oosthuizen, T., (2004). In marketing across cultures: Are you enlightening the world or are you speaking in tongues. *Massachusetts Institute of Technology, 20*(2), 61–72.

Orth, U. R., Koenig, H. F., & Firbasova, Z., (2007). Cross-national differences in consumer response to the framing of advertising messages: An exploratory comparison from Central Europe. *European Journal of Marketing, 41*(3/4), 327–348.

Owersloot, H. and Duncan T. (2008), 'Integrated Marketing Communication', London McGraw Hill, pp. 69-71.

Parle, A., (2017). *Frooti Fizz.* Retrieved from: https://parleagro.com/brand/frooti-fizz (Accessed on 28 August 2019).

Stewart, D. W., & Pavlou, P. A., (2002). From consumer response to active consumer: Measuring the effectiveness of interactive media. *Journal of the Academy of Marketing Science, 30*(4), 376–396.

Tian, K., & Borges, L., (2011). Cross-cultural issues in marketing communications: Anthropological perspective of international business. *International Journal of China Marketing, 2*(1), 110–126.

Vicks, (2018). *Vicks Does it In a Cool Way*. Retrieved from: https://www.vicks.co.in/en-in (Accessed on 28 August 2019).

Welspun India, (2014). *How Can Welspun India Do Well*. Retrieved from: http://www.welspunindia.com/index.php (Accessed on 28 August 2019).

CHAPTER 6

Internet of Things (IoT) and the Digital Business Environment: A Standpoint Inclusive Cyber Space, Cyber Crimes, and Cybersecurity

ANAND NAYYAR,[1] RUDRA RAMESHWAR,[2] and ARUN SOLANKI[3]

[1]*Duy Tan University, Graduate School, Viet Nam, Room No: 102, 254 Nguyen Lan Vinh Street, Da Nang, Vietnam, NA, Da Nang–550000, Vietnam, E-mail: anandnayyar@duytan.edu.vn*

[2]*Thapar Institute of Engineering and Technology, L.M. Thapar School of Management, Thapar School of Management, Thapar University, Patiala, NA, Patiala, Punjab–147004, India, E-mail: rudrarameshwar@gmail.com*

[3]*Gautam Buddha University, School of Information and Communication Technology, School of ICT, Noida, NA, Noida, U.P.–201312, India, E-mail: ymca.arun@gmail.com*

ABSTRACT

Internet of Things (IoT) has a strong backbone support via enhanced developments in the area of RFID, smart sensors, communication technologies, and internet protocols. The primary objective is to integrate smart sensors to work collaboratively without any sort of human intervention to deliver the best of the class applications and services. With the revolution via Industry 4.0, Smart Phones, Machine Learning, Deep Learning, and Smart Sensors, in the coming years, IoT is expected to bridge up all diverse technologies to enable new applications via connecting physical objectives to make the intelligent decision making. The chapter outlines an in-depth overview of IoT-Technical details, enabling technologies followed by an in-depth technical overview of protocols enabling IoT-device to device communication. The chapter enlists cybersecurity issues surrounding IoT to enable researchers and developers to speed up research in these areas along with an overview of various simulation tools available for carrying out research on IoT.

6.1 INTRODUCTION

6.1.1 INTERNET OF THINGS (IOT)

Well is quite known to everyone that in modern world advanced sensors and related devices used in hosting other devices such as mobile phones, monitors used at hospitals are connected in a closed envelope of cyberspace or cyber-physical system; simply to measure and monitor time, location of human beings and automobile movements, machine vibrations, humidity in the atmosphere, precipitation, and temperature, etc. (Lohr, 2012).

The Internet of Things (Camarinha-Matos et al., 2013) in short, mostly professional world called IoT, it confines new domain and endlessly produces data streams across the globe with its geographical impression or footprints from interconnected mobile devices, laptops, computers, actuators, sensors, RFID tags, etc. (Michael and Miller, 2013; Van-Den-Dam, 2013). However, in a broader spectrum, Big Data generated from these devices categorized in IoT, which contains well-off Spatio-temporal information-data. Presently in the age of technology development, an advancement in the area of IoT and Big Data Analytics, Advance Technologies presents an array of modern applications including better product-line management, supplementary effectual and appropriate unlawful examination, enhancing farming yield output (Jiang et al., 2009; Hori et al., 2010; Xing et al., 2010; Bo and Wang, 2011), and increasing speed of the expansion of smart cities (Belissent, 2010; Schaffers et al., 2011; Gubbi et al., 2013) with new architecture (Balakrishna, 2012; Mitton et al., 2012; Theodoridis et al., 2013; Jin et al., 2014), world-class infrastructure and communication network step-up.

In the age of Social Media and a high-speed internet connection, various social networks are active such as Facebook, Twitter, LinkedIn; Blogs generate Big Data, responsible for the transformation of the social sciences area in and around. However, during the time of writing, many social media users, particularly in case of Twitter, users across the globe roughly or on an average 6000+ tweet per second, fairly which corresponds to 500+ million tweets /day and around 200+ billion tweets/year, according to internet Live Stats-2016 database. Similarly, the big data count follows in the case of Facebook, LinkedIn, Blogs, WhatsApp, and other social media platforms.

Nonetheless, Social Scientists, Economists, Political Scientists, and other scholars use Big Data mining methods to analyze social variable and their interactions regarding research parameters in vibrant areas such as health records, phone logs, government records and other digital traces (Boyd and Crawford, 2012). Even as such data mining methods promote governments and social studies (Grimmer, 2015) increases big data acquaintances, it is still demanding to rapidly haul out spatiotemporal patterns from big data social affairs too, for example, it helps in the prediction of criminal activity (Heffner, 2014) on the continuous evaluation basis, scrutinize emerging public health threats and provide more effective intervention (Lampos and Cristianini, 2010; Jalali et al., 2012) solutions to ongoing threats in the concerned areas.

6.1.2 DIGITAL BUSINESS ENVIRONMENT

Digital business environment is altering customer expectations and business models. It is changing how businesses converse, conduct, and network with customers as well as with other business such as partners and suppliers. It is increasing the pace of business. And it is changing industries and the competitive landscape through a social web monitoring system based on IoT, Big Data, Information Framework, Networking Technologies, and Leveraging Cloud Computing."Digital business is blurring the digital and physical worlds. According to Forbes, Digital Business is Everyone's Business. Moreover, it agreed to support in an exceptional meeting of person, business strategies, and other belongings that interrupts accessible business models – even those born of the internet and e-business eras." According to Jorge Lopez, a 'distinguished analyst' with Gartner, Digital business is shaping the future-and businesses must evolve their business ecosystems and technology to remain competitive.

"If digital disruption can be viewed as a wave sweeping over industries, most are in the crest of that wave or soon will be," according to research from HBR, an overwhelming majority (80%) of the executives in a *Harvard Business Review* (HBR) study thinks "their industry will be disrupted" by global digital business trends, according to HBR, competing in 2020: Winners and Losers in the Digital Economy, April 2017.

"Digital technology platforms are the building blocks for a digital business. In case of Gartner's Top 10 Strategic Technology Trends for 2017, January 2017, every organization [must] have some mix of digital technology platforms: Information systems, customer experience, analytics, and intelligence, the Internet of Things and business ecosystems," said the HBR study authors in Gartner. One critical point in keeping pace with the evolving digital environment is ensuring the evolution of the management and transfer of business-critical data. In the digital era, constantly evolving forms of digital data must be synchronized and shared- and done so quickly, efficiently, and securely-with consumers, partners, and other businesses.

6.2 BACKGROUND

6.2.1 CYBERSECURITY REGULATIONS IMPROVEMENT

We ought to see a proceeding enhancement within the pertinent regulations as connected to cybersecurity. The fast-moving and dynamic nature of cybersecurity outpaces regulation which is far too clumsy and slow to be of any advantage and might really ruin security by building a culture of compliance with regulations and a wrong sense of security against enemies who are clever, motivated, and agile.

6.2.2 DATA THEFT TURNING INTO DATA MANIPULATION

We can anticipate seeing attackers changing their strategy from website hacking and pure data theft to attacking data integrity itself. This sort of attack, in comparison to a straightforward theft of data, will serve to cause long-term, reputational harm to people or groups by getting individuals to question the integrity of the data in question.

6.3 CYBERSECURITY AND INTERNET OF THINGS (IOT)

6.3.1 CYBERSECURITY AND INTERNET OF THINGS (IOT): CHALLENGES AND ISSUES

The rapid development of smart IoT devices (Rose et al., 2015) has opened up a number of ways for hackers to invade user privacy. The IoT is making our lives easier while leaving us less secure. Whenever a major technological innovation comes along, hackers find ways to exploit it. The world of IoT includes a huge variety of wire and wireless devices like smartphones, personal computers, PDAs, laptops, tablets, and other handheld embedded devices. IoT has also been known as Cyber-Physical Systems (CPS), M2M (Machine to Machine), and Simply Industrial Internet (SII) and connected devices. The generic topology of the IoT seen in layers to incorporate the Datacenter, Gateway, IoT Gadgets and Sensors. The IoT devices used sensors and wireless communication network to communicate with each other and transfer information to the centralized system. These days the embracing rate of the IoT devices is very high, increasingly gadgets are connected via the internet. In the current scenario, these devices are targeted by attackers and intruders. A report found that 70% of the IoT devices are very easy to attack (Hossain et al., 2015) (Figure 6.1).

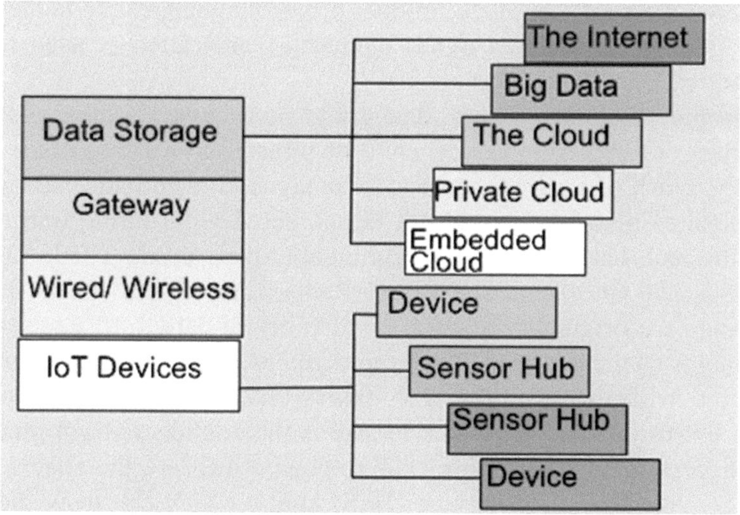

FIGURE 6.1 (See color insert.) IoT topology.

There are many critical challenges and issues regarding the security of IoT devices. This section covers these issues and suggests the probable solution. Generally, IoT has used Perception, Network, Middle, and Application layer in data transmission and processing (Suo et al., 2012). Each layer has its own importance.

Network layer transmits the gathered information obtained from the perception layer. This information is sent by using existing communication networks like the Internet, Mobile Network, or any other kind of reliable network (Yang et al., 2012). The middle layer comprises of information processing systems that take automated activities based on the results of processed data received from the network layer. This layer also links the system with the database, which is having information about the collected data. The middle layer is a service-oriented layer which ensures the same service type as the connected IoT devices (Khan, 2012). The application layer is founded on TCP and UDP to solve the communication challenges and used by various applications like Smart Home, Smart Environment, Smart Transportation and Smart Hospital, etc. (Rong and Tao, 2013).

The major security goal of IoT is Data Confidentiality, Data Integrity, and Data Availability; whereas Data confidentiality ensures that there will be no outer inference. It ensures that only trusted communication will be taken place, and no information will be leaked to outsiders. Two steps verification like Gmail or the encryption process may be used to achieve data confidentiality. These techniques provide authentication to the trusted user only. In the current trend, biometric verification is used for user authentication (Miorandi et al., 2012).

During data transmission, data could be accessed or changed by the intruders or cybercriminals or could be affected by various other aspects like the crash of a server or an electromagnetic disturbance. Data Integrity ensures that data should not be tampered with during transmission by intruders, hackers, or cybercriminals (Atzori et al., 2010). Different methods like Checksum and Cyclic Redundancy Check (CRC) are used to ensure the originality of data. Availability of data to trusted resources is a major challenge in IoT. IoT protocol ensures that every authenticate resource will always access the data as required. It may be possible some entrusted resource wants to access the confidential information so it is necessary to provide firewalls to countermeasure the attacks on the services like Denial-of-Service (DoS) attack which can deny the availability of data (Reddy et al., 2017).

1. Cybersecurity Lacks Control Environment and Tools for IoT Devices: In cyberspace, commercial products are at high risk. Currently, there is no up to the mark security measures, and even no tool exists to remove the existing risk. As the consumer is not aware of this risk, so consumer demand in the cybersecurity area remains low. Especially in consumer products as many types of attacks, such as denial-of-service attacks, do not affect the people who own the equipment. If media or social sites aware the consumer about risk in cyberspace, then vulnerabilities might raise consumer awareness of the impacts of inadequate cybersecurity, and with it the demands they make of manufacturers.

There is a need for "trusted IoT label" for IoT products that meet certain security requirements. This label will be aware of the specification of IoT product with risk, and it will clearly deliver more energy-efficient products to the consumer. The government should take some initiative through legislation and regulation. These rules will decrease the risk of IoT products. The labeling of IoT Products under these rules places greater responsibility on the manufacturers and importers of products, and this has generally worked well (Appazov, 2014). A feasible cybersecurity structure shall aim at the evolution of the adequate cybersecurity environment for IoT devices. Therefore, it shall include national and international cooperative efforts to develop standards, methodologies, procedures, and processes that align policy comprising business, health, entertainment, education, and technology approaches to address cybersecurity issues (Teplinsky, 2013). The private sector will play as significant a role in the implementation of the policy as does the public sector. The policy on cybersecurity shall be informed by the adequate understanding of the cyber-vulnerability threat on the part of the policy development.

6.3.2 WHY ATTACKERS WILL CONTINUE TO TARGET CONSUMER DEVICES?

In the era of the IoT, the smart devices are the part of every aspect of our lives which include homes, offices, cars, and even our bodies. As internet upgraded to IP4 to IPv6 and Wi-Fi network advances, IoT is growing at a very fast speed, and researchers estimate that by 2020, the number of active wireless connected devices will exceed 40 billion (ABI Research, 2014). As the number of active connected devices increases the chance of hacking the data will be increased and targeted by the cybercriminals.

The IoT companies are constantly making smart-devices at low cost to customers. As a result, companies have to lower down the security of an IoT device in favor of its cost, size, and low energy consumption. These devices are different from each other in their application area.

Nowadays, in the market, there are lots of smart-gadgets with poor security services available to a remote attacker. Consumers never think that their smart device is not venerable and these devices can be hacked in a matter of seconds. The large-scale gathering of IoT devices and user-generated content opens user to the risk of data misuse and abuse. If proper security measures have not taken then it becomes disasters to the IoT users. In the basic IoT environment, shows simple information that is easy to understand for possible vulnerable components.

By analyzing the IoT environment, there is a possibility to define possible attack vectors used by attackers. The environment consists of a basic sensor which collects measurable data. The transmission of data from the sensor to the server gives many loopholes for attackers. Attackers used these loopholes and hacked or modify the IoT environment.

This section will cover the loopholes in IoT devices which are used by attackers to hack the consumer devices:

- Users are not using the secure web/desktop/mobile applications which can provide proper authentication and authorization.
- Two-step authentications are not used by the IoT device security.
- Hackers used a Cross-Site Scripting (XSS), SQL injection, and Buffer over Flow (BoF) vulnerabilities to hack the IoT devices.
- Simple passwords like '123,' 'password,' the name of the user, date of birth, etc. lead to hacking of IoT device.
- Not using Captcha to secure IoT devices.
- Brute force attacks are done by the hackers.
- Companies are not providing the security updates including details on security fixes, the impact of the vulnerability, etc.
- Consumers are not using encryption for communication.
- Vendors or customers are providing sensitive data like name, phone number, DOB, etc., publicly through different social media platforms.
- Vendors are not testing the security of IoT devices when new features are added.

Following are some proposed solution to overcome the security issues in IoT devices:

- The user always updates their device security from a trusted source.
- User updates device regularly.
- Users always read the manual and user instructions provided by the vendor.
- 2-way authentication always provides the enhanced security to the device.
- Use strong, unique, and complex passwords at least 8–20 characters with a mixture of letters, numbers, and special characters.
- The user always ensures to take backups in a secure place.
- Stop unwanted services used by the devices.

DTH service is one of the examples of IoT device. What happens if hackers hack the personal information of a user and use this information to hack the DTH service? There will be a big impact on DTH business which may lead to the following consequences:

- Customer personal information can be used for social engineering or any other advanced attacks.
- DTH service may be interrupted.
- The consumer will not trust DTH service providers.
- Consumer account, including recharge money can be lost.
- Data are theft by the hackers, including favorite channels, recording, etc.

The IoT is forcing many business leaders to reassess their approaches to cyber risk management.

6.3.3 PROTECTION OF IOT DEVICES IN CYBER SPACE

The first question comes in the consumer mind why we need to protect IoT device data. The reason is to protect one's own intellectual property against other competitors. This is a way of preventing others to compromise one's own environment and use it for malicious purposes (Sagedhi, 2017). The IoT device security in cyberspace is presented with the triangle.

This triangle visualizes the essential steps desired to obtain a high level of security consciousness with IoT devices (Sagedhi, 2017).

The triangle will be visualized from bottom to top i.e., green part is the target. The first two gray steps, i.e., meaningfulness/usefulness and functional correctness are the starting point of the journey to a higher security level. These two levels show that an IoT device with internet without any security features can have duplicate MACs or IP addresses. The security in these three steps is like unique passwords; the latest software and firmware versions secured communication protocol such as HTTPS securing web front-end and security architecture. These three steps in brown color are the first implemented security procedures and therefore they aware consumer. The next three levels with yellow present a higher level of security implemented in an IoT environment. The practical implementations with these levels are like swarm attestation, IoT sentinel and the implementation of automatic vulnerability and attack detection (Figure 6.2).

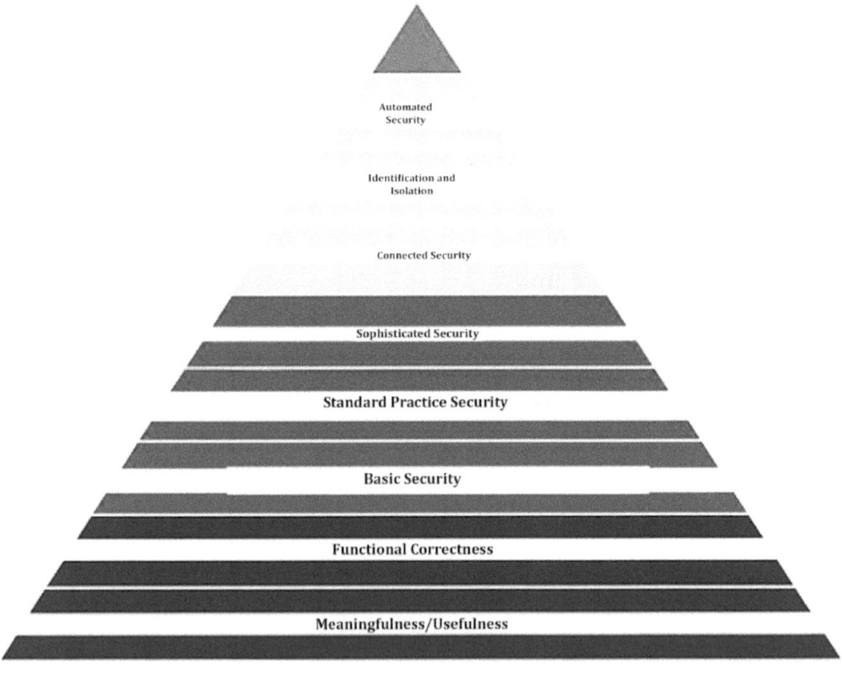

FIGURE 6.2 (See color insert.) Cybersecurity awareness triangle.

The top green triangle is the level IoT device should strive to whenever it is possible and reasonable as well as economically wise.

Within the current situation, there are more IoT gadgets than people on Earth, engineers, and developers got to embrace a new paradigm for cybersecurity within the IoT age. The rapid increase of IoT has produced big data. Smart sensors are gathering statistics, which are used in the machine learning algorithm and can be used to attract the customers for any business. The sheer volume of this data generated from IoT devices is extremely large. This large volume of data is a security threat if it is not secure. To make it secure, organizations/users have to do the following things:

- The organization has to prepare an inventory of all IoT usage because it's impossible to defend the unknown.
- The user has to change the default user id and password. Attackers know these platforms and their defaults. The new password should be strong.
- Passwords should be changed regularly.
- The user should disable unnecessary remote administration and features.
- Do not allow untrusted access to the device from the Internet.
- Don't empower universal plug and play on IoT devices.
- Use secure protocols like HTTPS and SSH for communications.
- Include IoT devices in regular vulnerability management programs.

1. **Defining Security Implementation:** To secure IoT devices in cyberspace, it is necessary to find possible attack vectors. These attack vectors might be against the IoT environment. Therefore, security administrator and system administrator need to secure the environment against the vulnerability of the IoT devices (Palmers, 2013). The methods for analyzing the environment could be similar to the zero-day vulnerability analysis (Figure 6.3).

The zero-day attack is a phenomenon that secures computer appliances before the vulnerability is known or patched against the vulnerable (Salonen, 2017).

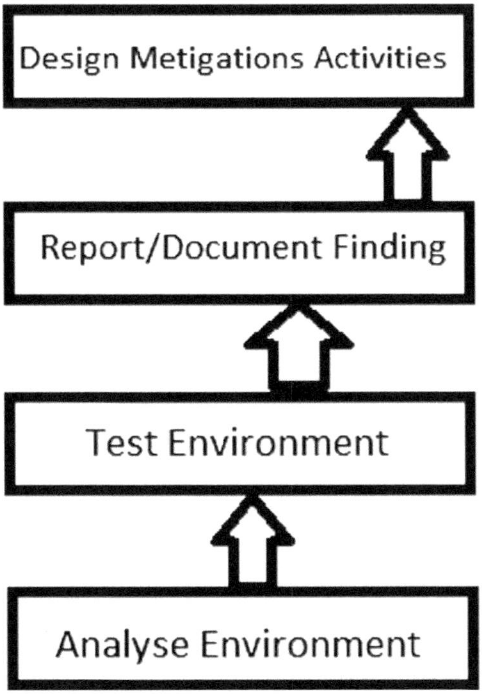

FIGURE 6.3 Security implementation in IoT environment.

6.4 SOLUTIONS AND RECOMMENDATIONS

6.4.1 PROTOCOLS FOR INTERNET OF THINGS (IOT)

6.4.1.1 INTRODUCTION

Internet of Things (IoT) plays a crucial role in varied application areas like Industry 4.0, healthcare, transportation, agriculture, autonomous vehicles and even disaster management. These days, IoT has also started improvising the quality of daily life of human beings by providing portable and smart gadgets, business operations in terms of logistics and manufacturing and future smart homes (Suryadevara and Mukhopadhyay, 2018). The most prominent example is Smart Home, where everything is controlled via Sensors like Temperature, Air Conditioning, Kitchen Home Appliances, Lighting, Security cameras, and even cleaning via Smart Robotics.

Seeing the current research interests of various researchers in IoT, many new protocols are evolving and getting standardized. Internet of Things (IoT) is expanding to a plethora of applications via hardware minimization, smart sensors and "Smart Objects," and many protocols are evolving to facilitate Machine to Machine communications. Because of the remote nature and requirement of wireless networking of smart objects, IoT systems have to cope up with various limitations in terms of unreliable, intermittent, and low bandwidth connections for network access (Javed et al., 2018). To facilitate communication between devices, lots of protocols are proposed for different layers of IoT: Data Link, Network Layer (Routing and Encapsulation), Session Layer, IoT Management and Security.

6.4.1.2 IOT ARCHITECTURE

Till date, there is no as such properly defined and proposed architecture for IoT, which is agreed universally. Internet of Things (IoT) architecture can be three-layer or five-layer (Mashal et al., 2015; Said and Masud, 2013; Wu et al., 2010). The three-layer architecture was proposed in the early days of IoT research. It contains three main layers: Perception layer, Network Layer and Application layer. The bottommost layer of IoT architecture is perception layer, which is primarily responsible for acquiring information from devices and convert into digital form. The network layer is primarily responsible for transmitting digital signals over the network and application layer is responsible for transferring digital signals to varied applications (Figure 6.4).

- **Perception Layer:** In the initial stage of Internet of Things (IoT), the major task of perception layer is to collect all the data from devices or objects sensed from the environment, i.e., Temperature, Humidity, Gases, etc. All the sensors operational in sensor network, performs the task of collecting, computing, processing, and transmitting varied types of data to the perception layer. The sensors and other objects in IoT, i.e., GPS Devices, IP Cameras, Actuators all communication among one another via specialized network technologies like ZigBee, Wi-Fi, and other short-range communication protocols. IoT connects lots of heterogeneous devices, so it is highly required to identify and connect every object or thing in a significant manner. Latest research technologies like 6LoWPAN are also used to connect the devices within the network without any hiccup.

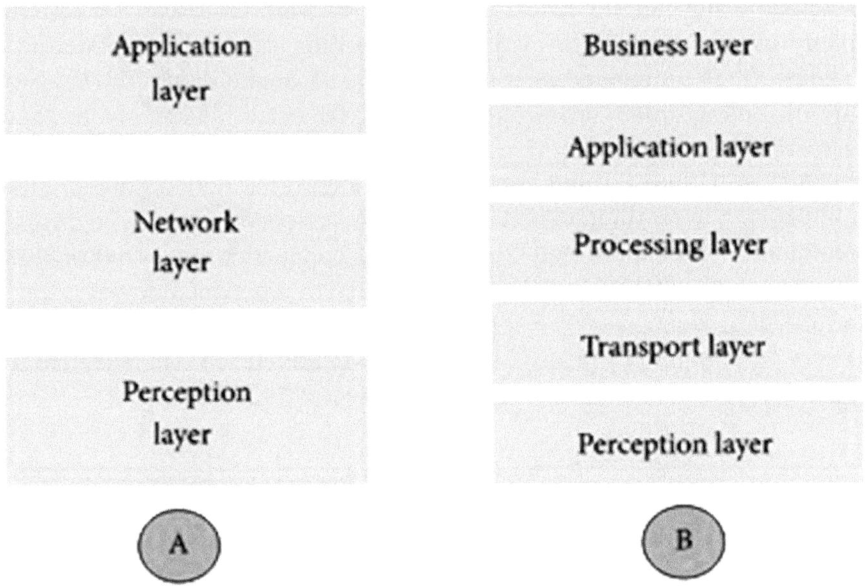

FIGURE 6.4 Three-layers and five-tier architecture for Internet of Things.
(*Source:* IoT Direct, (2017). *All About Internet of Things*. Retrieved from: http://iotdirect.blogspot.com/.)

- **Network Layer:** It is regarded as backbone for IoT architecture and transmits all the data from perception layer to the application layer in a highly secured manner. The network layer is regarded as the delivery layer to collect all information acquired from perception layer and deliver to application layer, i.e., Servers, and Application Softwares. Lots of research is being conducted till date to propose routing protocols, and sophisticated communication technologies in the network layer to improvise core network operations. Unique addressing efficient routing ensures seamless integration of devices comprising an IoT network. All types of Wired, Wireless Technologies-Wi-Fi, Bluetooth, RFID, 6LoWPAN ensure unique addressing and break-free connectivity in IoT network.
- **Application Layer:** The topmost layer of IoT architecture is the application layer providing a bridge between applications and users. It provides application-specific services to the end-user. It facilitates Intelligent IoT based real-time solutions like Remote Health Monitoring, Autonomous Transportation, Disaster Monitoring, Industrial

Manufacturing and many more. This layer handles all the global management of IoT applications (Mashal et al., 2015).

The IoT architecture comprising three layers (Perception, Network, and Application) no doubt have created a strong foundation of IoT, but it is not sufficient enough to accommodate the advanced concepts of Internet of Things. In order to facilitate, more advanced research, IoT architecture is facilitated with new architecture comprising five layers-Perception, Transport, Processing, Application, and Business layer (Mashal et al., 2015; Said and Masud, 2013; Wu et al., 2010; Khan et al., 2012) -

- **Transport Layer:** It performs the task of sensor data transfer from perception layer to processing layer and vice versa using network communication standards like Wi-Fi, LAN, Bluetooth, RFID, NFC, and 6LoWPAN.
- **Processing Layer:** It is also termed as middleware layer. It performs the tasks of storing, analyzing, and processing tons of data acquired from the transport layer. Various ICT technologies, i.e., Big Data, Cloud Computing, Data Processing Modules all forms processing layer.
- **Business Layer:** The entire management of whole IoT ecosystem is performed by business layer which includes all types of IoT Applications, models as well as ensuring privacy and security of end-users.

6.4.1.3 IOT PROTOCOL STACK

The most basic protocol considered for facilitating all sorts of IoT based communication operations was TCP/IP, as it is also regarded as a baseline protocol for doing all sorts of computer network related operations. But the utilization of IPv6 was restricted in IoT, as all the IoT enabled devices are low powered and require limited bandwidth. Many Interest groups have laid their efforts to define some IoT based standards to facilitate the IoT development like World Wide Web Consortium (W3C), Institute of Electrical and Electronics Engineers (IEEE), Internet Engineering Task Force (IETF) and EPC Global.

The following sections highlight various Network Communication protocols for IoT for various layers-Data Link, Network Layer (Routing + Encapsulation) and Session Layer Protocols (Salman and Jain, 2015; Silva et al., 2018) (Figure 6.5).

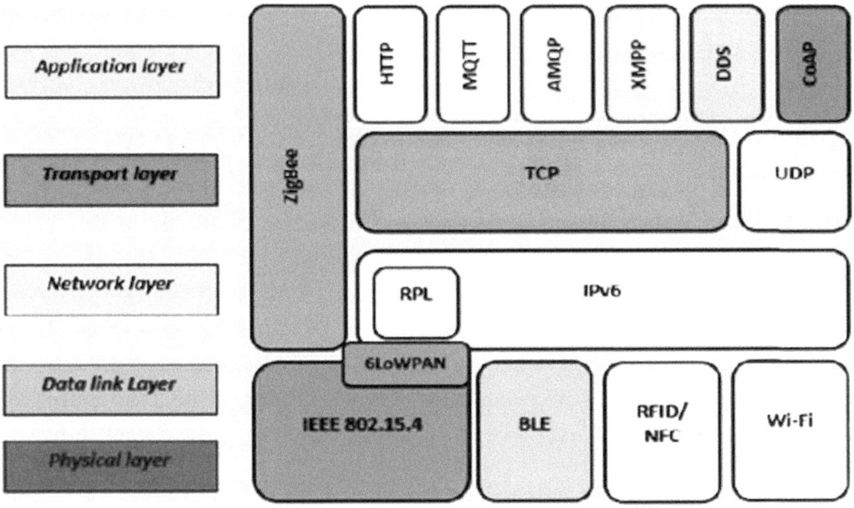

FIGURE 6.5 (See color insert.) IoT communication protocols.

6.4.1.4 IOT DATA LINK LAYER PROTOCOLS

The IoT data link layer protocols combine Physical and MAC Layer protocols. The following are the Data Link Layer protocols for Internet of Things (IoT):

- **IEEE 802.15.4:** This protocol was designed by the IEEE 802.15 PAN working group (Kivinen and Kinney, 2017). The protocol was designed for low-power, short-range, low-bandwidth as well as low-cost IoT based devices (LAN/MAN Standards Committee, 2003). It specifies physical layer and MAC for LR-WPANs. It defines 16 channels ranging between 2.4 GHz-2.48 GHz, where every change is 2 MHz wide and separated by 5 MHz from each other. There are two types of devices supported by IEEE 802.15.4:
  - **Fully Functional Device (FFD):** For creating, coordinating, and maintaining the network and communicating with all types of devices in the network;
  - **Restricted Functional Device (RFD):** Devices have limited access and can only communicate with the coordinator.

IEEE 802.15.4 was further extended to IEEE 802.15.4e to facilitate low power communication. The protocol defines the way how a schedule is executed via the MAC layer, which can either be centralized or decentralized. The protocol can facilitate star or peer to peer topologies (LAN/MAN Standards Committee, 2003). IEEE 802.15.4 facilitates packet size of 127 bytes and communication speed is restricted to 250 Kbps. There is inbuilt redundancy in IEEE 802.15.4 making the communication robust, detect losses and retransmission of all packet loss. The protocol fully supports 16-bit link addresses to decrease the header size, overloading of communication and memory requirements.

- **IEEE 802.11 AH:** To facilitate efficient Machine to Machine communications, a new standard protocol was proposed, i.e., IEEE 802.11ah by IEEE 80211ah task force (Adame et al., 2014). The standard was facilitated to solve M2M networks requirements like limited power devices, long transmission range, small data messages, low data rates and non-critical delay. IEEE 802.11ah operates over unlicensed radio bands depending on country-wise regulations, i.e., 863–868 MHz in Europe, 902–928 MHz in the US, and 916.5–927.5 MHz in Japan. It makes use of OFDM-based waveform comprising 32 or 64 carriers and supports BPSK, QPSK, and 256-QAM modulations. IEEE 802.11ah supports three types of stations: Traffic Indication Map (TIM); non-TIM stations and unscheduled stations. TIM stations only listen to access points beacons to transmit or receive data. Non-TIM stations don't listen to any sort of beacon for data transmission. They directly negotiate with Access Points to acquire transmission time allocated in a Periodic Restricted Access Window (PRAW).
- **WirelessHART:** It is IoT data link protocol based on Open Systems Interconnection model (OSI) and adopts IEEE 802.15.4–2006 standard as physical layer, and uses Time Division Multiple Access (TDMA) in MAC for channel access and reducing collisions (Hassan et al., 2017). WirelessHART, globally operates on standard 2.4 GHZ ISM bank using 15 different channels, i.e., 11–26. It encrypts messages and ascertains integrity to make reliable data transmission. Supports mesh, star, or cluster topology to facilitate reliable communication as compared to Wi-Fi especially in industrial environments. It employs redundant routing at network layer. WirelessHART includes features like: Self-organization, robust, Simple Implementation, and Interoperability with other HART devices, highly energy-efficient, scalable, and self – healing (Figure 6.6).

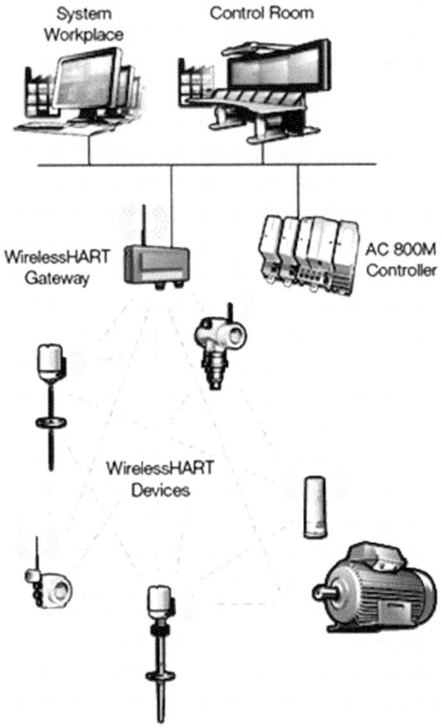

FIGURE 6.6 (See color insert.) WirelessHART working.
(*Source:* ABB, (2018). *WirelessHART Information and FAQs*. Retrieved from: https://new.abb.com/products/measurement-products/wireless-products-and-solutions/highlights/wirelesshart-information-and-faqs.)

- **Z-Wave:** This was designed and developed by Zensys in Denmark, is a low-power MAC protocol based on ZigBee for home automation and was used for IoT based communication especially for smart homes (Alliance, 2013, 2015; Said and Masud, 2013; Wu et al., 2010). Z-Wave operates at 908.42 MHz in the US and supports MESH topology. It can support up to 232 nodes and supports interoperable operations. It makes use of GFSK and Manchester channel encoding. In a Z-Wave network, the three main devices are required to comprise a network-Central Station, Network Controller and End-user devices. Every Z-Wave network has unique Network ID and every device has Node ID. The Network ID remains common for all nodes to associate with 1 Z-Wave logical network. It's 4 Bytes in size and nodes with different Network ID's can't communicate among each other. Node ID is unique for end-user device and is only 1-byte of size. Z-Wave supports mesh topology with primary and secondary controllers (Figure 6.7).

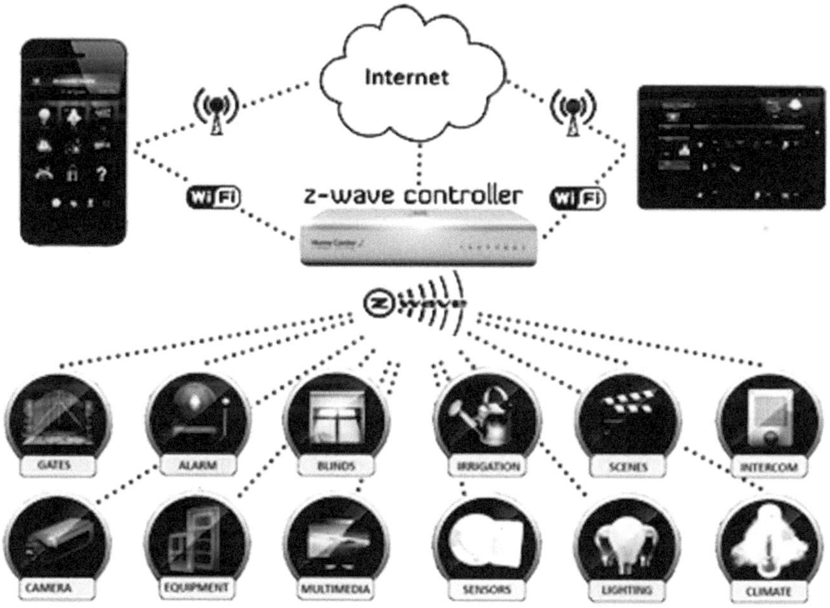

FIGURE 6.7 (See color insert.) Z-wave network.
(*Source:* SmarterHOME, (2016). *Hierarchy of the Z-Wave System.* Retrieved from: https://smarterhome.sk/en/blog/hierarchy-of-the-z-wave-system_39.html.)

- **Bluetooth Low Energy (BLE):** It also termed as "Bluetooth Smart" is part of Bluetooth 4.0 specification (Panwar and Misra, 2017; Tosi et al., 2017). It is a short-range communication protocol and is widely adapted in-vehicle networking as well as Machine to Machine communication. BLE is regarded as a best alternative wireless solution as compared to existing standards like IEEE 802.11b, ZigBee, ANT+ and Bluetooth 3.0. It operates in standard 2.4 GHz ISM band ranging from 2.4–2.5 GHz and is divided into 40 channels having frequencies 2402 + k x2 MHz, where k = 0, 1, 2, 3…39. It makes use of CSMA/CA for all sorts of collision detection and ACK for reliable message transmission. Its basic network comprises of Master and Slave called "Piconet."

- **HomePlug:** HomePlug AV is IoT based data link protocol designed by HomePlug Powerline Alliance especially for home automation products (Homeplug Alliance Kit, 2018). It operates at speed of 200 Mbps and frequency range of 2 to 28 MHz. HomePlug Power line Alliance has designed three main specifications: HomePlug AV, HomePlug Green PHY, and HomePlug AV2.
  - **HomePlug 1.0:** It was the first HomePlug specification with a speed of 14 Mbps. It makes use of CSMA/CA to transport data from 46 to 1500 bytes long from encapsulated IEEE 802.3 frames as MAC Service Data Units.
  - **HomePlug AV:** It was designed for applications like HDTV and VoIP. It operates at speed of 200 Mbps in Physical Layer and 80 Mbps in MAC Later. It uses OFDM carriers spaced at 24.414 kHz, with carriers from 2 to 20 MHz.
  - **HomePlug AV2:** It is based on IEEE 1901 standard and fully interoperable with HomePlug AV and HomePlug Green PHY devices. It supports PHY rates of 1300 Mbps with full MIMO. It was designed for emerging applications for consumers like HDTV, Online Gaming, Home Storage Systems, and VoIP. It offers 1 Gbps speed at Physical Layer and 600 Mbps at MAC Layer.
  - **HomePlug Green PHY:** It was basically proposed for home appliance connectivity like HVAC (Heating, Ventilation, and Air Conditioning) as well as Smart Meters to facilitate smart grid applications with data transmission speed of 10 Mbps. HomePlug Green PHY is designed for low power, cheap cost and small devices facilitate reliable connectivity during data transmission.
- **ITU-T G.9959:** G.9959 is MAC layer protocol proposed by International Telecommunication Unit (ITU) especially for low-cost, bandwidth, and reliable wireless communication of IoT based devices (Gomez et al., 2017). G.9959 based networks operate mostly in unlicensed frequency bands depending on the countries, i.e., US-908.4 MHz, 860.4 MHz in Europe at data rates ranging from 9.6 Kbps to 100 Kbps. G.9959 networks has two dissimilar nodes: Control Nodes and End Device nodes. All the commands are initiated by control nodes and these commands are executed by end-device nodes. G.9959 networks, supports mesh topology and protocol supports max 4 hops between end

nodes and max 232 nodes in a single network. The average communication distance between two nodes can be 100 feet with throughput of 40 Kbps.

- **DASH7:** DASH7 Alliance Protocol is wireless protocol originated from ISO 18000–7, an active RFID standard and operates on different ISM bands, i.e., 433, 868 and 915 MHz (Ayoub et al., 2018; Weyn et al., 2013, 2015). It was designed especially for IoT device to device communication. It specifies all layers of the OSI model, making it very easy to implement. DASH7 protocol is highly scalable, supports long-range wireless communication in encrypted format and supports higher data rate. It is based on Master/Slave architecture and is highly suitable for lightweight and transitive traffic. DASH7 protocol has 4 classes of devices-Blinker Device-Simplex device, can perform only data transmission but cannot act as receiver; Endpoint-A low power device that can perform both roles of data transmission and receiving. As the device supports wake-up events which gives the device the capability to receive a request and response can be transmitted; Gateway-connects DASH7 network to outside network; Sub controller-makes use of wake on scan cycles like typical endpoint devices. DASH7 has two foreground network protocols: D7A Network Protocol (D7ANP) and D7A DataStream Protocol (D7ADP).
- **DECT/ULE Protocol:** DECT (Digital Enhanced Cordless Telecommunications/Ultra Low Energy) is a wireless communication universal European standard designed for Sensor Communications and Smart home automation (Das and Havinga, 2012; Gomez and Paradells, 2010). The main features of DECT/ULE are being ultra-low-power and consume less energy as compared to 802.11 standards and have wider coverage range as compared to IEEE 802.15 and BLE. DECT/ULE follows star topology, i.e., all devices are connected to each other, and any device can communicate with any device. DECT/ULE support FDMA, TDMA, and time division multiplexing and faceless issues with regard to interferences and collisions.

6.4.1.5 IOT-NETWORK LAYER PROTOCOLS

The main task of Network layer protocols is to route the packets from one IoT device to another, i.e., from source IoT device to destination IoT device. Network layer protocols are further categorized into two categories: Routing Protocols and Encapsulation Protocols. The following are the Network Layer Routing Protocols:

- **RPL (Routing Protocol for Low-Power and Lossy Networks):** IETF Routing Over Low-power and Lossy links (ROLL) working group proposed a new distance vector protocol for IoT networks based on IPv6 for resource-constrained nodes called RPL (Winter, 2012; Vasseur, 2011). It was primarily designed to perform minimum routing via constructing a robust topology over lossy links. All sorts of network models like Point to Point, Point to Multipoint and Multipoint to point are supported by RPL protocol. RPL protocol constructs Destination Oriented Directed Acyclic Graph (DODAG) which demonstrates all the nodes routing directions. In DODAG, every node has complete information regarding its parent node, but no information regarding child nodes. When a node wants to transmit a message, it sends Destination Advertisement Object (DAO) to its parent nodes, it is transmitted to the Root node and the root nodes transmit is based on the destination node. If a new node wants to join the existing network, it transmits a DODAG Information Solicitation (DIS) message and the acknowledgment is sent by the root node via acknowledgment message called DAO-ACk. In the entire network, the only root node has complete information of DODAG, so all communications can only take place via the root node.
- **CORPL or Cognitive RPL Protocol:** This was proposed for Cognitive Radio Enabled AMI Networks (Aijaz et al., 2015). The protocol upgrades the conventional RPL protocol to solve different routing issues in Cognitive radio environments by securing primary users as well as fulfilling utility prerequisites of the auxiliary network. In order to solve routing issues, CORPL protocol makes use of opportunistic forwarding approach to route the packets by making use of multiple forwarder nodes and implementing a coordinating scheme to choose the best possible next hop to forward the packet. The approach enhances overall throughput, reduces end-to-end delay and brings overall reliability in the network. The two techniques are utilized by CORPL for enhancing performance in the network because of spectrum sensing. In the first technique, the performance is improvised via collecting sensing schedule information of all the neighboring nodes. The second technique improvises the performance by reducing the sensing time of spectrum. The performance evaluation of CORPL protocol was done in MATLAB with square region of side

1000 meters which is occupied by 9 PU transmitters and results state that CORPL protocol has enhanced network reliability by reducing interferences by up to 50% and also reduced the deadline violation probability of delay-sensitive traffic.

- **CARP Protocol:** CARP (Channel Aware Routing Protocol) is a multi-hop routing protocol designed for Underwater WSN networks (Basagni et al., 2015). Highly efficient for IoT as only small data size packets are transmitted. The protocol takes into consideration the Link Quality, which was calculated on the basis of previous network transmissions from neighboring nodes to determine the forwarding nodes. CARP Protocol operates as:
  - **Network Initialization:** In this, a source node transmits a HELLO packet in the network which contains the UNIQUE ID of the node as well as Hop Count. When the packet is received, every node checks hop count variable, whether greater than the hop count carried by packet by 1+ value. If it is greater, the value is updated and the packet is transmitted otherwise dropped.
  - **Data Forwarding:** When a source node has many packets, it broadcasts PING packet to search for suitable delay in neighboring nodes. If the packet pairs>1 then packets are transmitted otherwise dropped. Simulation-based results demonstrate that CARP protocol is best in PDR, Delay, and minimizes energy as compared to FBR and EFlood protocols.

6.4.1.6 NETWORK LAYER ENCAPSULATION PROTOCOLS

- **6LoWPAN:** Low power Wireless Personal Area Network (6LoWPAN) is regarded as one of the most important network layer encapsulation protocol on which different types of IoT communications can be done (Gomes et al., 2018). It encapsulates IPv6 long headers in IEEE 802.15.4 small data size packets of max size of 127 bytes. 6LoWPAN protocol was designed in 2007 by IETF 6LoWPAN working group. It is primarily responsible for performing re-ordering and fragmenting of IPv6 packets, compressing protocol stack headers, enabling stateless addressing and creating mesh routing to maintain efficient consistency with packet transmission to upper layers. Considering IoT, 6LoWPAN is highly efficient in terms of supporting different length addresses, efficient bandwidth, topologies, i.e., Star or Mesh, low power utilization, low cost, scalability, mobility, and wakeup mode.6LoWPAN has four types of headers: No 6LoWPAN (00); Dispatch (01); Mesh-Addressing (10); Fragmentation (11).

- **6Lo:** In 2013, IPv6 over networks of resource-constrained nodes (6Lo) Working Group proposed 6Lo (Yushev et al., 2016). Even though two different protocols were developed, i.e., 6LoWPAN and 6TiSCH based on IEEE 802.15.4 and IEEE 802.15.4e, but still more data links are required, so 6Lo was proposed.6Lo defines the specifications for IPv6 over constrained networks as follows:
    - Limitations in terms of memory, power, and processing capabilities.
    - Energy Optimization and efficient bandwidth utilization.
    - Hard upper bounds on the state, code space, and processing cycles.
    - Missing of some Layer 2 services like connectivity and Broadcast.

The 6Lo has produced many standards like IPv6-Over-BLE, IPv6-Over – Z-Wave and many are in progress like IPv6-over-DECT Ultra Low Energy, IPv6-over-BACNET Master-Slave/Token-Passing networks, IPv6-over-NFC, IPv6 over IEEE 802.11 AH, IPv6 over WIA-PA (Wireless networks for Industrial Automation-Process Automation).

- **6TiSCH:** IETF, new working group was formulated in November 2013 to allow IPv6 to pass through a Time Slotted Channel Hopping (TSCH) mode of IEEE 802.15.4E data links (Dujovne et al., 2014). It proposes a new concept called "Channel Distribution Usage (CDU)" which is regarded as a matrix of cells comprising of frequencies available in columns and time slots available for all sorts of network scheduling operations in rows. 6TiSCH is regarded as novel solution for MESH type networks that makes use of deterministic slotted channels to avoid all sorts of attenuations, collisions, optimize energy consumption and maintain load balancing of data. It was designed to provide efficient delivery and reduce jitter and latency in WSN networks like WirelessHART and ISA 100.11a. 6TiSCH makes use of "On-The-Fly (OTF) Scheduling" in which every node keeps a close eye on the packet transmission queue. If the queue gets filled up, OTF determines that there is not enough outbound bandwidth and fires additional timeslots with neighboring nodes. It supports different scheduling approaches like-Distributed, Centralized, and Hybrid.

6.4.1.7 SESSION LAYER PROTOCOLS

• **Message Queuing Telemetry Transport (MQTT):** MQTT was outlined by IBM in 1999 and is ISO Standard: ISO/IEC PRF 20922 publish-subscribe based messaging protocol (Light, 2017). It was designed to connect all sorts of embedded devices and networks with middleware and applications by using TCP protocol for transport layer. MQTT is highly simple to implement and very suitable for those devices having limited bandwidth and operate under unreliable network. MQTT protocol is advantageous in terms of throughput, but lacks latency. MQTT protocol is highly successful for implementation, where system has three main components: Publishers, Subscribers, and Brokers. Considering IoT, publishers are defined as "Sensors" connected to all brokers to transmit the data and go to sleep mode, when not in action. Subscribers are termed as "Applications" or "Sensor Data" to connect to brokers to inform when at any point of time new data is available. The brokers perform the task of data classification and send to subscribers according to their requirements (Figure 6.8).

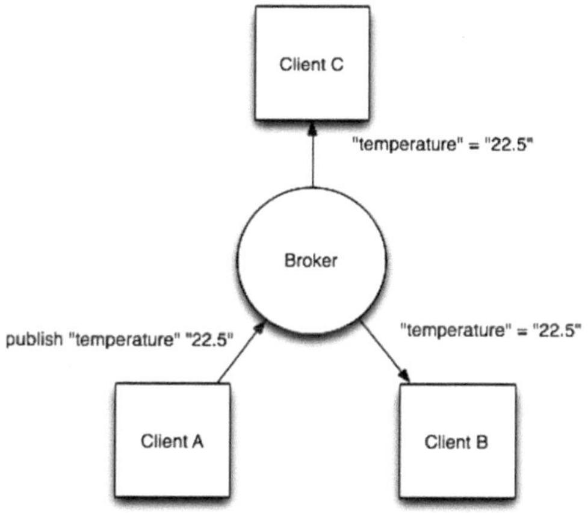

FIGURE 6.8 MQTT architecture.
(*Source:* Jabby, (2014). *MQTT and CoAP, IoT Protocols.* Retrieved from: https://www.eclipse.org/community/eclipse_newsletter/2014/february/article2.php.)

MQTT has three QoS levels:

- **Fire and Forget:** Transmit the message and forget about it, i.e., No receipt of acknowledgment of any message transmitted from source to the destination device.

- o **Delivered at Least Once:** Message is delivered at least once for every delivery and acknowledgement is sent to the source device.
- o **Delivery Exactly Once:** In order to assure that the message is delivered to the destination node, 4-way handshake procedure is applied (Figure 6.9).

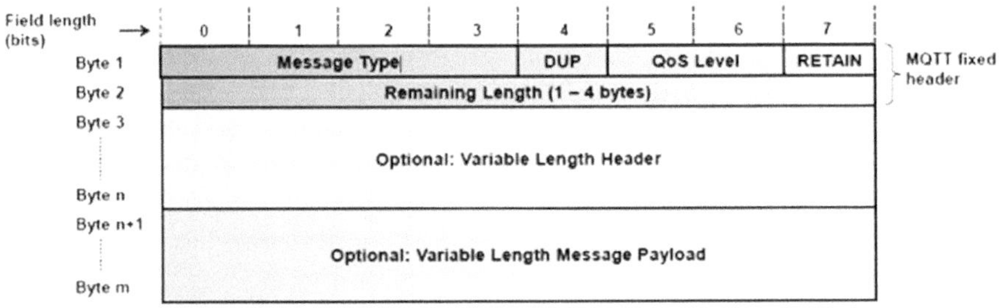

FIGURE 6.9 MQTT message format.
(*Source:* Living Mobile, (2015). *MQTT Message Format.* Retrieved from: http://ritzinmobileworld.blogspot.com/2015/05/mqtt-message-format-part-i.html)

It consists of fixed message header variable header and payload. Header comprise of 4 bytes, length is 1–4 bytes.

- **Secure MQTT (SMQTT):** SMQTT (Secure MQTT) is regarded as an extension of MQTT protocol, which combing standardized MQTT with lightweight Attribute-Based Encryption (ABE) over elliptic curves (Singh et al., 2015). The main reason behind adapting ABE algorithm is because of its unique design which broadcasts encryption and with one encryption a single message can be transferred to multiple users. SMQTT protocol inherits security features via addition of new MQTT publish message 'Spublish' with reserved message type '0000.' SMQTT has three main entities:
    - o **Publisher Device:** Performs the task of publishing the data under given topic.
    - o **Subscriber Device:** Receives data under same topic via broker.
    - o **PKG or Broker:** Third-party highly trusted.

  SMQTT works in four main stages:
    - o **Setup Phase:** In this, all devices (Publisher and Subscriber) register with PKG by giving a UNIQUE ID along with associated attributes. The master Secret Keyset is generated by PKG as

per CP/KP-ABE scheme and all parameters are published with a universal set.

- **Encrypt Phase:** As soon as data is published, it gets encrypted, published by broker which sends the data to the subscribers and data gets decrypted at the subscriber end using same master secret key. Broker transmits SUBACK to all receiving devices.
- **Publish Phase:** All the encrypted data is embedded as Payload by publisher in SPublish command and the SPublish packet is transmitted back to Broker. It's the duty of the broker to respond to PUBACK Packet. Publisher on receiving PUBACK acknowledges with reply by transmitting PUBREL packet. Broker then broadcasts the message to all subscribers and deletes the data and transmits PUBCOMP packet to sender.
- **Decrypt Phase:** The message is decrypted using private attribute keys by verifying whether the access policy is satisfied or not.

• **XMPP (Extensible Messaging and Presence Protocol):** Extensible Messaging and Presence Protocol (XMPP) is an XML based communication protocol for message-oriented middleware (Ozturk, 2010; Saint-Andre, 2011). The protocol was outlined by Jabber Open source community in 1999 for maintaining contact list, presence information, and real-time instant messaging and was named "Jabber." It was later standardized by IETF in 2002 by creating XMPP working group. XMPP utilizes XML text format for facilitating person-to-person communication and unlike MQTT, it runs over TCP. Considering IoT, XMPP facilitates device addressing in a highly simple way. XMPP supports both architectures: Publish/Subscribe and Request/Response, facilitating developers with enough flexibility and scalability to choose the required architecture. XMPP gives a system for messaging over a network, which encourages tons of applications as compared to conventional messaging and data dispersion (Figure 6.10). The following are the key XMPP technologies:

- **Core:** Information with regard to core XMPP technologies for XML streaming.
- **Jingle:** SIP compatible multimedia signaling for file transfer, video, voice, and other applications.
- **Multi-User Chat:** Flexible, multi-party communications.
- **PubSub:** All sorts of alerts and real-time notifications with regard to data synchronization, etc.
- **BOSH:** An HTTP binding for XMPP traffic.

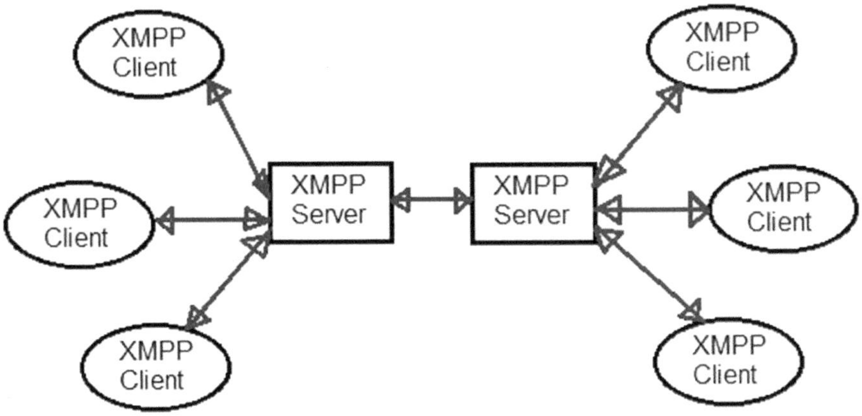

FIGURE 6.10 XMPP protocol – working.
(*Source:* rfwireless World, (2018). *XMPP*. Retrieved from: http://www.rfwireless-world.com/IoT/XMPP-protocol.html.)

- **CoAP (Constrained Application Protocol):** CoAP was proposed by the IETF and is regarded as a service layer protocol designed for utilization in resource-constrained internet devices like WSN sensor nodes (Shelby et al., 2014; Jan et al., 2016; Bormann et al., 2012, 2018; CoAP, 2016). CoAP supports easy translation to HTTP for simple integration with web and also provides specialized requirements in terms of less overhead, multicasting support, simplicity towards implementation. The protocol provides lightweight RESTful (HTTP) interface which acts as a standard interface between HTTP client and servers. The protocol is planned to encourage Machine-to-Machine (M2M) applications like smart energy and smart building, automation, etc. CoAP utilizes two types of messages: Requests and Responses via simple, binary, and base header format. In terms of security, CoAP utilizes UDP protocol as default and also supports DTLS for high-level communication security for message exchange. CoAP is sub-divided into two layers: Message Sub-layer and Request/Response sub-layer. The message layer performs the task of reliable end to end transfer of messages between nodes and the other layer deals with all sorts of REST communications. It incorporates a CoAP Version (V), Transaction Type (T), OC (Option Count), Code, and Message ID (M ID) in the header, and rest it contains Token, Options, and Payload (Figure 6.11).

```
 0                   1                   2                   3
 0 1 2 3 4 5 6 7 8 9 0 1 2 3 4 5 6 7 8 9 0 1 2 3 4 5 6 7 8 9 0 1
+-+-+-+-+-+-+-+-+-+-+-+-+-+-+-+-+-+-+-+-+-+-+-+-+-+-+-+-+-+-+-+-+
|Ver| T |  TKL  |      Code     |          Message ID           |
+-+-+-+-+-+-+-+-+-+-+-+-+-+-+-+-+-+-+-+-+-+-+-+-+-+-+-+-+-+-+-+-+
|   Token (if any, TKL bytes) ...
+-+-+-+-+-+-+-+-+-+-+-+-+-+-+-+-+-+-+-+-+-+-+-+-+-+-+-+-+-+-+-+-+
|   Options (if any) ...
+-+-+-+-+-+-+-+-+-+-+-+-+-+-+-+-+-+-+-+-+-+-+-+-+-+-+-+-+-+-+-+-+
|1 1 1 1 1 1 1 1|    Payload (if any) ...
+-+-+-+-+-+-+-+-+-+-+-+-+-+-+-+-+-+-+-+-+-+-+-+-+-+-+-+-+-+-+-+-+
```

FIGURE 6.11 CoAP protocol message format.
(*Source:* Shelby, Z., Hartke, K., Bormann, C. (2014). The Constrained Application Protocol (CoAP). Internet Engineering Task Force (IETF). : https://tools.ietf.org/html/rfc7252)

CoAP protocol has four messaging modes:

- **Confirmable Message (CON):** It requires a response, i.e., Positive or Negative Acknowledgement. In case of non-receipt of any acknowledgment, retransmissions are done till all attempts are done. Retransmissions make use of non-linear and exponential strategy between attempts.
- **Non-Confirmable Message (NON):** It doesn't require any sort of Acknowledgement, but the message requires Message ID for supervising in case of retransmission. If the message fails to be processed at the recipient side, server responds with RST.
- **Piggy-Backed:** It is used for Client/Server communication, when the server responds directly to the client after receiving the message, i.e., Acknowledgement message for successful receipt of the message. ACK message also contain response message, for failure response, ACK contains failure response code.
- **Separate Response:** It is only used when the response from the server comes in a message entirely different from acknowledgment and server takes some time to transmit the message.

CoAP protocol makes use of different messages: GET, PUT, PUSH, DELETE to perform operations like create, update, select, and delete.

- **AMQP (Advanced Message Queuing Protocol):** AMQP is another open standard protocol for IoT typically designed for message-oriented environments (Vinoski, 2006; Naik, 2017; Luzuriaga et al., 2015). AMQP supports reliable communication via directives in terms of at-most-once, at-least-once and exactly once-delivery. Unlike MQTT, it makes use of TCP for message exchange. In AMQP protocol, all types of exchanges are handled by two main components: Exchanges and message queues (Figure 6.12).

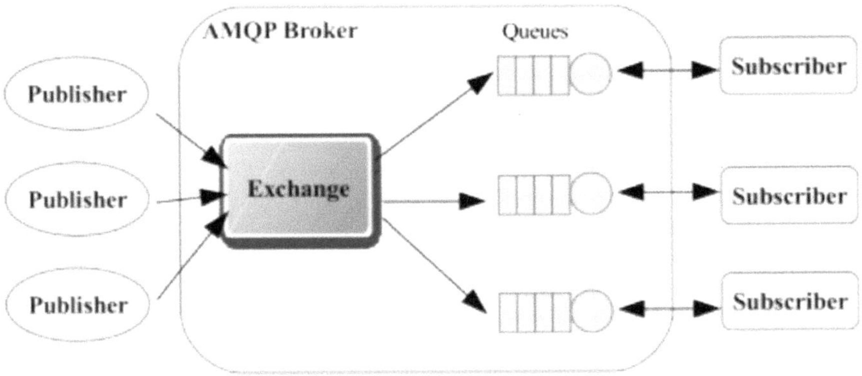

FIGURE 6.12 AMQP protocol working.
(*Source:* Electronics for You, (2018). *AMQP*. Retrieved from: https://iot.electronicsforu.com/research-articles/internet-things-protocols-landscape/2/.)

The exchanges perform the task of message routing and all messages are stored in message queue before they are routed to the receiver. It is mostly found in business messaging. It defines "devices" as mobile units communicating at back-end data centers.

- **DDS (Data Distribution Service):** This was proposed by Object Management Group (OMG) for facilitating Machine to Machine communication and is regarded as yet another publishes/subscribes protocol (Pardo-Castellote et al., 2005). Unlike MQTT, AMQP, and XMPP, DDS is a highly reliable and secured protocol as it makes use of SSL and DTLS. DDS has two levels of Interfaces-
  - **Data-Centric Publish-Subscribe (DCPS):** DCPS, whose main task is to assure efficient delivery of information to the designated receivers.
  - **Data-Local Reconstruction Layer (DLRL):** DLRL, which facilitates easy integration with application layer.

DDS has 23 different levels of QoS like Security, Urgency, Priority, Durability, Reliability, etc.

6.4.2 SIMULATION TOOLS FOR IOT-INTERNET OF THINGS

In order to implement newly designed protocols and IoT devices in real-time, everything has to undergo a thorough stage of evaluation and testing to determine the required capability and performance of the protocol and device. The testing and evaluation is carried out by the variety of simulation tools. Example: Testing a large number of IoT nodes is not practical in the real world during the initial stages of design and exploration because of financial and operational constraints, especially when the protocol reliability, security, and utility are unknown. So, to overcome all these challenges, a Simulation tool is the only way out. A standardized research in IoT starts with idea formulation and ends with real-world implementation comprising both virtual and real entities. In order to test virtual entities, a simulation and testbed is essential, but for real-world live scenarios are required for final testing before launching it finally for the end-user.

In this section of the chapter, we explore some simulators especially designed for IoT which can lay a strong foundation for researchers to perform novel research, propose novel designs cum architectures and protocols to extend IoT technology.

- **Bevywise IoT Simulator:** It is free IoT simulation suite which facilitates researchers to design, test, and emulate real-time IoT Devices, middleware, and management solutions (Bevywise, 2018). The simulator can simulate few to thousands of IoT devices under single box. It facilitates simulation of Smart Buildings, Smart Cities, Smart Manufacturing, Smart Farming, Healthcare devices and many more. Features:
    - Simple and Intuitive GUI Interface for easy operations.
    - Fully support MQTT protocol.
    - Powerful Python programming support to program user-customized messages for real-time transmission.
    - Support for WILL topic and retain messages.
    - REST-based APT for full simulation integration.
    - Highly scalable, robust, and throughput oriented.
- **IoTIFY:** It is a cloud-based simulation platform for IoT facilitating the user to design production-ready IoT and blockchain in web browser (IoTIFY, 2018). It facilitates the user to design and develop powerful simulations with regard to blockchain wallets, autonomous vehicles, smart cities, smart meters, smart hospitals and many more.

Features:

- o Easy design of IoT device models using JavaScript-based templates.
- o Directs the virtual IoT devices to all types of cloud-Amazon AWS, Azure, IBM Bluemix, and Google Cloud.
- o Supports tons of protocols-MQTT, CoAP, UDP, TCP, DTLS, TLS, HTTP, etc.
- o Facilitates generation of specific tokens based for devices for performing all sorts of registration and provisioning.
- o Facilitates users to perform research on tons of varied network conditions to determine throughput and latency.

- **CupCarbon U-One:** CupCarbon is a discrete wireless sensor network and IoT simulator, which is multi-agent and performs geolocations-based simulations (CupCarbon, 2017; Mehdi et al., 2014). It enables to users to run simulations and keep close eye on varied events changing over time. It is well-facilitated simulator for IoT and can simulate smart cities along with Unmanned Aerial Vehicles (UAVs) and detailed street-level topology and maps belonging to real-world. It offers users to create paths, open or congested roads and agent's assignments to make them mobile. Trajectories are generated in two ways: Manual or Automatic.

Cup carbon simulator has for modules:

- o **Agent Module:** Includes devices and events to simulate WSN and configure simulator for performing simulations.
- o **OpenStreetMap:** Deploying wireless sensors on the map.
- o **WiSen Simulator Module:** Simulate WSN and connected to all agents of the simulator.
- o **Solver Module:** Integrates all sorts of algorithms like Routing, Coverage, etc.

Features:

- o Interactive and highly simple GUI interface for users to perform all sorts of simulations based on JavaFX.
- o Simple script language based on SenScript which also facilitates interactive commands.
- o 2D/3D Visualization with OpenStreetMap.
- o Facilitates Automation and repeated tasks using CupCarbon Scripts.
- o Arduino/XBee code generation.
- o Integrates various radio modules and standards like 802.15.4, ZigBee, Wi-Fi, and LoRa.

- Intelligent Mobility models for running comprehensive simulations.

- **SimpleIoTSimulator:** It creates a powerful and highly interactive work environment for end-users to simulate hundreds or thousands of sensors and gateways on a single computer (SimpleSoft, 2018). The simulator supports tons of common IoT protocols like HTTP, HTTPS, MQTT, MQTT-SN, CoAP, etc. The simulator can easily adapt to changing customer requirements and create powerful replica of real-time scenarios. The simulator supports both IPv4 and IPv6 protocols to simulating sensors in constrained environments. Features are:
  - Built-in learner utilities to make SimpleIoTSimulator to learn more IoT scenarios.
  - Powerful enough to simulate thousands of sensor nodes and gateways to single simulator.
  - Easy generation of user-based scripts to generate real-time scenarios, modifications, and dynamic changing properties.
  - Realistic simulations with easy to use GUI interface.
  - Supports almost every IoT standard protocol along with IPv4 and IPv6.
  - Supports other management protocols like SNMP, Telnet, SSH, etc.
  - Real-time sensor data modifications as per changing operational scenarios.

- **MIMIC IoT Simulator:** MIMIC IoT simulator provides a strong real-world replica of operational test environment for managing all types of IoT based sensors, gateways, and devices to test the IoT environment (Gambit Communications, 2018). The simulator is fully competent to simulate various scenarios like: Industry 4.0, Smart Factories, Smart Cities and Agriculture. It is regarded as fully loaded simulator and bundles almost types of applications, Middleware's, Brokers, Load Balancers, MQTT gateways and clients. MIMIC IoT Simulator has the following main components:
  - **MQTT Simulator:** It is fully equipped to simulate almost 1,00,000 MQTT v3.1 and MQTT 3.1.1 based sensors and devices. Every device has a unique IP Address, Port, and Address. Once the simulation scenario is created and started, every device can

respond to all sorts of MQTT requests to and fro from a broker or application from the network.

- **HTTP/REST Simulator:** All the sensors can communicate using HTTP/HTTPS/REST. Every device is fully efficient to respond to REST requests to any application from anywhere in work.
- **CoAP Simulator:** MIMIC can simulate tons of CoAP enabled devices managed by one or more clients.
- **Modbus Simulator:** It facilitates the creation of Modbus slave devices to communicate with Modbus Master Server using TCP protocol. Every slave has a unique IP address that can request to server and respond accordingly.
- **MIMIC Shell:** Main controller of all activities of MIMIC commands using commands and even equipped with GUI interface.

In addition to above, MIMIC also allows the users to simulate SNMP v1, SNMPv2c, SNMPv3, NetFlow, sFlow, IPMI, Cisco IOS, Juniper JUNOS, Telnet/SSH based devices.

Features:

- Facilitates creation of dynamic scenarios of Smart Cities, Industry 4.0 and other IoT based environments.
- Tons of IoT sensors can be simulated with an almost exact replica of real-world.
- Facilitates reliable testing and scalability of sensors.
- Handles all sorts of connections and scenarios with regard to heterogeneous environments.

- **Cooja:** It is regarded as Contiki network emulator-An extensible Java-based simulator for emulating Tmote Sky and other nodes (Mehmood, 2017; Nayyar and Singh, 2015; Thiruveedula, 2017; Bagula and Erasmus, 2015). It is basically an operating system for simulating resource-constrained and networked low-power IoT devices. It facilitates multitasking and contains a fully equipped TCP/IP suite. It consists of a powerful built-in suite for simulating and constructing powerful network scenarios. Contiki supports IPv4; IPv6 along with various IoT based protocols like: 6LoWPAN, RPL, CoAP. Cooja supports three different levels of simulation -

- **Networking/Application Level:** Cooja facilitates the users to exchange all sorts of simulator modules like device drivers or radio medium modules.
- **Operating System Level:** Cooja simulates OS via execution of native operating system code. All sorts of user processes are executed to alter Contiki core functionality.
- **Machine Code Instruction Set Level:** It facilitates new nodes creation under varied structures as compared to standard nodes.

Features:

- Highly memory efficient and contains a defined set of mechanisms for memory allocation like Memory block allocation memb, managed memory allocator mmem, and C memory allocator malloc.
- Fully operational IP network stack with almost all standard protocols of networks TCP, UDP, and HTTP along with IoT protocol stack.
- Fully support low-power devices simulation and gives a precision estimation of nodes power utilization.
- Fully support dynamic loading and modules linking at simulation run-time. The module load faster and link standard ELF files.
- Supports platforms like 8051, MSP430, ARM, and AVR devices.
- Supports commands via optional shell.

6.5 CONCLUSION

In view of the above arguments that support an exclusive discussion on IoT and Digital Business Environment inclusive Cyber Space, Cyber Crimes and Cybersecurity paradigm. Even though there was evidence of such increased concerns about the cybercrime, cyber threat, risk approaching from cyberspace along with the idea of serious cyberattacks from time to time against decisive infrastructure perceived by top-notch agencies globally in digital business. Regarding social and sustainable business, positive benefits from cyberspace are essential under ethical circumstances and codes; however, acknowledges the threats related to cyberspace. Moreover, it highlights cyber world importance, even though more concerned is still focused on connecting the opportunities connected to cyberspace and minimizes the threats in today's competitive world.

KEYWORDS

- **cybersecurity**
- **digital business environment**
- **internet of things (IoT)**
- **IoT protocols**
- **IoT security**
- **message queuing telemetry transport (MQTT)**
- **simulation tools**

REFERENCES

Abera, T., Asokan, N., Davi, L., Koushanfar, F., Paverd, A., Sadeghi, A. R., & Tsudik, G. (2016, June). *Things, trouble, trust: on building trust in IoT systems*. In Proceedings of the 53rd Annual Design Automation Conference (p. 121). ACM.

ABI Research, (2014). Retrieved from: https://www.abiresearch.com/press/the-internet-of-things-will-drive-wireless-connect (Accessed on 28 August 2019).

Adame, T., Bel, A., Bellalta, B., Barcelo, J., & Oliver, M., (2014). IEEE 802.11 AH: The Wi-Fi approach for M2M communications. *IEEE Wireless Communications, 21*(6), 144–152.

Aijaz, A., Su, H., & Aghvami, A. H., (2015). CORPL: A routing protocol for cognitive radio enabled AMI networks. *IEEE Trans. Smart Grid, 6*(1), 477–485.

Alliance, Z. W., (2013). *About Z-Wave Technology*. Z-Wave Alliance, Tech. Rep. https://z-wavealliance.org/about_z-wave_technology/ (Accessed: December 2018)

Alliance, Z. W., (2015). Z-wave. Retrieved from: https://www.z-wave.com/ (Accessed: December 2018)

Appazov, A., (2014). *Legal Aspects of Cybersecurity*. University of Copenhagen. Retrieved from: http://justitsministeriet.dk/sites/default/files/media/Arbejdsomraader/Forskning/Forskningspuljen/Legal_Aspects_of_Cybersecurity.pdf (Accessed on 28 August 2019).

Atzori, L., Iera, A., & Morabito, G., (2010). *The Internet of Things: A Survey* (pp. 2787–2805). In Computer Networks.

Ayoub, W., Mroue, M., Nouvel, F., Samhat, A. E., & Prévotet, J. C., (2018). Towards IP over LPWANs technologies: LoRa WAN, DASH7, NB-IoT. In: *Digital Information, Networking, and Wireless Communications (DINWC), Sixth International Conference* (pp. 43–47). IEEE.

Bagula, B., & Erasmus, Z., (2015). *IoT Emulation with Cooja*. In ICTP-IoT workshop.

Balakrishna, C., (2012). Enabling technologies for smart city services and applications. In: *6th International Conference on Next Generation Mobile Applications, Services and Technologies (NGMAST)* (pp. 223–227).

Basagni, S., Petrioli, C., Petroccia, R., & Spaccini, D., (2015). CARP: A channel-aware routing protocol for underwater acoustic wireless networks. *Ad Hoc Networks, 34*, 92–104.

Belissent, J., (2010). *Getting Clever About Smart Cities: New Opportunities Require New Business Models*.

Bevywise, (2018). Retrieved from: https://www.bevywise.com (Accessed on 28 August 2019).

Bo, Y., & Wang, H., (2011). The application of cloud computing and the internet of things in agriculture and forestry. In: *International Joint Conference on Service Sciences (IJCSS)*, 168–172.

Bormann, C., Castellani, A. P., & Shelby, Z., (2012). Coap: An application protocol for billions of tiny internet nodes. *IEEE Internet Computing, 16*(2), 62–67.

Bormann, C., Lemay, S., Tschofenig, H., Hartke, K., Silverajan, B., & Raymor, B., (2018). *CoAP (Constrained Application Protocol) Over TCP, TLS, and WebSockets (No. RFC 8323)*.

Boyd, D., & Crawford, K., (2012). Critical questions for big data: Provocations for a cultural, technological, and scholarly phenomenon. *Information, Communication & Society 15*(5), 662–679.

Bryant R, Katz RH and Lazowska ED (2008) *Big-data computing: Creating revolutionary breakthroughs in commerce, science and society*. In: Computing Research Initiatives for the 21st Century, Computing Research Association, Ver. 8. Available at: http://www.cra.org/ccc/docs/init/Big_Data.pdf

Camarinha-Matos, L. M., Tomic, S., & Graça, P., (2013). Technological innovation for the internet of things: 4th IFIP WG 5.5/SOCOLNET. *Doctoral Conference on Computing, Electrical and Industrial Systems, DoCEIS, Costa de Caparica, Portugal, Proceedings* (Vol. 394). Springer.

Chen, H., Chiang, R. H., & Storey, V. C., (2012). Business intelligence and analytics: From big data to big impact. *MIS Quarterly, 36*(4), 1165–1188.

CoAP, (2016). Retrieved from: http://coap.technology (Accessed on 28 August 2019).

CupCarbon., (2017). *Cup Carbon U-One 3.8: A Smart City & IoT Wireless Sensor Network Simulator*. Retrieved from: http://www.cupcarbon.com (Accessed on 28 August 2019).

Das, K., & Havinga, P., (2012). Evaluation of DECT-ULE for robust communication in dense wireless sensor networks. In: *Internet of Things (IoT), 3rd International Conference* (pp. 183–190). IEEE.

Duan, L., Street, W. N., & Xu, E., (2011). Healthcare information systems: Data mining methods in the creation of a clinical recommender system. *Enterprise Information Systems, 5*(2), 169–181.

Dujovne, D., Watteyne, T., Vilajosana, X., & Thubert, P., (2014). 6TiSCH: Deterministic IP-enabled industrial internet (of things). *IEEE Communications Magazine, 52*(12), 36–41.

Färber, F., Cha, S. K., Primsch, J., Bornhövd, C., Sigg, S., & Lehner, W., (2012). SAP HANA database: Data management for modern business applications. *ACM Sigmod Record, 40*(4), 45–51.

Gambit Communications, (2018). *MIMIC IoT Simulator*. Retrieved from: https://www.gambitcomm.com/site/iot_simulator.php (Accessed on 28 August 2019).

Gölzer, P., Cato, P., & Amberg, M. (2015, May). Data Processing Requirements of Industry 4.0-Use Cases for Big Data Applications. In ECIS.

Gomes, T., Salgado, F., Pinto, S., Cabral, J., & Tavares, A., (2018). A 6LoWPAN accelerator for internet of things endpoint devices. *IEEE Internet of Things Journal, 5*(1), 371–377.

Gomez, C., & Paradells, J., (2010). Wireless home automation networks: A survey of architectures and technologies. *IEEE Communications Magazine, 48*(6).

Gómez, C., Paradells, J., Bormann, C., & Crowcroft, J., (2017). From 6LoWPAN to 6Lo: Expanding the universe of IPv6-supported technologies for the internet of things. *IEEE Communications Magazine, 55*(12), 148–155.

Gopalkrishnan, V., Steier, D., Lewis, H., & Guszcza, J., (2012). Big data, big business: Bridging the gap. In: *Proceedings of the 1st International Workshop on Big Data, Streams and Heterogeneous Source Mining: Algorithms, Systems, Programming Models and Applications* (pp. 7–11). ACM.

Grimmer, J., (2015). We are all social scientists now: How big data, machine learning, and causal inference work together. *PS: Political Science & Politics, 48*(1), 80–83.

Gubbi, J., Buyya, R., Marusic, S., & Palaniswami, M., (2013). Internet of things (IoT): A vision, architectural elements, and future directions. *Future Generation Computer Systems, 29*(7), 1645–1660.

Gui, Z., Yu, M., Yang, C., Jiang, Y., Chen, S., Xia, J., & Huang, Q., (2016). Developing sub-domain allocation algorithms based on spatial and communicational constraints to accelerate dust storm simulation. *PloS One, 11*(4), e0152250.

Hassan, S. M., Ibrahim, R., Bingi, K., Chung, T. D., & Saad, N., (2017). Application of wireless technology for control: A wireless HART perspective. *Procedia Computer Science, 105*, 240–247.

Heffner, J., (2014). Predictive policing. *GEO: Geoconnexion Internal, 13*(7), 20–23.

Homeplug Alliance Kit, (2018). Retrieved from: http://www.homeplug.org/media/filer_public/79/fb/79fbf901-d875–475f-9abc-8a272329f4e2/homeplug_press_kit_091114.pdf (Accessed on 28 August 2019).

Hori, M., Kawashima, E., & Yamazaki, T., (2010). Application of cloud computing to agriculture and prospects in other fields. *Fujitsu Scientific & Technical Journal, 46*(4), 446–454.

Hossain, M. M., Fotouhi, M., & Hasan, R., (2015). Towards an analysis of security issues, challenges, and open problems in the internet of things. In: *Services (SERVICES), IEEE World Congress on* (pp. 21–28).

Hsu, M. H., (2008). A personalized English learning recommender system for ESL students. *Expert Systems with Applications, 34*(1), 683–688.

Indeck, R. S., & Indeck, D. M., (2012). *U.S. Patent No. 8,156,101*. Washington, DC: U.S. Patent and Trademark Office.

IoTIFY, (2018). Retrieved from: https://iotify.io (Accessed on 28 August 2019).

Jalali, A., Olabode, O. A., & Bell, C. M., (2012). Leveraging cloud computing to address public health disparities: An analysis of the SPHPS. *Online Journal of Public Health Informatics, 4*(3), 1–7.

Jan, S. R., Khan, F., Ullah, F., Azim, N., & Tahir, M., (2016). Using CoAP protocol for resource observation in IoT. *International Journal of Emerging Technology in Computer Science & Electronics, ISSN,* 0976–1353.

Javed, F., Afzal, M. K., Sharif, M., & Kim, B. S. (2018). *Internet of Things (IoT) operating systems support, networking technologies, applications, and challenges: A comparative review*. IEEE Communications Surveys & Tutorials, 20(3), 2062-2100.

Jiang, S., Fang, L., & Huang, X., (2009). An idea of special cloud computing in forest pests' control. In: *Cloud Computing* (pp. 615–620). Springer Berlin Heidelberg.

Jin, J., Gubbi, J., Marusic, S., & Palaniswami, M., (2014). An information framework for creating a smart city through internet of things. *IEEE Internet of Things Journal, 1*(2), 112–121.

Kagermann, H., Helbig, J., Hellinger, A., & Wahlster, W., (2013). *Recommendations for Implementing the Strategic Initiative INDUSTRIE 4.0: Securing the Future of German Manufacturing Industry*. Final Report of the Industry 4.0 Working Group. Forschungsunion.

Khan, R., Khan, S. U., Zaheer, R., & Khan, S., (2012). Future internet: The internet of things architecture, possible applications and key challenges. In: *Proceedings of the 10th International Conference on Frontiers of Information Technology (FIT)* (pp. 257–260).

Kivinen, T., & Kinney, P., (2017). *IEEE 802.15*. 4 Information element for the IETF (No. RFC 8137).

Krogh, B. H. (2008). Cyber Physical Systems: The Need for New Models and Design Paradigms. Presentation Report.

Lampos, V., & Cristianini, N., (2010). Tracking the flu pandemic by monitoring the social web. In: *2010 2nd International Workshop on Cognitive Information Processing (CIP)* (pp. 411–416). IEEE.

LAN/MAN Standards Committee, (2003). *Part 15.4: Wireless Medium Access Control (MAC) and Physical Layer (PHY) Specifications for Low-Rate Wireless Personal Area Networks (LR-WPANs)*. IEEE Computer Society.

Light, (2017). Mosquitto: server and client implementation of the MQTT protocol. *Journal of Open Source Software, 2*(13), 265, doi:10.21105/joss.00265.

Liu, X., Zhao, M., Li, S., Zhang, F., & Trappe, W., (2017). A security framework for the internet of things in the future internet architecture. *Future Internet, 9*, 1–28; doi: 10.3390/fi9030027.

Lohr, S., (2012). *The Age of Big Data* (p. 11). New York Times.

Luzuriaga, J. E., Perez, M., Boronat, P., Cano, J. C., Calafate, C., & Manzoni, P., (2015). A comparative evaluation of AMQP and MQTT protocols over unstable and mobile networks. In: *Consumer Communications and Networking Conference (CCNC)* (pp. 931–936). 12th Annual IEEE.

Mashal, I., Alsaryrah, O., Chung, T. Y., Yang, C. Z., Kuo, W. H., & Agrawal, D. P., (2015). Choices for interaction with things on internet and underlying issues. *Ad. Hoc. Networks, 28*, 68–90.

Mehdi, K., Lounis, M., Bounceur, A., & Kechadi, T., (2014). Cupcarbon: A multi-agent and discrete event wireless sensor network design and simulation tool. In: *7th International ICST Conference on Simulation Tools and Techniques* (pp. 126–131). Lisbon, Portugal. Institute for Computer Science, Social Informatics and Telecommunications Engineering (ICST).

Mehmood, T., (2017). *COOJA Network Simulator: Exploring the Infinite Possible Ways to Compute the Performance Metrics of IOT Based Smart Devices to Understand the Working of IOT Based Compression & Routing Protocols*. arXiv preprint arXiv:1712.08303.

Michael, K., & Miller, K., (2013). Big data: New opportunities and new challenges. *Computer, 46*(6), 22–24.
Miller, L., (2016). *IoT Security for Dummies, INSIDE Secure Edition.* Retrieved from: https://www.insidesecure.com/content/download/909/10066/file/IOT%2BSecurity%2Bfor%2BDummies.pdf (Accessed on 28 August 2019).
Miorandi, D., Sicari, S., Pellegrini, F. D., & Chlamtac, I., (2012). Internet of things: Vision, applications and research challenges. *In Ad Hoc Networks*, pp. 1497–1516.
Mitton, N., Papavassiliou, S., Puliafito, A., & Trivedi, K. S., (2012). Combining cloud and sensors in a smart city environment. *EURASIP Journal on Wireless Communications and Networking, 1*, 247.
Moniruzzaman, A. B. M., & Hossain, S. A., (2013). Nosql database: New era of databases for big data analytics-classification, characteristics and comparison. *International Journal of Database Theory and Application, 6*(4), 1–14.
Naik, N., (2017). Choice of effective messaging protocols for IoT systems: MQTT, CoAP, AMQP and HTTP. In: *Systems Engineering Symposium (ISSE)* (pp. 1–7). IEEE.
Nayyar, A., & Singh, R., (2015). A comprehensive review of simulation tools for wireless sensor networks (WSNs). *Journal of Wireless Networking and Communications, 5*(1), 19–47.
O'Donovan, P., Leahy, K., Bruton, K., & O'Sullivan, D. T., (2015). Big data in manufacturing: A systematic mapping study. *Journal of Big Data, 2*(1), 1–22.
Ozturk, O., (2010). Introduction to XMPP protocol and developing online collaboration applications using open source software and libraries. In: *Collaborative Technologies and Systems (CTS), International Symposium* (pp. 21–25). IEEE.
Palmers, T., (2013). Implementing a vulnerability management process. *SANS Institute InfoSec Reading Room.* Retrieved from: https://www.sans.org/reading-room/whitepapers/threats/implementing-vulnerability-management-process-34180 (Accessed on 28 August 2019).
Panwar, G., & Misra, S., (2017). Inside bluetooth low energy. *IEEE Wireless Communications, 24*(4), 2, 3.
Pardo-Castellote G, Farabaugh B, Warren R: An Introduction to DDS and Data-Centric Communications. Real-Time Innovations 2005.
Reddy, K. M., Kumar, K. N., & Vardhan, V. H., (2017). Analysis on securing the data over the internet of things (IoT). *International Journal of Scientific & Engineering Research, 8*(5), 24–29.
Rose, K., Eldridge, S., & Chapin, L., (2015). *The Internet of Things (IoT): An Overview-Understanding the Issues and Challenges of a More Connected World.* Internet Society. October 2015. Available at: https://www.internetsociety.org/wp-content/uploads/2017/08/ISOC-IoT-Overview-20151221-en.pdf
Said, O., & Masud, M., (2013). Towards internet of things: Survey and future vision. *International Journal of Computer Networks, 5*(1), pp. 1–17.
Saint-Andre, P., (2011). *Extensible Messaging and Presence Protocol (XMPP): Core (No. RFC 6120).*
Salman, T., & Jain, R., (2015). Networking protocols and standards for internet of things. *Internet of Things and Data Analytics Handbook*, 215–238.

Salonen, J., (2017). Tietoturvaratkaisut (SSCP), TestOut Security Pro, web-based online training. *Helsinki Metropolia University of Applied Sciences*. Retrieved from: https://cdn.testout.com/client-v5–1–10–495/startlabsim.html? (Accessed on 28 August 2019).

Schaffers, H., Komninos, N., Pallot, M., Trousse, B., Nilsson, M., & Oliveira, A., (2011). Smart cities and the future internet: Towards cooperation frameworks for open innovation. *Future Internet Assembly, 6656*: 431–446.

Sequeira, H., Carreira, P., Goldschmidt, T., & Vorst, P., (2014). Energy cloud: Real-time cloud-native energy management system to monitor and analyze energy consumption in multiple industrial sites. In: *Proceedings of the IEEE/ACM 7th International Conference on Utility and Cloud Computing* (pp. 529–534). IEEE Computer Society.

Shelby, Z., Hartke, K., & Bormann, C., (2014). *The Constrained Application Protocol.* https://tools.ietf.org/html/rfc7252 *(CoAP) (No. RFC),* 7252).

Shi, Y. R., & Hou, T. (2013). Internet of Things key technologies and architectures research in information processing. In Applied Mechanics and Materials (Vol. 347, pp. 2511–2515). Trans Tech Publications.

Silva, B. N., Khan, M., & Han, K., (2018). Internet of things: A comprehensive review of enabling technologies, architecture, and challenges. *IETE Technical Review, 35*(2), 205–220.

SimpleSoft, (2018). *Simple IoT Simulator*. Retrieved from: https://www.smplsft.com/SimpleIoTSimulator.html (Accessed on 28 August 2019).

Singh, M., Rajan, M. A., Shivraj, V. L., & Balamuralidhar, P., (2015). Secure mqtt for Internet of Things (IoT). In: *Communication Systems and Network Technologies (CSNT), 5th International Conference* (pp. 746–751). IEEE.

Suo, H., Wan, J., Zou, C., & Liu, J., (2012). Security in the internet of things: A review. In: *Computer Science and Electronics Engineering (ICCSEE)* (pp. 648–651).

Suryadevara, N. K., & Mukhopadhyay, S. C., (2018). *Internet of Things: A Review and Future Perspective*. Reliance. Available at: https://www.europeanbusinessreview.com/internet-review-future-perspective/

Teplinsky, M., (2013). Fiddling on the Roof: Recent Developments in Cybersecurity. 2 American University Business Law Review.

Theodoridis E, Mylonas G, Chatzigiannakis I. *Developing an IoT smart city framework.* In: Proceedings of 4th International Conference on Information Intelligence Systems and Applications, Piraeus, 2013. 1–6.

Thiruveedula, B., (2017). *IoT Network Testbed Simulation with Cooja*. Retrieved from: https://blog.imaginea.com/iot-network-testbed-simulation-with-cooja (Accessed on 28 August 2019).

Tosi, J., Taffoni, F., Santacatterina, M., Sannino, R., & Formica, D., (2017). Performance evaluation of Bluetooth low energy: A systematic review. *Sensors, 17*(12), 2898.

Van-Den-Dam, R., (2013). Internet of things: The foundational infrastructure for a smarter planet. In: *Internet of Things, Smart Spaces, and Next Generation Networking* (1–12). Springer Berlin Heidelberg.

Vasseur, J. P., Agarwal, N., Hui, J., Shelby, Z., Bertrand, P., and Chauvenet, C., RPL: The IP routing protocol designed for low power and lossy netwroks. Internet Protocol for Smart Object (IPSO) Alliance, White Paper, Apr. 2011.

Vinoski, S. (2006). Advanced message queuing protocol. *IEEE Internet Computing*, (6), 87–89.

Weyn, M., Ergeerts, G., Berkvens, R., Wojciechowski, B., & Tabakov, Y., (2015). DASH7 alliance protocol 1.0: Low-power, mid-range sensor and actuator communication. In: *Conference on Standards for Communications and Networking (CSCN)* (pp. 54–59). IEEE.

Weyn, M., Ergeerts, G., Wante, L., Vercauteren, C., & Hellinckx, P., (2013). Survey of the dash7 alliance protocol for 433 MHz wireless sensor communication. *International Journal of Distributed Sensor Networks*, *9*(12), p. 870430. Retrieved from: https://doi.org/10.1155/2013/870430 (Accessed on 28 August 2019).

Winter, T., (2012). *Routing Protocol for Low-Power and Lossy Networks*. Technical report, rfc 6550, 6551, 6552. Internet Engineering Task Force (IETF), Fremont, CA, USA.

Wu, M., Lu, T. J., Ling, F. Y., Sun, J., & Du, H. Y., (2010). Research on the architecture of internet of things. In: *Proceedings of the 3rd International Conference on Advanced Computer Theory and Engineering (ICACTE '10)* (Vol. 5, pp. 484–487). IEEE, Chengdu, China.

Xing, F. X., Pang, J., & Zhang, B. Q., (2010). Application about the objects internet technology in the modern agricultural production. *Agricultural Technology & Equipment, 8,* 16–17.

Yang, X., Li, Z., Geng, Z., & Zhang, H., (2012). A multilayer security model for internet of things. In: *Communications in Computer and Information Science* (Vol. 312, pp. 388–393).

Yushev, A., Sikora, A., & Sebastian, E. J., (2016). Open source 6Lo protocol stack for wireless embedded systems. In: *Wireless Telecommunications Symposium (WTS)* (pp. 1–7). IEEE.

Zhang, Y., (2011). Technology framework of the internet of things and its application. In: *Electrical and Control Engineering (ICECE)* (pp. 4109–4112).

CHAPTER 7

Big Data: Analytics, Technologies, and Applications

E. SWEETLINE PRIYA

Madras Christian College, Department of Computer Application (BCA), F1 B Block, Shahin Villas, Rajiv Nagar 4th Street, Kovilampakkam, Chennai, Tamilnadu – 600117, India,
E-mail: sweetlinepriya.edwin@gmail.com

ABSTRACT

Big data refers to the massive quantity of data that are excessively large in size; hence, the traditional processing tools cannot manage the same. It is high volume, high velocity, and high variety of information that require advanced technologies to process. This chapter presents a brief introduction to big data, big data analytics, the various categories of big data, and the characteristics that determine the big data, how different technologies are involved in big data analytics and a brief introduction on the tools involved handling big data. This chapter also explains the applications of big data in various fields like health care, banking, marketing, politics, meteorological department, finance, and so on.

7.1 INTRODUCTION

We dwell in the data era. In this digital world, the data is stored electronically, and the size of the stored data is estimated at 4.4 Zettabytes in 2013. It is forecasted that the growth would be reaching a size of 44 Zettabytes by the year 2020 (White, 2015). A Zettabyte is equivalent to 1021 bytes or is equal to one thousand Exabyte, one million Petabytes, or one billion Terabytes. In the digital world, a huge volume of data is generated by means of different internet applications. These data can be collected from a real-time transaction, tweets, Facebook posts, photo uploads, mobile messages, purchase information on e-shopping, which is very huge in volume and is difficult to be processed. Such sort of data is called as 'Big Data.'

The term 'big data' denotes the collection of massive amount of data which are growing exponentially over time. In other words, it is any sort of data that is very huge and complex and is tedious to be handled by traditional data management tools. Also, it is tough to process and store the data in an efficient manner. Below are some of the examples of big data:

- Social media like Facebook that generates/produces and stores petabytes of new data in formats such as text, image, audio, video, and message exchanges, etc.
- YouTube portal that stores large volume (typically in MBs and GBs) of videos.
- Patient's details recorded in health care.
- Web blogs with different stories and comments.
- E-shopping cart with various products, purchases, and customer information.
- Bank transactions.
- Stock exchange information.

7.2 CATEGORIES OF BIG DATA

As per Guru99 (2018), big data is wildly classified into three main categories:

• **Structured Data:** Any kind of data that can be accessed, processed, and kept in a number of fixed columns/fields of a table is termed as structured data. Relational database management system (RDBMS) is one of the examples of structured data. Different techniques in computer science field have achieved bigger success in dealing with data that has a stable format and well known in advance and also ease to derive useful information out of it. A database can hold any number of tables (Table 7.1).

TABLE 7.1 A 'Student' Table

Reg. No.	Last Name	First Name	Age	Gender	Date of Birth
10111	Lal	Manohar	20	Male	27/4/1998
10112	Edwin	Priya	21	Female	2/7/1997
10113	Jenner	Nancy	22	Female	13/2/1996
10115	John	Thomas	21	Male	5/4/1997

- **Unstructured Data:** Any data that has unknown structure is categorized as unstructured data. In addition to its unknown form, it contains several challenges in processing and deriving values out of it. Example of unstructured data is, data source containing a combination of text files, audio, video, images, etc. It often includes multimedia content. Facebook updates of individuals, weather forecasting content, the data store in a flight, a twitter account with different posts and comments, etc. Even though they possess a rigid structure, still they are considered as unstructured as the values/data they contain cannot be fit properly into a database.
- **Semi-Structured Data:** In addition to structured data and unstructured, there exists a third category data called 'semi-structured' data. It is nothing but the data that contains both the forms (structured and unstructured) of data. Example for such data is the data stored in XML and JSON format.

Example for XML file format is given below:

```
<library>
<books>
<book>
<name>Java Complete Reference</name>
<author>Herbert Schildt</author>
<edition>9</edition>
<price>800</price>
</book>
<book>
<name>Database System Concepts</name>
<author>Abraham Silberschatz</author>
<edition>6</edition>
<price>750</price>
</book>
```

```
<book>
<name> Hadoop the Definitive Guide </name>
<author> Tom White </author>
<edition> 4 </edition>
<price> 900 </price>
</book>
</books>
</library>
```

The same content of data in JSON format is given below:

```
{
“books”: [
{
“name”:”Java Complete Reference,”
“author”: “Herbert Schildt,”
“edition”: “9,”
“price”: “800”
},
{
“name”: “Database System Concepts,”
“author”: “Abraham Silberschatz,”
“edition”: “6,”
“price”: “750”
},
{
“name”: “Hadoop the Definitive Guide”,
“author”: “Tom White,”
“edition”: “4”,
“price”: “900”
}
]
}
```

7.3 FEATURES OF BIG DATA

As per Sadiku et al., (2016), the 3Vs which depicts the features of big data are as follows (Figure 7.1):

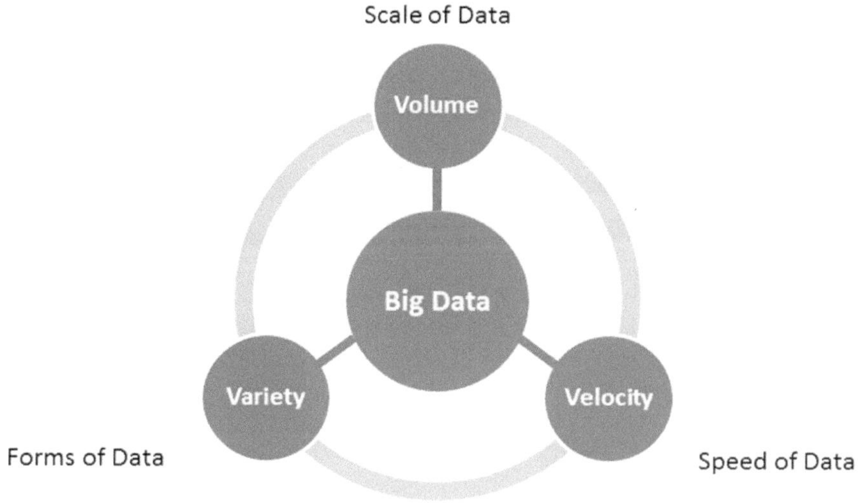

FIGURE 7.1 Various features of big data.

- **Volume:** A decade ago, organizations typically measured the stored data in TeraBytes (TB). But now they require data storage infrastructure in Peta Bytes (PB). As big data deals with profusion of information and because the size of the data plays a crucial role in determining value out of it, volume is considered as one important feature. Nowadays data is accumulated in Gigabytes and Petabytes. With the expansion of this data, we should think about re-evaluating the tools and architectures built up to handle data.
- **Velocity:** It generally refers to speed. Here in big data we speak about the speed of making of data and the response after processing the data. How fast the data is produced and managed to come across the target is the real challenge for many of the organizations.
- **Variety:** It means the different forms of data (text, image, audio, video, etc). Today data arises in all these kind of forms through internet and social network and hence is one among the features of big data.

Other than Velocity, Variety, and Volume there exists other metrics to be counted in when talking about big data. They are Variability and Complexity.

7.4 BIG DATA ANALYTICS AND TECHNOLOGIES USED

Big data analytics is the process of gathering, organizing, and analyzing large quantity of data sets to derive useful pattern and figures from the data set (Toshniwal et al., 2015). By using big data analytics, data-driven organizations and industries can increase the profit and sales. It also helps in forecasting the future, take productive decisions and hence to withstand in the competitive world. Data scientists use big data analytics to mine useful information which cannot be identified by traditional business intellectual tools. The list of newer technologies involved in the field of big data analytics are discussed in the following subsections..

7.4.1 APACHE HADOOP

Apache Hadoop was constructed by Doug Cutting, the founder of "Apache Lucene," the extensively used search library for indexing of texts. Apache Hadoop is originated from "Apache Nutch," a web search engine and is an open-source. The name Hadoop is not an abbreviation; it is a made-up name. Doug Cutting, the founder named it as Hadoop behind the name of his kid's stuffed toy elephant. Hadoop is modeled after Google's MapReduce (Morais, 2015). Hadoop is intended to process and manage massive volume of information by connecting number of commodity computers (nodes) parallel to work together. The group of these nodes which is connected together is called a cluster. To take benefit of parallel processing provided by Hadoop, the programmer has to express the queries as MapReduce jobs.

7.4.1.1 MAPREDUCE

MapReduce is a programming framework aimed at distributed computing which was created by the company "Google." It uses divide and conquers method to break down complex big data problems into tiny junks of work and process them in parallel (Sagiroglu and Sinanc, 2013). MapReduce works by splitting the process into two different phases namely map phase and the reduce phase. For each of the phases gets key-value pairs in place of the input and output. In map phase, map function extracts the data in a form that the reduce phase can process and perform its function on the data. The output of map function is processed by the help of MapReduce framework, before it is directed to the reduce function. The job of reduce function is to collect, summarize, filter or transform the data by iterating through the key-value data pairs. Finally the results got from reduce phase is being written.

There are five independent entities involved in the MapReduce job process at the highest level which are listed below:

- **The Client:** It submits MapReduce jobs.
- **YARN Resource Manager (RM):** It manages resource management and allocation in the cluster.
- **YARN Node Managers:** They launch and keep track of the compute containers on the machines in the cluster.
- **The MapReduce Application Master (AM):** It coordinates the tasks that are running the MapReduce jobs. The application master (AM) and the containers that hold MapReduce tasks are scheduled by resource manager (RM) and are managed by the node managers.
- **HDFS:** The distributed file system which is used for sharing job/task files among nodes that exist in the cluster.

7.4.1.2 HADOOP DISTRIBUTED FILE SYSTEM (HDFS)

HDFS is the file system, here in Hadoop, which stands for Hadoop Distributed File System (HDFS). When the data is large in quantity, the storage capacity of a single machine is not enough. Hence it is essential to partition the data across multiple machines. File system that manages data storage across multiple systems in a network is called as distributed file system. HDFS splits big data into stores and blocks across multiple nodes in a cluster. This also replicates data (storing multiple copies of data) in the cluster and therefore provides high availability of data in case of loss of data in a single node. Let us see in detail of the HDFS concepts (Apache Hadoop, 2018a).

- **Cluster:** A cluster is considered to a single logical unit that consists of multiple computers (called as nodes) that are associated together through network.
- **Blocks:** A block is nothing but the minimum amount of data that a disk can read and write. A disk is divided into several blocks. HDFS too

has the concept of blocks, but here we talk about larger units like 128MB by default. HDFS blocks are very large in size compared to disk blocks and the purpose for this is to minimize the rate of seeks. The applications write their data once but they read it one or more times which require these reads to be fulfilled at streaming speeds. HDFC supports the semantics called write-once-read-many on files.

- **NameNodes and DataNodes:** A HDFS cluster consists of two types of nodes: NameNode and DataNode which operate as master-slave pattern. NameNode is a master server/node that manages the namespace of file system and controls the access to the files by clients. DataNode is the slave that exists one per each node in a cluster. The files are stored as one or more blocks in DataNode. The NameNode executes the file system tasks like opening, closing, and renaming of files and directories. The NameNode also manages the mapping of the blocks of data into DataNodes. The client's read and writes request can be handled by the DataNodes. Also, based on the instructions from NameNode they do creation, deletion, and replication of blocks.
- **Block Caching:** Generally a DataNode reads various blocks from disk, but in case if the files are repeatedly accessed, the corresponding blocks are cached in the memory of DataNode, in a block cache. Block is cached for the increased read performance from the clients.
- **Staging:** NameNode is not reached immediately when a client raise a request to create file. Initially the file data is been cached by HDFS client in a temporary local file. All the application's writes happen to this local file only. But when the local file's size reaches the HDFS block size, the client reaches the NameNode. Then the NameNode allocates a data block for the file after it inserts the file name into the file system hierarchy and it responds to the client request with the identity of the DataNode and the destination data block. Then the client moves the data block from the temporary local file to the specified DataNode. Once the file is closed, the left overrun-moved data in the temporary local file is transferred to the DataNode. After transfer, the client communicates to the NameNode to inform that the file is been closed. At this point in time, the NameNode commits the creation operation of actual file into a permanent store. This approach has been implemented for many applications that get executed on HDFS.
- **Replication Pipelining:** As we discussed in previous section, when the client writes data into HDFS, first it is been written to the temporary

local file. Suppose HDFS data file has a replication factor of 3. When the local file reaches the maximum block size of HDFS, the client requests for and retrieves the list of DataNodes from the NameNode. This list comprises the DataNodes which will host the replica of that particular data block. Then the client forwards the data block to the first DataNode. Out of the three DataNodes, the first one gets the data in small portions (like 4 KB), writes each of the portions to the local storage and passes that written portion to the second DataNode. Similarly the second DataNode gets the data block from first DataNode, writes the portion to its local storage unit and forwards the same to the third one. Thus, the DataNode receives the data block from the previous one in the pipeline and at the same forwards the data block to the next one in the pipeline. In this way the transferred data is been pipelined in the list of data blocks.

7.4.1.3 YET ANOTHER RESOURCE NEGOTIATOR (YARN)

It assigns CPU, storage, memory, and other resources to applications that get executed on a Hadoop cluster. The first version of Hadoop could run MapReduce applications alone. But YARN supports other applications like SPARK to run on Hadoop as well, which now opens up a wide range of possibilities. The central occupation of YARN is to part up resource management and job planning/observing functionalities into particular daemons (Apache Hadoop, 2018b). Hence it is required to have a global RM for each AM. An application can be a single job or multiple jobs. Data computation framework is formed by RM and Node Manager together. RM oversees and plans the resources among the running applications in the framework. Node manager is liable for monitoring the resources (like CPU, disk, network, storage) and reporting the same to RM. AM works in par with Node Manager and RM to run and monitor the tasks.

7.4.2 SPARK

Apache Spark is a cluster computing framework for performing fast distributed computing and data processing for large-scale with the help of in-memory primitives. MapReduce methodology is not used as execution engine in Spark; instead Spark uses its own distributed runtime environment for executing tasks on a cluster. Spark lets the users to load data blocks into primary memory and to query it repeatedly. Hence it is suitable for online and iterative processing. In addition to spark several projects have been developed recently like Spark Core, Spark Streaming, Spark SQL, Spark GraphX, etc. Originally, Apache Spark was hosted by UC Berkeley R&D lab (which is nowadays known as AMPLab) in 2009. Later, in 2010 it became open source under BSD license. By 2014, it turned out to be the top-level Apache project (Data Flair, 2016).

Initially Hadoop MapReduce was used to execute batch processing. Other tools (like Apache Storm, Apache Tez/Apache Impala and Apache Giraph) were used for stream processing, interactive processing and graph processing respectively. Hence there was no powerful tool in industry to manage and process data in both real-time and batch mode. Also there was a requirement that one tool can perform all in memory-processing and hence Apache Spark came into picture which is an open source engine that offers real-time batch processing, stream processing, interactive, and graph processing. Spark is closely integrated with Hadoop system. It can run on YARN and can work with Hadoop file formats and storage backend like HDFS.

7.4.3 APACHE HIVE

Apache hive is used to build query typescripts (like SQL) to perform MapReduce operations. It is a data warehousing and an ETL (Extract, Transform, and Load) tool developed on top of HDFS. Hive makes the job simple for the following operations like:

- Data encapsulation;
- Ad-hoc queries;
- Analyzing huge quantity of datasets.

In Hive, databases, and tables are designed first and then data is fed into these tables. Hive is mainly designed for querying and managing the structured data that is stored in tables.

7.4.3.1 HIVE DATA TYPES

Hive supports two types of data types namely primitive and complex (Confluence, 2017).

1. Primitive Data Type:

Integers:

- TINYINT (1 byte).
- SMALLINT (2 bytes).
- INT (4 bytes).
- BIGINT (8 bytes).

Boolean:

- BOOLEAN (TRUE/FALSE).

Binary:

- BINARY (Sequence of bytes).

Float:

- FLOAT (Single precision).
- DOUBLE (Double precision).

Fixed Point Numbers:

- DECIMAL (Number with fractions in user-defined precision)

String:

- STRING (Arrangement of number of characters in a particular scale and character set)
- CHAR (Sequence of characters in a specific scale and character set with a defined length)
- VARCHAR (Sequence of characters in a specified scale and character set with a maximum length)

Date and Time:

- TIMESTAMP
- DATE

2. Complex Data Types:

Structs: The elements instructs data type can be accessed with dot (.) operator. For example in STRUCT c {a INT; b INT}, a can be accessed by the expression "c.a"

Maps: The elements in the map are constructed with a key-value pair.

Arrays: The elements in the array is of the same type and are stored in contiguous (adjacent) memory locations

7.4.3.2 OPERATORS

Hive supports all the basic operators like:

- Relational Operators (=,!=, <, >, >=, <=, etc.)
- Arithmetic Operators (+, -, *, /,%&, |, ^, etc.)
- Logical Operators (AND, OR, NOT, etc.)

7.4.4 APACHE PIG

Pig is a high-level data-flow language for exploring very large datasets. With Pig, MapReduce programs can be generated and compiled on the fly. It creates the scripts that can run on Apache Hadoop. To write data analysis programs/scripts, Pig offers a high-level language called Pig Latin. It is the replacement for the programmers who do not know the java programming language. Pig Latin is a procedural language. It consists of different scalar data types like int, long, float, double, char array, byte array. Pig Latin provides the following features (Apache Hadoop, 2018c):

- **Ease of Programming:** Complex logics can be converted to programs and it can be easily maintained.
- **Optimization Opportunities:** The manner by which tasks are coded permits the system to enhance the execution automatically. This allows the user to focus on programming and semantics rather than efficiency.
- **Extensibility:** It is possible for the users to write their own logical functions to do reading, writing, and processing.

As discussed early, In order to analyze data, programmers have to write programs use Pig Latin. All the code written is converted to map and reduce tasks internally. Apache Pig contains a component called Pig Engine which accepts the scripts that have been written using Latin and converts them to MapReduce tasks/jobs. Generally it is been used by the programmers who finds difficulty with java programming language. They can use Pig in the place of java to write MapReduce tasks which may have complex logic in java. Pig uses multi-query approach, which in turn reduces the length of the code. For example, 5 lines of code in Pig Latin are equal to 50 lines of code in Java (MapReduce program). Hence it can increase the productivity. It is a SQL like language and is easy to be trained and write the same. Apache Pig supports data operations like join, filter, group, sort, etc. Both Apache Pig and Apache Hive are used in creation of MapReduce tasks/jobs. Both of them run on HDFS in a similar way (Table 7.2).

TABLE 7.2 Features of Apache Pig and Apache Hive

Apache Pig	Apache Hive
Pig Latin is the high-level language used by Pig which is originally created by Yahoo.	HiveQL is the language used by Hive which is originally created by Facebook.
Pig Latin is procedural and it supports pipeline paradigm.	HiveQL is declarative.
Apache Pig supports structured, semi-structured and unstructured data.	Hive is used for handling structured data.

(*Source:* Tutorials Point, (2018). *Apache Pig-Overview*. Retrieved from: www.tutorialspoint.com/apache_pig/apache_pig_overview.htm.)

7.4.5 NOT ONLY SQL (NOSQL)

NoSQL stands for Not Only SQL (NOSQL). NoSQL is found to store and handle huge quantity of data. In traditional database systems, the programmer or the database administrator has the responsibility of defining schema and creating table before inserting values to the table. But in NoSQL data can be recorded and updated on the fly. Few advantages of NoSQL are, it is easy to scale, much faster in the database operations and is able to hold larger quantity of data. Several NoSql databases are available in market today. Some of them are MongoDB, Hbase, Cassandra, Neo4j, CouchDB, GemFire, Redis, Mnesia, and Hazel cast.

7.4.6 SQOOP

Apache Sqoop is designed to transfer bulk data among Hadoop and structured data stores like relational database systems or mainframes. With Sqoop it is possible to import/export data from/to relational databases and Hadoop HDFS efficiently. Sqoop can work with several databases such as Oracle, MySQL, Teradata, Netezza, Postgres, and so on. The input to Sqoop tool is either table from database or mainframe datasets (Apache Sqoop, 2018). For databases, Sqoop will read the tables' row-by-row and feed into HDFS. For mainframes, Sqoop will read records from each dataset and feed them to HDFS. The outcome of this import process is the collection of files that keeps a replica of the imported datasets or tables. Sqoop uses MapReduce methodology to import data which is done in parallel. Hence, the output will be in multiple files in the form of delimited text files.

For exporting files from Hadoop, Sqoop will read a set of all delimited text files from HDFS, parse them into records, and insert/add them as rows in tables of target database (Figure 7.2).

FIGURE 7.2 Workflow of sqoop.
(*Source:* Tutorials Point, (2018). *Sqoop-Introduction.* Retrieved from: www.tutorialspoint.com/sqoop/sqoop_introduction.htm.)

7.4.7 OOZIE

Apache Oozieis a work-flow scheduling system of Hadoop jobs that can be managed and executed on distributed environment. Parallel execution of multiple Hadoop jobs is possible with Oozie tool. Oozie workflow jobs are executed as Directed Acyclic Graphs (DAGs) of actions (Apache Oozie, 2018). Based on constant time intervals or availability of data Oozie work-flow jobs are triggered. Oozie work system can be integrated with Hadoop jobs such as Java map-reduce, Pig, Sqoop, Hive or system-specific jobs like shell scripts. Oozie is a trustworthy, extensible, and scalable system. Oozie has two components namely workflow engine and coordinator engine. The primary responsibility of workflow engine is to run the stored workflows of Hadoop jobs. Coordinator engine takes care of the workflow jobs and executes the same based on time schedules specified in advance and based on data availability (Figure 7.3).

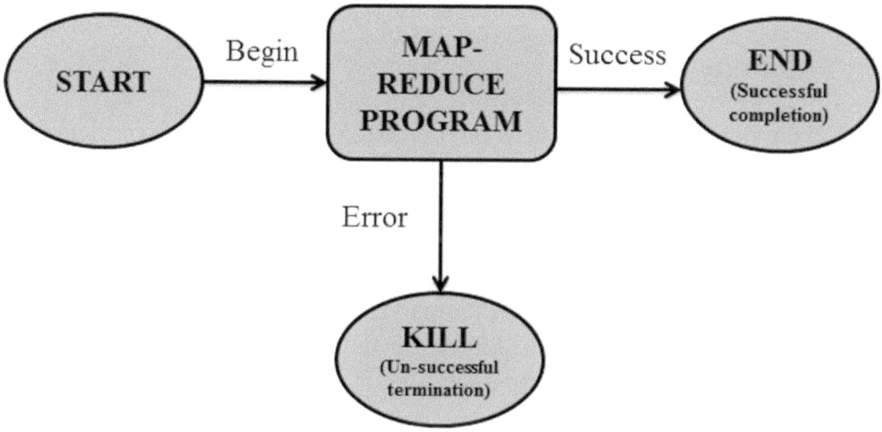

FIGURE 7.3 Sample workflow of oozie.
(*Source:* Guru99. (2018). *Learn OOZIE in 5 Minutes-Hadoop Tutorial*. Retrieved from: https://www.guru99.com/learn-oozie-in-5-minutes.html.)

Oozie workflow has start, action, end, and error nodes. Start node indicates the start of Oozie workflow jobs/tasks. Action node represents the workflow tasks. End node represents the completion of the job. Error node designates the occurrence of various errors and the matching error messages will be printed accordingly and a proper action will be triggered (sending email about the error).

7.5 BIG DATA APPLICATIONS

By managing big data,

- Business can derive useful value out of it.
- For better customer service/satisfaction.
- For increasing productivity by knowing the customer interests and personalized fashion.
- For better operational efficiency.
- Identification of risk to product/service if any.
- Prediction of failure.
- To analyze market and trading opportunities.

- In the field health care, to manage the patient details and to predict the disease and find the causes of diseases with the already information available.

We will see in detail of how the big data technologies are implemented in different fields:

- **Telecom Industry:** To increase customer satisfaction and to retain customer for a long time, big data plays a vital role. Call detail records that includes all the information like when the call is been made, to whom, the time duration, price for the call made, customer service call logs, web, messages received by the customer are the few examples of data that is been accessible to the telecom operators. Handling such data and bringing useful information about customer is a challenge. But tools like Hadoop can help in mining the data to enrich the quality of service provided and to make the customer experience better with particular service providers. It is also useful in promoting their novel products to the customer based on their necessity and to reduce the operational costs.
- **Politics:** Big data is proven to be very useful in mining the user data. A report (Illing, 2018) says, in 2016 US presidential election, a British data analytics company named "Cambridge Analytica" has helped Donald Trump team to win the election. This firm has used Facebook and Twitter as the source to find the individual voter's likes and the dislikes and in-tern targeted them with emotionally charged content. With just the personal information it actually changed the manner the people voted. See how the big data influences the world!
- **Healthcare:** With collection of several data like diseases, symptoms, treatments, therapies from the patients, big data is capable of providing better understanding on the diseases and the best treatment to be taken (Big Data Value Association, 2016). Nowadays, the data collected is shared globally which helps the health providers among various health care centers to lead to the shared decision making of physicians and the right pathway of treatments. This is highly helpful in case of rare and uncommon diseases. Also the analytics helps finding the root cause of the particular disease and to afford awareness on the same.
- **Marketing:** With big data, companies can do personalized marketing by analyzing each customer's individual buying habits. With the bill generated for the products bought by the customer, a correlation can be derived based on the products that have been purchased together (example

bread and jam are purchased together. Other products like mobile phone, mobile cover, and mobile screen scratch card are often purchased in same transaction). So that in shop as well in online shopping websites, these products can be shown together which in turn increase the sales. The retail legends like Walmart gathers the bills collected from various branches of the company to check the products purchased frequently by the customers to retain more stock on the same and to place these products in a more visible way hence can increase business.

- **Banking:** In banking and financial sectors fraud detection can be done with the support of big data analytics. By applying the analytics and machine learning, it is possible to differentiate the fraudulent business transactions from the genuine business transactions. Hence the analytic systems can recommend prompt actions, for example, obstructing of fraudulent and suspect transactions before it could produce major business impact.
- **Meteorology:** Often in various cities, the temperature for a day is captured at different timings. These temperature values captured are compared with other temperature values for the same day and can find the highest temperature of the day at a particular city.
- **Agriculture:** A biotechnology firm uses the sensor data to improve the crop efficiency (Sravanthi and Reddy, 2015). By planting test crops and observing the crops periodically, how plants reacts to several changes of environmental conditions (atmosphere moisture level, temperature, water levels, soil composition) is been captured. With growth and output of the crops, it is possible to determine the best environmental condition for particular crop.
- **Finance:** Big data analytics and machine learning can be very useful in examining the complex stock market values and moves and helps in making global financial decisions (Mukherjee and Shaw, 2016). Also it helps business to increase profit by forecasting the stock market. Many financial companies have now started adopting big data technologies and policies to gain an economical edge.
- **Social Media:** It is the computer-based technology that allows individuals to post new ideas, communicate with their friends, family, and colleagues over internet. Facebook is considered as one among the biggest social media in today's world. Generally in Facebook, photos, and videos are being shared and in addition to that people nowadays started liking pages and posts shared by other people. The pages can be selling pages

and posts can consist of advertisements. Data scientists and analytics team uses this likes and shares to find the people's interest over product or a shop or a place and can provide more customized advertisements when a customer logins into Facebook. Every day more than 5TB of multimedia data is added, hence analyzing them is possible with powerful tools like Hadoop. Hadoop can also contribute to filter spam emails. A sample set of spam emails are examined and the frequently used set of words are found out. This set of spam words are verified against the new emails coming in and can decide whether it is spam mail or not and an action can be proceeded based on the decision.

7.6 CONCLUSION

There is no doubt that the way internet applications deal with data has been increased significantly over the past decade. With more data collected it is the real big challenge to analyze the data and derive useful information out of it. We also need to think about the performance and scalability of the database management system. Hence big data is the boom in the information technology field to manage such datasets when comparing traditional database management systems. By using different technologies of big data various fields can improve the profit and can sustain in the competitive world.

KEYWORDS

- **Apache Hadoop**
- **Apache hive**
- **Apache pig**
- **big data**
- **big data analytics**
- **NoSql**
- **Apache Oozie**
- **Apache Spark**
- **Apache Sqoop**

REFERENCES

Apache Hadoop, (2018a). *HDFS Architecture Guide*. Retrieved from: hadoop.apache.org/docs/r1.2.1/hdfs_design.html (Accessed on 28 August 2019).

Apache Hadoop, (2018b). *Apache Hadoop YARN*. Retrieved from: https://hadoop.apache.org/docs/r2.7.2/hadoop-yarn/hadoop-yarn-site/YARN.html (Accessed on 28 August 2019).

Apache Hadoop, (2018c). *Welcome to Apache Pig*. Retrieved from: https://pig.apache.org (Accessed on 28 August 2019).

Apache Oozie, (2018). *Apache Oozie Workflow Scheduler for Hadoop*. Retrieved from: http://oozie.apache.org (Accessed on 28 August 2019).

Apache Sqoop, (2018). *Sqoop User Guide (v1.4.6)*. Retrieved from: https://sqoop.apache.org/docs/1.4.6/SqoopUserGuide.html (Accessed on 28 August 2019).

Big Data Value Association, (2016). *Big Data Technologies in Healthcare*.

Confluence, (2017). *Apache Hive*. Retrieved from: https://cwiki.apache.org/confluence/display/Hive/Tutorial (Accessed on 28 August 2019).

Data Flair, (2016). *What is Spark-Apache Spark Tutorial for Beginners*. Retrieved from: https://data-flair.training/blogs/what-is-spark (Accessed on 28 August 2019).

Guru99, (2018). *Introduction to BIG DATA: Types, Characteristics & Benefits*. Retrieved from: www.guru99.com/what-is-big-data.html (Accessed on 28 August 2019).

Illing, S., (2018). *Cambridge Analytica, the Shady Data Firm That Might be a Key Trump-Russia Link, Explained*. Retrieved from: https://www.vox.com/policy-and-politics/2017/10/16/15657512/Cambridge-analytica-FaceBook-Alexander-nix-Christopher-Wylie (Accessed on 28 August 2019).

Morais, T. S., (2015). Survey on frameworks for distributed computing: Hadoop, spark and storm. *Proceedings of the 10th Doctoral Symposium in Informatics Engineering-DSIE, 15*, pp. 95–105.

Mukherjee, S., & Shaw, R., (2016). Big data-concepts, applications, challenges and future scope. *International Journal of Advanced Research in Computer and Communication Engineering, 5*(2), pp. 66–74.

Sadiku, N. O. M., Mahamadou, T., & Musa, S. M., (2016). Big data: An introduction for engineers. *Journal of Scientific and Engineering Research*, pp. 106–108.

Sagiroglu, S., & Sinanc, D., (2013). Big data: A review. *IEEE*, pp. 42–47.

Sravanthi, K., & Reddy, T. S., (2015). Applications of big data in various fields. *International Journal of Computer Science and Information Technologies, 6*(5), pp. 4629–4632.

Toshniwal, R., Dastidar, K. G., & Nath, A., (2015). Big data security issues and challenges. *International Journal of Innovative Research in Advanced Engineering (IJIRAE), ISSN: 2349–2163, 2*(2), pp. 15–20.

White, T., (2015). *Hadoop: The Definitive Guide* (4th edn.). O'Reilly Media.

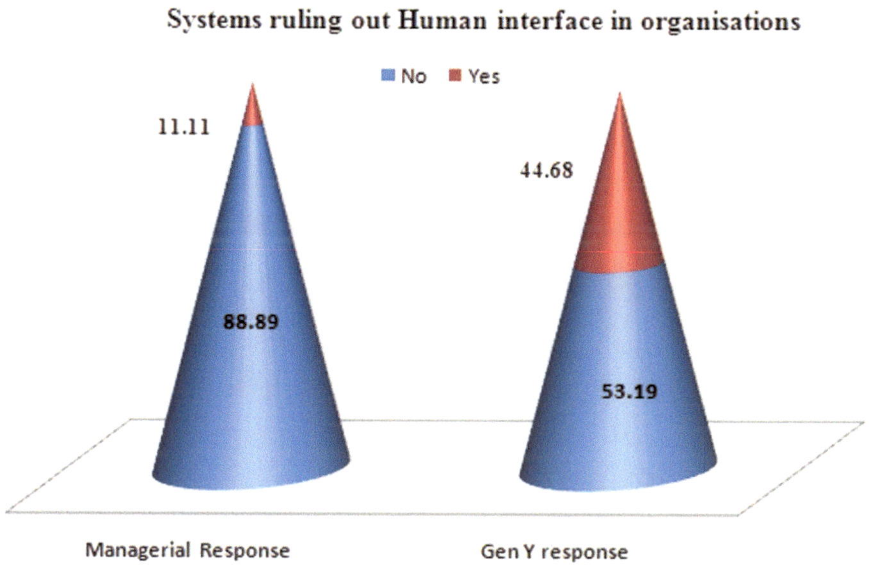

FIGURE 1.5 Acceptance of AI replacing human interface in India.

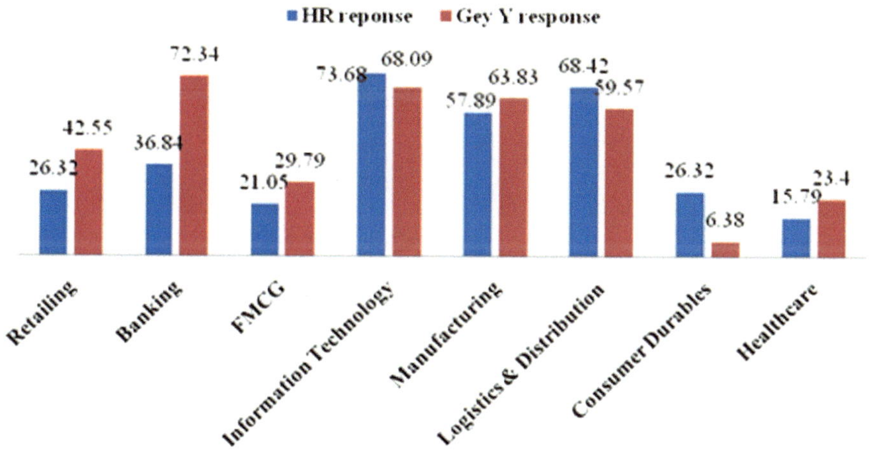

FIGURE 1.6 Possibility of AI replacing human interface in various sectors.

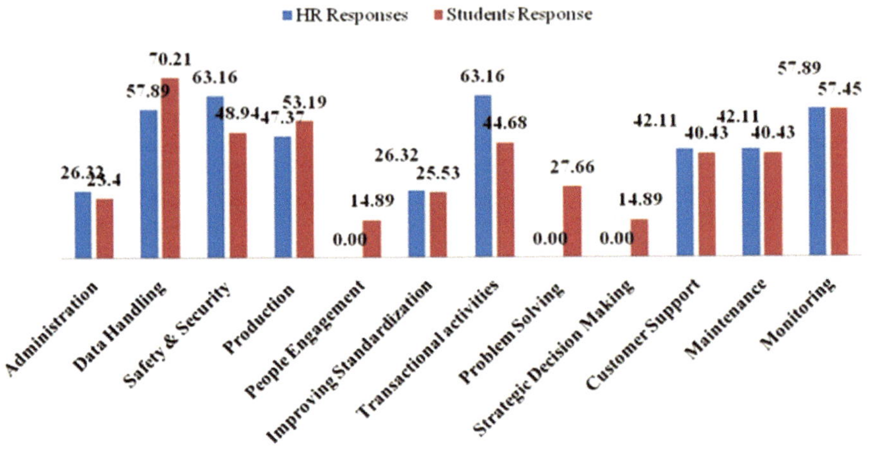

FIGURE 1.7 Jobs that can be replaced by AI.

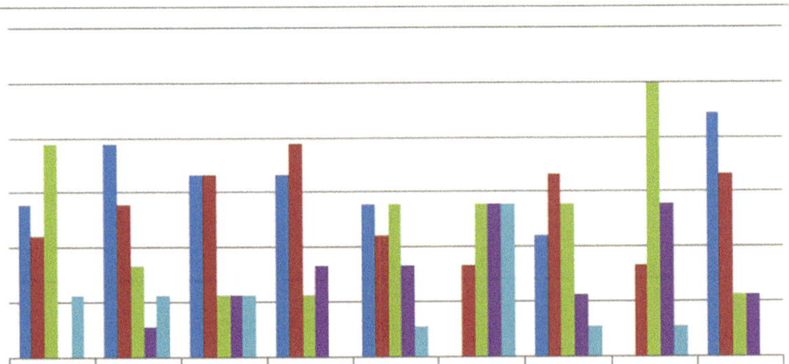

	Objective observation / Evaluation	Mentoring	Flexibility	People Orientation	Business logic	Focus on productivity	Intutive Capabilities	Control	Team Responsibility
Not Much	27.78	38.89	33.33	33.33	27.78	0.00	22.22	0.00	44.44
Little	22.22	27.78	33.33	38.89	22.22	16.67	33.33	16.67	33.33
Somewhat	38.89	16.67	11.11	11.11	27.78	27.78	27.78	50.00	11.11
Much	0.00	5.56	11.11	16.67	16.67	27.78	11.11	27.78	11.11
A great deal	11.11	11.11	11.11	0.00	5.56	27.78	5.56	5.56	0.00

FIGURE 1.8 Capabilities of robotics -HR viewpoint.

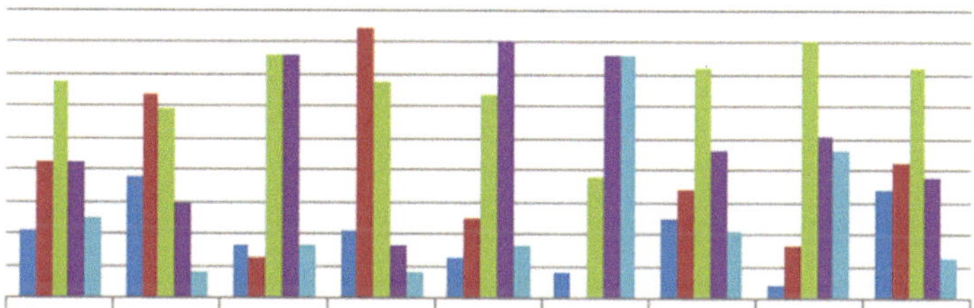

	Objective Evaluation	Mentoring	Flexibility	People Orientation	Business Logistics	Focus on Productivity	Intutive Capability	Control	Team Responsibility
Not Much	10.64	19.15	8.51	10.64	6.38	4.26	12.77	2.13	17.02
Little	21.28	31.91	6.38	42.55	12.77	0.00	17.02	8.51	21.28
Somewhat	34.04	29.79	38.30	34.04	31.91	19.15	36.17	40.43	36.17
Much	21.28	14.89	38.30	8.51	40.43	38.30	23.40	25.53	19.15
A Great Deal	12.77	4.26	8.51	4.26	8.51	38.30	10.64	23.40	6.38

FIGURE 1.9 Capabilities of robotics -gen Y viewpoint.

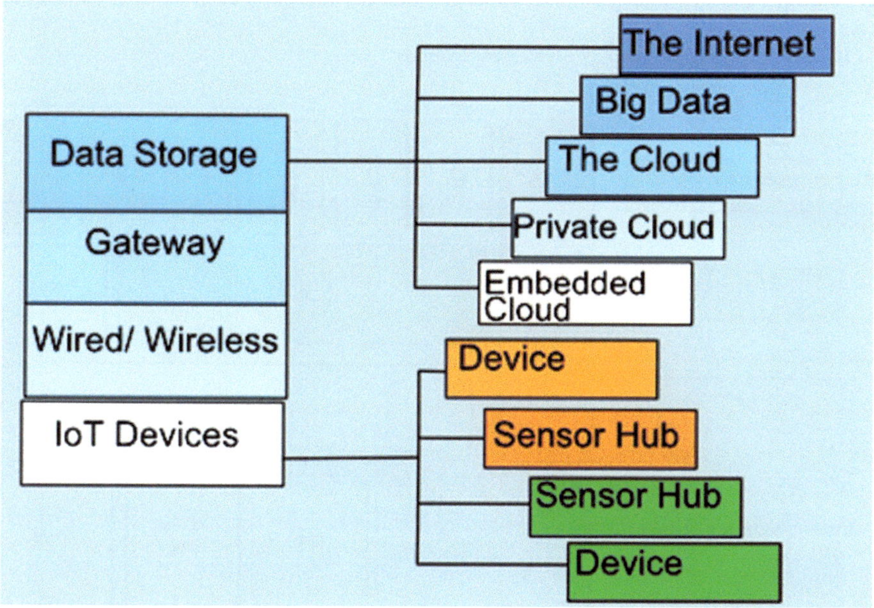

FIGURE 6.1 IoT topology.

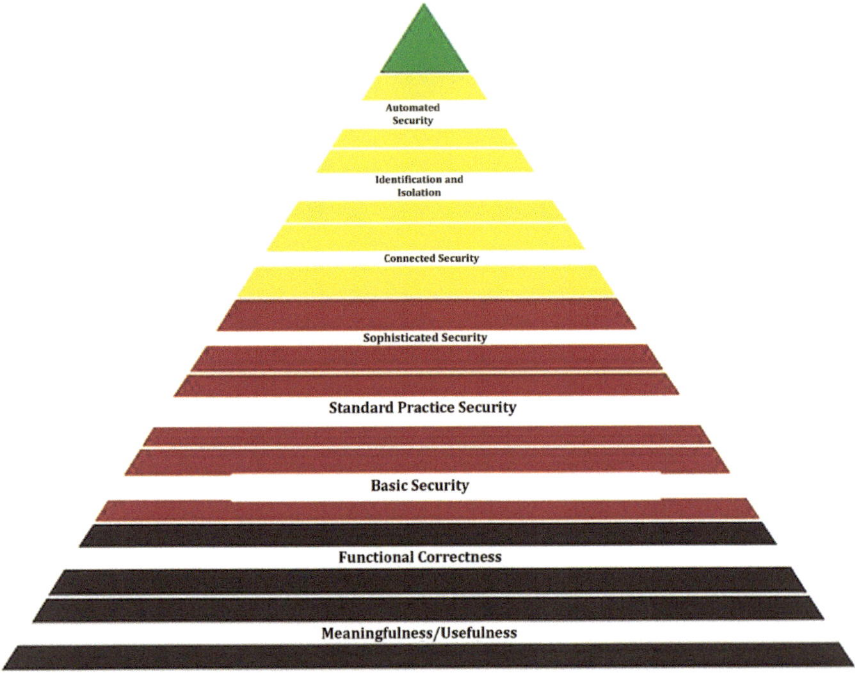

FIGURE 6.2 Cybersecurity awareness triangle.

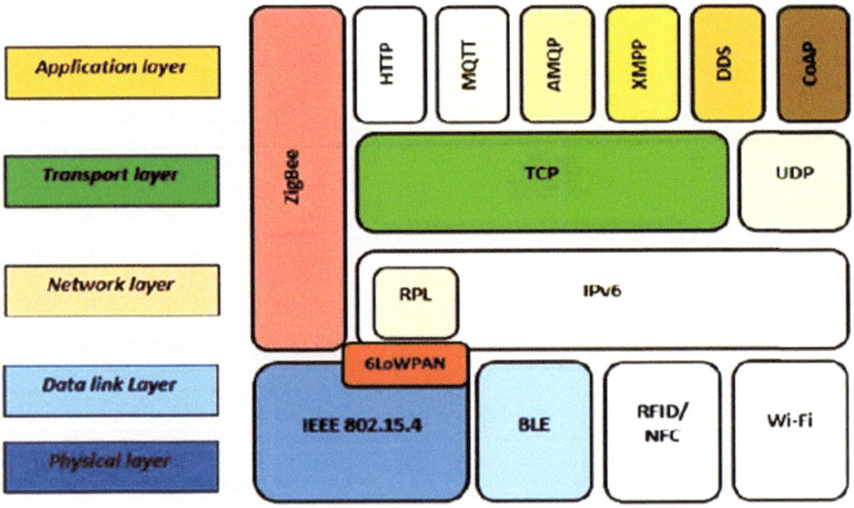

FIGURE 6.5 IoT communication protocols.

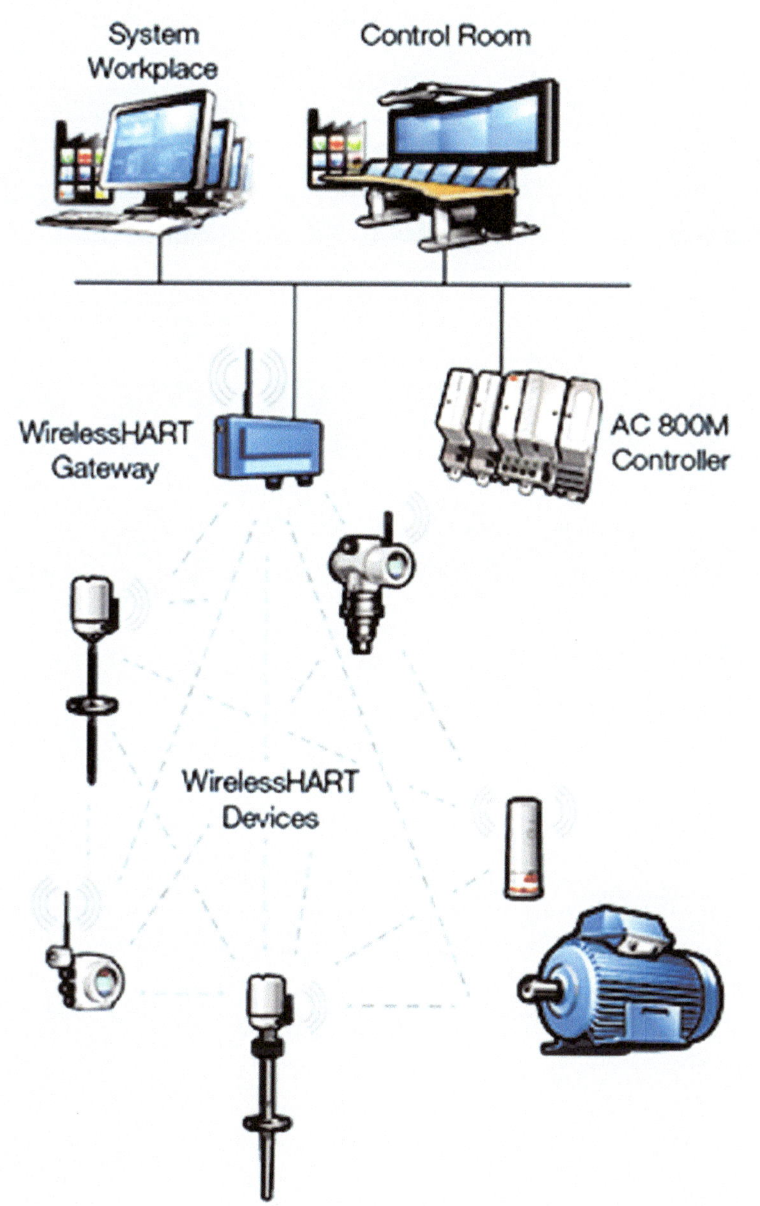

FIGURE 6.6 WirelessHART working.
(*Source:* ABB, (2018). *WirelessHART Information and FAQs*. Retrieved from: https://new.abb.com/products/measurement-products/wireless-products-and-solutions/highlights/wirelesshart-information-and-faqs.)

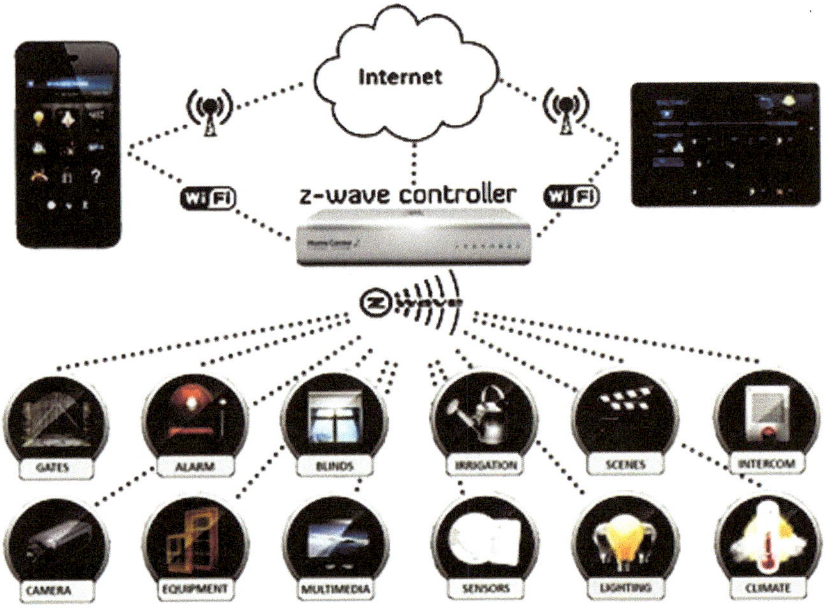

FIGURE 6.7 Z-wave network.
(*Source:* SmarterHOME, (2016). *Hierarchy of the Z-Wave System*. Retrieved from: https://smarterhome.sk/en/blog/hierarchy-of-the-z-wave-system_39.html.)

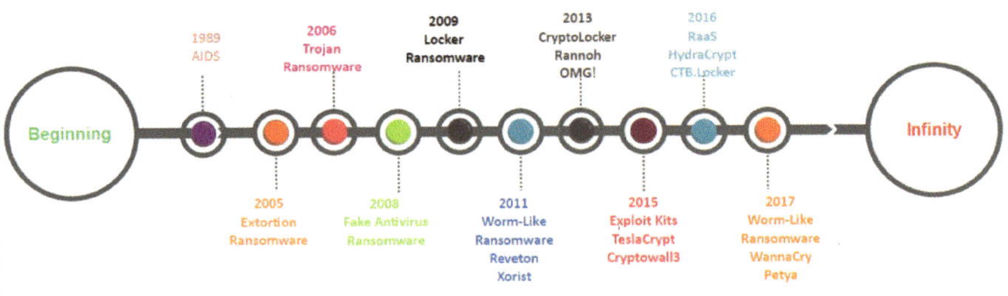

FIGURE 9.1 Ransomware timeline.

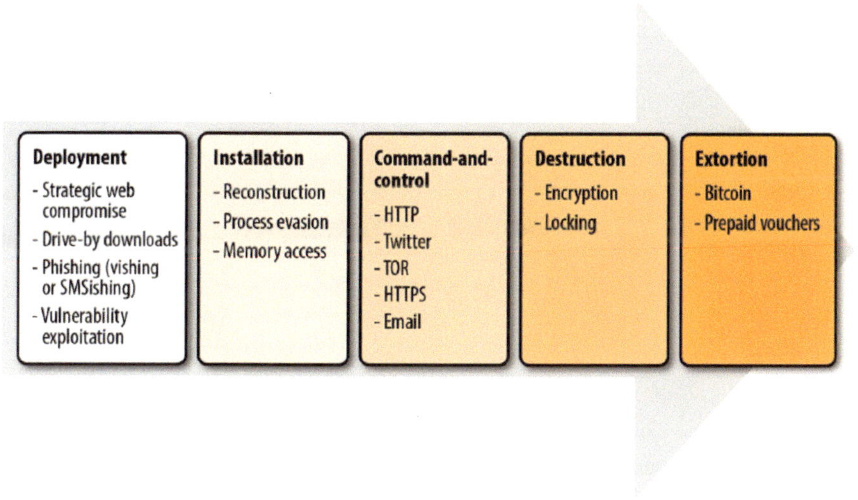

FIGURE 9.3 Stages of compromise.
(*Source:* Liska and Gallo, (2017). *Ransomware: Defending Against Digital Extortion* (First Edit). California: O'Reilly Media, Inc.)

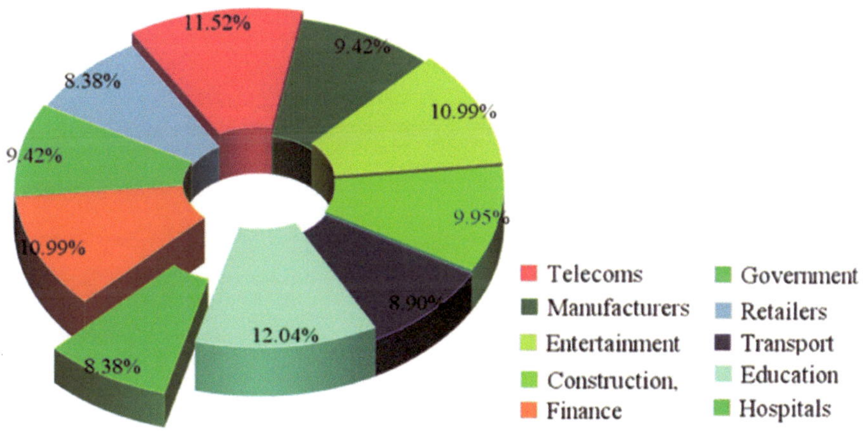

FIGURE 9.4 Ransomware attack targets.
(*Source:* Kaspersky, (2017a). *Kaspersky Security Bulletin: Kaspersky Lab Threat Predictions for 2018*. Retrieved from: https://goo.gl/nbXMo6.)

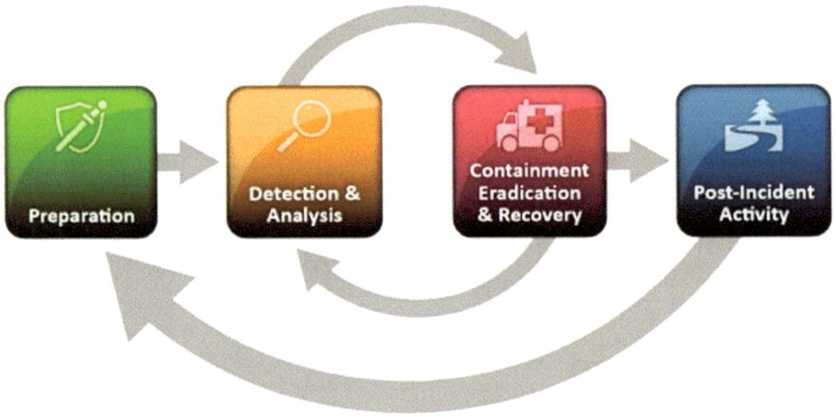

FIGURE 9.6 Incident response plan.
(*Source:* NIST, (2014). *Framework for Improving Critical Infrastructure Cybersecurity*. US: National Institute of Standards and Technology.)

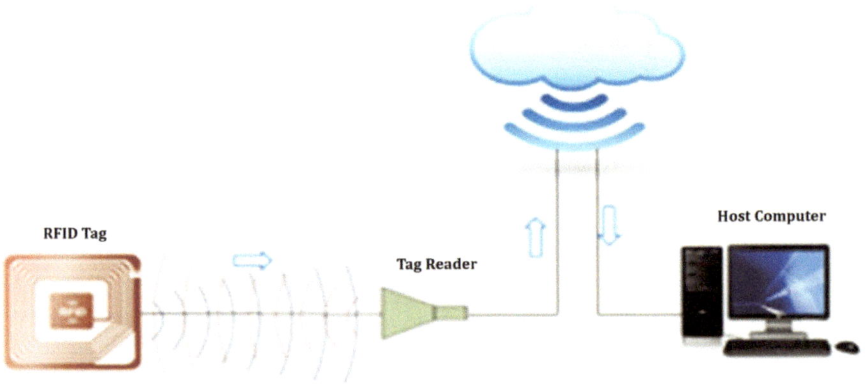

FIGURE 11.2 RFID block reader diagram.
(*Source:* Asghar, (2015))

PART II
Cyber Threat and Cybersecurity

CHAPTER 8

Cyber Threat Detection and Secured Alternative Technology: A Survey

S. ANITHA and HEMANTH G. KUMAR

ACS College of Engineering, Biomedical Engineering, #207, ACS College of Engineering, Kambipura, Mysore Road, Bangalore, Karnataka – 560079, India

E-mails: dranithasammilan@gmail.com (S. Anitha), hemumanju@gmail.com (Hemanth G. Kumar)

ABSTRACT

Cybersecurity plays a vital role in securing the information, data, etc., may be medical of in various other fields. It may act as a guardian during hazards like drive failures or power outages and other attacks. The second one includes everyone from initial threat makers to hackers and groups of criminals who are a threat to the society and also the enterprises. Reviews on various threats in different fields of business and the various existing methods to overcome the said threats have been discussed. Internet of things coupled with integration of internetworks is changing the landscape including utilities. The proposed survey model is based on outcome of the surveys carried on internet banking in other countries. The survey focused on users' practices in internet banking. The proposed model bridges the gap by widening the responsibility of banks to reduce cybersecurity risks for users.

8.1 INTRODUCTION

Cybersecurity is the practice of ensuring the integrity, confidentiality, and availability of information. It may represent the ability to defend against and recover from accidents like hard drive failures or power outages, and from attacks by adversaries. A very serious warning to the enterprise as it includes everyone from script kiddies to hackers and groups of criminals, who are efficient in executing advanced persistent threats etc. is posed by the latter one. A review of various threats in different fields of business and the various existing methods to outclass the said threats has been discussed. A survey on a few existing techniques has been performed in this chapter. The cyber-world is broadening everyday rapidly and a large number of people is getting connected to this world, lead in generating a huge amount of data known as Big Data. Cybercriminals are increasing drastically in number in parallel with the cyber world. To work against the criminals in cyber and protect the interest of common civilians, Big Data Analytics takes the lead. Big data is huge in both quantity and quality and can be thoroughly used to analyze various patterns and behavior anomalies which can avoid threats or any forthcoming attack. This proactive and analytical approach will reduce Cyber Crimes and also get the knowledge out of that data which was not observed before. This paper projects some of the key characteristics of big data, architecture, categories of cyber-crimes, and big data analytic methods that can be used. The advantages of added association must be taken against the risk of increased exposure to cyberattacks (Bayardo, 1998), the decision to upgrade traditional power grid architectures using smart grid technologies are decided by electric sector stakeholders.

Risk of adding connections to a network and hiring a number of cyber defense teams, with the objective to help decision-makers formally assess tradeoffs and set priorities given limited resources. New Bayes-adaptive network security model based on a theory of the "Multi-Armed Bandits" (MAB) problem is a central to this method. The network protector is facing an attack on uncertainty network nodes according to Poisson distributed rates here, replacing designs with opportunities for unsuccessful success as in classic MAB problem. A strong view of investing cyber investments, evaluating how network protectors can properly fit cyber interactions numbers is shown in this new method, which is called "Multi-Node Bandits." Enforcing hiring by supporting the defense purposes and collecting information from those who often respond to cyber violations after it happens; this technique finds an approach in effect. This approach is used to investigate the power by considering how to integrate the need for feedback on their smart grid network, in conjunction with identifying both the level of communication and equality. A series of distribution systems related to cybersecurity. Throwing up new cybersecurity challenges, these devices provide valuable information that can be used to improve reliability and reduce operating costs. Internet banking has become one of the fastest and easiest ways of banking. The threat of cybersecurity attacks set a great challenge for the Internet banking and electronic commerce industries.

A discussion on one more existing method is been done here, which was due to the outsourcing of the power a numbers of utility customers as in December 2015, a confirmed targeted cyberattacks for many electrical services in Eastern Europe. This event reflects on the Distribution System Operators (DSOs) systems are at stake and are at great risk of monitoring. With many newly integrated programs in the DSO's systems and online vulnerable information, there is a great risk of exit due to cyberattacks.

To help reduce the risk and construction of a cyberattack surface, DSOs should use the security organization and take active safety procedures. This study will explain one of these processes that must be kept strictly. Reducing the risk and expectation of disability on utilities grids is not just the only one in the latest technology. DSOs today are experiencing the difficulties of implementing a patch management system. The proposed options will also be processed for a smart patch management, which includes the risk detection process of the DSO system components and patch correction tips. Therefore, cybersecurity is not just a matter of 'getting' a secure cyber status. High protection of cyber threats requires analytical, complete, and ongoing analysis of the many changes and disaster risk management processes. Therefore, it is important for DSOs to think how their cybersecurity strategies will be changing over time. This is about making the DSO system more secure and ready to protect the system from possible known threats in a planned and iterative manner. Strengthening a strong defense against cyberattacks requires continuous action. Cybersecurity deployment first focused on being compliant to standards and regulations. This approach is very beneficial for the whole industry due to the increased level of awareness about cybersecurity, even if it may have led to oversimplifications.

As the electrical grid is getting too complex with distributed energy resources (DER) integration and feeder automation, a novel approach is required more oriented towards risk management. As an illustration, the same type of substation can feed a residential area or a critical infrastructure, with significantly different impacts during an attack. The transfer of cybersecurity to this leads to a different way. This also implies a change in the traditional boundaries like substation fences comprising DER, presumes, etc. This study discusses the new trends introduced in the latest standards and presents an approach for deploying security in-depth and improving the security of the grid. A survey has briefed up on cybersecurity, moving towards risk management. Internet of things coupled with integration of Internetworks is changing the landscape, including utilities. Bigger relays are installed now in utilities are right now installing huge numbers of present-day relays in their substations, not as it were to replace legacy defensive transfers but moreover for metering and hardware monitoring. Reliability can be enhanced and operating costs can be minimized as these devices provide valuable information in the latest cybersecurity threat. Internet banking has become one of the fastest and easiest ways to carry out banking transactions.

Threats to cybersecurity attacks pose a major challenge to Internet banks and trade unions. A technique has been presented, wherein first the analysis in detail of the cybersecurity of Internet Banking in three emerging countries and then introduces a novel model to minimize the cybersecurity risk to fill the gap between banks and customers. The said model is based on results of surveys conducted on Internet banking in various countries. The survey focused on users' practices in Internet banking. The proposed model draws the gap growth of banks to reduce the risk of cybersecurity for users (Agrawal and Ramakrishna, 1994).

8.2 BACKGROUND

8.2.1 CYBERSECURITY

Cybersecurity is sometimes said to be the safety of technical technologies in any way processes, which are used to protect networks, devices (any electronic device types that cannot create data or may work with IoT), programs, and data (formal, informal) on attacks, damages, or unauthorized accesses. The continuous discussion of the indication is from starting the traditional approach cybersecurity had to focus on resources on the main features of the system for well-known. The most popular threats but now this is a change, and the organizations now uses much of the flexibility, which is possible because of the fastest and constant changes in safety risks.

Domain cybersecurity is available at:

- Information protection;
- Network security;
- Security of the app;
- Operational safety.

8.2.2 *NEED OF CYBERSECURITY*

It is known that the internet user database is growing exponentially. Everyday new users are connecting with internet platform in some way or the other. According to some expert's predictions, by 2020, there will be more than 200 billion connected things that will come into existence. We are applying technology everywhere. Technology nowadays plays a major part in our lives. Nowadays, it isn't just the number of cybersecurity threats are increasing; the degree to those threats and attacks are also increasing in their nature. According to UK consultancy, the net loss in previous 12 months was up to $280 billion and adding to that the Forbes published an article stating that by 2019 it will become $2.1 trillion (Treglia and Delia, 2017). That is undoubtedly a large sum of loss for economy just because of lack of cybersecurity.

8.3 MAIN FOCUS OF THE CHAPTER

8.3.1 *BIG DATA: AN INTRODUCTION*

As mentioned above, the rapid growth and user clarification, as a result in a huge amount of data created by users. This growing data is far higher than both quantities and quality. Every day, a new source of upcoming data is added regularly.

8.3.1.1 3 V'S IN BIG DATA

Big Data is commonly associated with 3 V's for a better understanding, these are:

1. **Volume:** It refers to the amount of data generated in any for media, text, files, etc, Bayardo, R. J. Jr., (1998). The volume here consists of files, records, transactions, archives, and logs;
2. **Velocity:** This here refers to the speed of processing of the data that is being generated. The velocity here consists of batch, unpredictable fast streams, and real-time; and
3. **Variety:** It means the different and miscellaneous types of data generated. The variety here consists of structured, unstructured, semi-structured, etc.

8.3.1.2 IMPORTANCE OF BIG DATA

The core of the big data is not how much data you have, rather what you can do with that data that can prove beneficial for everyone. With the help of data analytics and algorithm big data can be used for following:

- Determining root causes of failures, issues, and threats in near-real-time. Big data can be analyzed to get to the root causes with the help of finding out unusual patterns by big data analytics.
- Recalculating the risks/threats portfolio in just a matter of minutes. With proper algorithms and analytics reports can be generated indicating the risks or threats.

8.3.1.3 REDEFINING CYBERSECURITY WITH BIG DATA ANALYTICS

Detecting fraudulent or unusual behavior by the means of proper analytics visualizations can be generated like box plots, graphs, and scatter plots can help easily identify fraudulent or unusual behavior (Dubbudu, 2017). One can get the idea of the intensity of this ever-growing huge amount of data just by some of these facts and figures. Nowadays where there is 8k capable cameras are available which takes up to 10 GB for a clip of just 2–3 minutes.

- Online media giant YouTube's users upload 72 hours of new media content every minute of the day.
- Every Minute more than 571 websites are being created and hosted.

- Call Detailed Records (CDR) are nowadays so huge and have so much data, as total minutes used by Indian consumer is approx. 1,339.7 billion minutes.
- Total data uploaded to Facebook daily is up to 100 TB.
- Total e-mail sent in one minute is up to 204 million.

So, this shows the intensity of the data that is being generated and also shows us the glimpse of what to expect from the future regarding this ever-growing huge amount of data. This data will become extremely profitable and useful if proper analysis and interpretation is done on this data which can bring out the facts, which in turn can help us take decisions which will prove very beneficial to us Javaid, A. Y., Devabhaktuni, V. K., & Alam, M. (2012). This is called "Data Mining." The first and initial definition of data mining states that there is a need to discover patterns and trends that are very much beyond simple analysis which can extract out knowledge from big data. Data mining is the practice of automatically searching large stores of data.

8.3.1.4 SOURCES OF DATA FOR CYBER CRIMES

Massive amount of data is being generated by the following:

- **Businesses E-mail:** Businesses rely on e-mail as a major source of communication now days. every minute nearly 204 million e-mails are being sent and received.
- **Database:** All the companies save their huge amount of user database on their servers Financial Transactions (Martins et al., 2014). After demonetization in India, the government is encouraging online banking, e-banking methods which have boosted 400–1000% increase in digital transactions. So, one can get the idea of how much transactions are being done right now, and also how much it's important to secure them.
- **Location-Based Data:** As now there is a huge integration of cloud with our electronics; thus, they are storing the location-based data of each user using those services, for every minute of the day. This amounts to a large sum of data generated daily. This data proves to be very helpful while investigation also. Call Detail Record (CDR)

and Internet Protocol Detail Record (IPDR) nowadays also shows a large amount of data just for a particular time period, which has been increased on a much larger scale from past.

- **Cloud Integration with Electronics:** Our systems and mobile electronics are connected with cloud, e.g., Google cloud for android, iCloud for apple products. These clouds keep all our data, including our location, photos, documents, contacts, and every day more and more data is being uploaded to cloud systems. If the security of these are breached or compromised it can prove to be very harmful for the user.
- **Social Media:** Arguably social media can be said to have been generating the most data now than ever. For example, in just one minute Skype is used for 1.4 million minutes of communication, 2.5 lakh tweets are being done in twitter, 350 GB of data on Facebook from posts and status updates.

8.3.1.5 PRESENT SCENARIO

Today's hackers have evolved from taking hacking as just a hobby to using it as a hardcore profession. These hackers are nowadays very well trained, backed, and very well-funded, thus are now more capable of more than ever before. Cybersecurity nowadays is needed the most than ever before, as the number of technology and internet consumers is increasing exponentially day by day, the risk of cyberattacks and cyber threats are also increasing. According to the report generated by National Crime Records Bureau (NCRB) for 2011 and 2015 for cyber crimes in India shows us the intensity of cyber crimes in India. These reported cases might not depict the true picture of the scenario of cyber crimes as in a country like India people try to avoid any types of reporting a crime or filing a complaint due to social and religious beliefs. Crimes are being committed in each sector whether it is business industries, banking, and economy industries, defense, healthcare sector etc. But the main target of today's criminals are the individuals of a nation be it an adult, child or senior people. These targets are being exploited by the means of various methods. Some of the vital ones are:

- **Social Media Crimes:** Social Media crimes are Criminal offenses committed via the means of Social Media platforms (e.g., software social apps) and which are aided by various forms of computer technology, which can harm an individual or society both mentally or physically.
- **E-mail Crimes:** These are the criminal offense committed via e-mails, e.g., Scam e-mails, threatening e-mails, incorrect e-mails, e-mail explosions, deleting e-mails, etc. The major motive behind most of the e-mail crimes are for financial gains.
- **E-Commerce Crime:** These are the criminal offenses which takes place by the means of e-commerce platform or on the e-commerce platform (Bank fraud, fraudulent pages, robbery shopping sites, cards, etc.).
- **Website Crimes:** These are the crimes in which attackers attack on the website to steal the data (SQL Injection), deface it, Distributed Denial of Service (DDoS), etc. India is now observing a significant rise in cyberattacks against financial and government sectors of 34% and 43% respectively which was previously 15% and 19%. If we talk about cybercrimes on a global scale, the picture is quite clear about the intensity of the cyber crimes occurring and the huge need of cybersecurity to counter these crimes. Few of the eye – opening facts regarding cybercrime on a global scale -
  - The global cost of cyber crimes will reach $2. 1 trillion, according to a brief up by IBM.
  - According to Identity Theft Resource Center (ITRC), More than 29 million records were available 858 postponement has been postponed.
  - Global spending to overcome crime in cyber top $80 billion this year.
  - The average cost involving thefts of assets for SMB's totaled $879,582 and another $955,429 to get back damages from them.

8.3.1.6 USE OF BIG DATA ANALYTICS IN CYBERSECURITY

The core or the backbone to this perspective of big data analytics in cyber-security is improved detection of cyber crimes and that is where the big data analytics and its domains come into play. National and international cybercrimes continue to rapidly grow so the Chief Information Security Officers (CISO) and other security researchers need to become better at anticipating and recognizing the pattern of behavior, in order to provide effective and efficient security management services (Qiu et al., 2015). According to a report by Ponemon Institute and IBM, organizations are 2.25 times more likely to seek a security-related incident within just hours or minutes in some cases when they integrated big data analytics to their security. Big data is able to provide the real-world insight into the current scenario and state of affairs and thus acts as an important element for the future. Now as the emergence of secure platforms like:

- Hadoop is a source which is open project. Apache project has this as a part; based on Java a framework that supports the maintenance and processing of a large amount of data it sets in a broadcast environment supported by Java applications.
- Splunk is an emerging industry-grade tool to collect and update huge sets of data and get the real and unseen insights/knowledge out of that data have made it easy for us to analyze, search, and monitor the data sets fairly easily.

8.3.1.7 TYPES OF ANALYTICS APPROACHES

- **Descriptive Analytics:** This can be told to be the starting stage of data processing which does a summarization of historical data to get out useful information.
- **Diagnostic Analytics:** This is a type of advanced analytics which gets to the bottom of the notion why certain thing happened with techniques like drill-down, data discovery, etc.
- **Predictive Analytics:** This analytics technique is used to determine or predict the future events on the basis of prior behaviors, by the techniques of data mining, modeling, etc.
- **Prescriptive Analytics:** This analytics approach deals with finding of best course of action to take after certain situation is dealt with using analytics approach. This in layman's terms is the remedy to take in consideration after certain situations.

8.3.1.8 *STEPS TO TURN MASSIVE AMOUNT OF DATA INTO MEANINGFUL INFORMATION*

- **Establishing the Context:** Behind the behavior of the data and how it is acting compared to its peers.
- **Providing the Meaning:** The good thing about analytics is that it does the main work for you. Data manually need not be looked into as it does the job for us. In addition, if we work some of the deep algorithms and gadgets can get well within knowledge and can give us some different patterns and ideas.
- **Making it Visible:** Now when insight is got and some patterns from the data by analytics,

Further use this information to do something that will benefit us or the company. Analytics should be integrated into the response process.

8.3.2 *BIG DATA ANALYTICS PROCEDURE*

- **Collection of Logs from Many Sources:** In this step, the collection of data takes places from Different sources. This collected data has variety of nature, some might be structured some might be unstructured. This collected data also needs to be stored for further processing.
- **Maintaining the Data Consistency:** In this step, the different nature data is taken and then preprocessing (the process of transforming raw data into understandable format) is done on it to make that data even. Different ways of data transfer are data cleaning. Data cleaning is the way of filling out unavailable information, removing noise from data and deleting unwanted value. Data Integration: Data is taken and put together with different representation to resolve the conflicts within the data sets.
- **Data Transformation:** The collected information is normalized and the redundancy from the data is removed and is generalized.
- **Data Reduction:** After the information is already installed it will reduce your availability. It is consolidated between the last data when the update system will take place in this data.

- **Perform Correlation:** In this process after the data is entered into the data warehouse the correlation and dependencies of one data set is checked with other data set to possibly find out that to which extent those sets fluctuate together.
- **Positive Correlation:** If the correlation amongst the data sets results to be positive then we can get the extent of those data sets decrease or increase in parallel they are directly proportional to each other.
- **Negative Correlation:** If the correlation between the data sets comes out to be negative then we can make out up to what extent to which one data set increases as other decreases they are inversely proportional to each other.

8.3.2.1 APPLYING VARIOUS INTELLIGENT ALGORITHMS AND ANALYTICS

Now after following above procedures, the data becomes ready to be analyzed. Different types of algorithms and analytics are applied regarding the type of that data (Alghazo and Kazimi, 2013). This analysis will give us the insights of the information which we would have noticed by just observation, and those are the insights which enable us to find the data and also possibly find out how to treat them (Agrawal and Ramakrishnan, 1994). Some examples of various algorithms are:

- **Apriori Algorithm:** It is an algorithm used for frequent itemset mining over transactional databases. It works on the basis of bringing the frequent data variables and extending them to larger and other data sets as far as they are frequent in nature.
- **Naive Bayes Classifier Algorithm:** This is a type of classification algorithm which is based on Bayes theorem with taking assumptions of independence among predictors. In other words, it is said to take the availability of an element in a class that does not agree with the existence of any other factor. Naïve Bayes model is easy to construct and comparatively work properly for large data sets.

8.3.2.2 FINDING OUT PATTERNS/BEHAVIOR ANOMALY TO DETECT ANY THREAT/HARM

Applying the algorithms to data sets produces Visualizations of the data in the form of the box plot, graphs, and scatter-plots, etc. which decides in very easy to find out any unusual behavior or any different pattern in the data sets just by observation. But Different ways can be used to find and analyze patterns or wrong behavior.

- **Clustering:** Integration is called integration or data integration in any way pattern or behavior is easily seen.
- **Classification:** The modeling model is also used in some cases to get a pattern again connecting, separation often involves two parts. Interchange variables and consolidation law the values of this behavior to obtain the predicted amount of the given annotation.

8.3.2.3 DATA MINING

Data mining is also an important factor when we talk about a great data analysis. Data mining is process or practice of analyzing existing collections, and making new information on that pre-existing data. Used to find a bigger problem (Bayardo, 1998):

- **Misuse Detection:** These are used to detect attacks or threats whose instances have not been observed yet. In this, predictive models are used which label datasets as 'normal' and 'intrusive.'
- **Anomaly Detection:** This helps in identifying the deviations from the normal behavior to find anomaly. This will be more helpful in finding out any unusual behavior on the particular data set and then the proper security measures can be done prior to any mishap.

8.3.2.3 CRITERIA FOR BETTER AND EFFECTIVE CYBERSECURITY

Solution using great data analytics:

- **Comprehensiveness:** It means to have a centralized particular view of all the relevant data and threat analysis and resolution.

- **High-Speed Gulping:** Data should be collected quickly, normal, and stored in it a great speed that makes it easy to reach.
- **Real-Time Processing:** Real-time procession must be in a condition to take place to manipulate streaming data feeds with related information as threat intelligence, geolocations, etc. for any investigation.
- **Efficiency:** This whole process should have efficiency both in time and cost. Analysis should be taking a little time to uninstall and lose the cost of keeping data and other infrastructure requirements.

8.3.3 KEY CHALLENGES

Despite having so many positive aspects of big data analytics, it's still not up to the mark in some areas as it is still in an evolving phase of technology. Some of the main challenges facing them are to be taken about large data analytics right now:

- **High Result Cost:** The building up and setup of these big data analytics infrastructures have very high financial demands. And to maintain ROI (Return on Investment) is fairly a difficult task for small scale projects.
- **Unavailability of Skilled Manpower:** As this field is still in its evolving phase, the specialized and skilled personnel for this are very less as compared to demand.
- **Difficulty in Data Integration:** As the data can and is coming for various sources so some of this data is structured and some of it is unstructured so integrating this data becomes fairly difficult task.
- **Handling Incoming Data:** As the data nowadays coming is of huge intensity and requires both speed and storage infrastructure, which sometimes becomes a challenge for data analysis.

8.4 SOLUTIONS AND RECOMMENDATIONS

8.4.1 PASSIVE TECHNOLOGIES

A discussion about random monitoring technologies such as the black net and the honeypot/sandbox will be available, including the accessibility of IT-related malware devices through black-hazardous black done. Additionally, the availability of cyberattacks with regard to use can be used for active cybersecurity responses as effective solutions. Technology is constantly responding to cyberattack under the wise city and connected conditions, future processing will be provided for use alert monitoring (Gasser, 1988).

8.4.2 CYBERSECURITY MOVING TOWARDS RISK MANAGEMENT?

A check on cybersecurity moving towards risk management has been presented here. Cybersecurity deployment first focused on being compliant to standards and regulations. This approach has been very beneficial for the entire industry because it has increased the height of awareness about cybersecurity even if it may have led to oversimplifications (such as focusing on transmission only, removing serial communications, defining security perimeters, etc. As the electrical grid is getting more and more complex with DER integration and feeder automation, a new approach is required more oriented towards risk management. As an example, the same type of substation can feed a residential area or a critical infrastructure, with significantly different impacts in the case of an attack. The deployment of the cybersecurity in this context leads to a different approach. One of them is the deployment of cybersecurity in depth. This also implies a change in the traditional boundaries like substation fences integrating DER, presumes, etc. This study describes the new trends introduced in the latest standards and provides an approach for deploying security in-depth and improving the security of the grid (Dubbudu, 2017). North American Electric Reliability Corporation-Infrastructure Protection (NERC-CIP) it has become a high level of cybersecurity, especially for all the power and service organization in North America. The CIP was made in 2006, and external examinations began in June 2007. It defines a set of accessibility controls for which Energy firms and services to be acquired obey. This includes creating and maintaining security, identifying, and implementing policy control electronic accessibility of detecting assets, keeping access control documents, and continuously monitoring and protecting electronic access to sensitive issues.

8.4.3 *THE FRAMEWORK OF NICE*

Discussion about the National Cybersecurity Education Initiative (NICE) which is an inter-agency try it by the National Institute of Standards and Technology (NIST). The agency focuses on the diagnosis, cyber safety education, and awareness, training, and professional development. NICE has increased with the cybersecurity application framework. If the monitoring process is used for this frame function emphasized the fulfillment of safe cyberinfrastructure as described in the context. Also, the frame is missing, including new emerging new technologies that increase the cybersecurity challenges threats. Students also need to have enough cybersecurity standards and procedures, which needs regular review. Researchers continue to show that the structures are not included abusive threats, and therefore, disaster management strategies need to be targeted. Also, the authors recommend that the cybercrime law does not exist in the area of management criminals. Finally, a successful safety strategy can work in partnership with the model business processes.

In terms of identifying thieves, people who get unauthorized personal access information such as social security, and financial account numbers are considered. Internet Spyware practices collect and sends personal information without the user's knowledge of "pork pigs" in downloaded applications for Spyware information. In addition to the above mentioned – discussed the problem, the company may face many other consequences if it fails to take steps to protect personal information and user information. The result includes loss of campus access computer network, confidential information, integration, and data acquisition of an important university, research on personal electronic issues, loss of public hope and giving opportunities, pursuits, internal conflicts action or termination of work (Qiu et al., 2015).

8.4.4 ***INTEGRITY ATTACKS AND ALLOY ANALYZER***

System integrity can be lost through two basic functions, the conversion of the existing information and the creation of new information. Convert data during travel or while storage is over conversion is the main purpose. There are various environmental events such as thunderstorms and thunderstorms lightning, which can lead to unwanted symptoms and the loss of the signal occurring. However, these natural events are unusual, and many communication processes and machines care for the problems caused by Jamming, signature integrity, and call/maintenance are three major stages of threatening the air. Jamming also aims as this has caused disturbing communication. Calling on wireless networks has much protective censorship proposed by researchers. For signal integrity, disturbance or increase the signal rate to the sound most smart by following the frequency of the sign, distance, etc. is the third-way tap or to pick up the worst type of attack to get started as necessary (Javaid et al., 2012). The first logic, by displaying structural obstacles in software processes. Discussed models for discussion using the Alloy can be studied at three levels of extinction: at a high level of output, targeted translation provides an easy way to quickly make a look at the system.

The second level of output is the introduction, and the low level of atomic release and the relationship, which are display titles of language semantics. The use of the Alloy ensures that the graphical structure is complicated. Basically, using algorithms and care features (such as low typing), a person can continue to extend an illustration frame to meet specific needs. For example, one can add a special type of system that introduces additional obstacles that do not interfere with other system types. Alloy Analyzer is a software tool that can be used to ensure that the system model meets whatever you want buildings. If the model is properly tested using the Alloy Analyzer, the model range must be clarified. In this activity, this involves clarifying a number of apps, data units; data channels, and more (Kulik et al., 2018).

8.5 CONCLUSION

This paper addresses the growing challenges of cybersecurity and relevant math support, followed by significant data changes, analytics types and other key algorithms that can it has been very helpful in terms of certain analytical procedures to deal with the threat of crime. This page also investigates and discuss how to use data analytics and what important things they will do keep in mind when you export useful information outside of the main data that can be seen great security, cybersecurity services. About Google Translate Community Mobile Google Privacy & Terms Help Send feedback. A survey on cybersecurity moving towards risk management has also been briefed.

KEYWORDS

- **big data**
- **cyber crimes**
- **cybersecurity**
- **distribution system operators**
- **internet banking**
- **smart grids multi-armed bands**
- **workspace framework**

REFERENCES

Agrawal, R., & Ramakrishnan, S., (1994). Fast algorithms for mining association rules. *Proceedings of the 20th International Conference on Very Large Data Bases, VLDB* (pp. 487–499). Santiago, Chile.

Alghazo, J. M., & Kazimi, Z., (2013). Social engineering in phishing attacks in the eastern province of Saudi Arabia. *Asian Journal of Information Technology, 12*(3), 91–98.

Bayardo, R. J. Jr., (1998). *Efficiently Mining Long Patterns from Databases*. ACM SIGMOD Record, pp. 85–92.

Dubbudu, R., (2017). *Most Number of Cyber Crimes Reported in Maharashtra & Uttar Pradesh*. Retrieved from: https://factly.in/cybercrimes-in-India-which-state-tops-the-chart (Accessed on 28 August 2019).

Gasser, M., (1988). *Building a Secure Computer System* (p. 3). Van Nostrand Reinhold. ISBN 0–442–23022–2.

Javaid, A. Y., Devabhaktuni, V. K., & Alam, M., (2012). *Cybersecurity Threat Analysis and Modeling of an Unmanned Aerial Vehicle System*, pp. 585–590.

Kulik, T., Tran-Jorgensen, P. W. V., Boudjadar, J., & Schultz, C., (2018). A framework for threat-driven cybersecurity verification of IoT systems. *IEEE International Conference on Software Testing, Verification and Validation Workshops*, pp. 89–97.

Martins, C., Oliveira, T., & Popovič, A., (2014). Understanding the internet banking adoption: A unified theory of acceptance and use of technology and perceived risk application. *International Journal of Information Management, 34*(1), 1–13.

Qiu, M., Gai, K., & Ali, M. L., (2015). *An Investigation on Cybersecurity Threats and Security Models*, pp. 307–311.

Treglia, J., & Delia, M., (2017). *Cybersecurity Inoculation*. Presented at NYS cybersecurity conference, Empire State Plaza Convention Center, Albany, NY, pp. 3–4

CHAPTER 9

Ransomware Evolution: Solving Ransomware Attack Challenges

MUHAMMAD UBALE KIRU and AMAN JANTAN

University of Science, School of Computer Science, Penang, Pulau Pinang–560079, Malaysia

E-mails: muhd.kiru@student.usm.my (Muhammad Ubale Kiru); aman@usm.my (Aman Jantan)

ABSTRACT

This research focuses on the world's most frightening cybersecurity threat known as Ransomware. Ransomware is popularly described by experts as scareware that makes data and resources on a victims' computer inaccessible and forces the victims to pay a ransom with bitcoins or through other means by frightening and intimidating them. Ransomware these days needs no introduction, the perpetrators behind ransomware have done more than enough damage to critical infrastructures and collected billions of dollars from victims across the world and are still collecting. Considering this, the research is vested upon uncovering the underlying mysteries behind the sudden growth and popularity of ransomware through the in-depth study of literature and efforts made by experts globally in understanding ransomware and how to fight and stop it. Moreover, the research seeks to bring together the collective professionals' views and recommendations on how to set up strategic defense-in-depth for fighting against ransomware.

9.1 INTRODUCTION

Ransomware is popularly described as a type of malware that makes a file on a victim's computer or device inaccessible and then demands the victim to pay ransom mostly in the form of Bitcoin or other means of payment to regain access to the hijacked system (Micro, 2017). Ransomware is usually considered by specialists as scareware, as it forces the victims to pay the ransom by frightening or intimidating them, especially looking at their nature and mode of operations, which are perhaps intimidating and scary (Micro, 2017). However, Liska and Gallo (2017) view ransomware as a new type of extortion, therefore, describes it as a criminal practice for obtaining something especially money or its equivalence from an individual or institution through coercion or threats. Ransomware is most commonly spread by hackers. But we know from experience that employees also sometimes contribute to the spread due to human error and or ignorance caused by lack of awareness (Fimin, 2017). Some of the common methods of spreading ransomware include exploiting system's known and unknown vulnerabilities or by visiting compromised sites or deep webs where all sorts of people with different intents come together with their goals; good and bad.

Studies suggest that the sudden rise of ransomware attacks recently is a signal to the world that ransomware has come back with full force in both complexity, impact, and size (Downs, Taylor, and Whiting, 2017). The year 2017 was the year history will never forget as per as internet security breach is concerned. It was the year in which the world saw some of the most dangerous attacks in the history of computing including WannaCry pandemic, Petya, NotPetya, Cerber, Cryptomix, Locky, CrySis, and many others. The aforementioned ransomware attacks were massive global ransomware attacks that mostly affect Windows operating systems (especially Windows 7 and other Windows OS before it) that have not been patched or secured. More importantly, the WannaCry attack became prominent following the leaked exploit kits which were stolen from the United States NSA by the infamous group known as 'Shadow brokers' which opens doors to other variants of ransomware to be created and eventually affected thousands of devices across the globe (Barracuda, 2017). These events led different social media observers and professionals in various domains to name 2017 as "the year of ransomware" (Cabaj et al., 2017).

The damage caused by ransomware did not catch much attention until recently when hundreds of companies and global security agencies across the world have begun to cry out (Brodsky, 2017). So far, the famous variant known as WannaCry had rapidly spread to around 200,000 to 300,000 machines in over 150 countries across the globe since its first appearance (Yaqoob et al., 2017), making it the world's largest attack in history if measured in terms of wide coverage, complexity, and impact. Earlier in 2016, the FBI reported that over $206 million was paid to ransomware criminals in the first quarter of 2016. In a another report by the United States Department of Justice, there are over 4000 ransomware attack reports per day, and that every month new variant of ransomware is being produced, which makes it more likely to increase with 100% by Q4 of 2018 (Harpur, 2017). Perhaps, the emergence of IoT devices has also contributed as well as fastened the spread of ransomware and the modern security challenges we are facing today (Yaqoob et al., 2017). The large availability of devices on the internet has open access to all perpetrators who have malicious intent to start a ransomware campaign.

The question many experts keep asking is why is ransomware so prevalent and unbeatable in every part of the world? The reason is simply that antivirus and anti-malware are no longer capable of detecting ransomware because modern ransomware use polymorphism to avoid detection. Secondly, the advent of Ransomware as a Service (RaaS) and Exploit kits in black markets make it even more difficult to deal with the situation. With RaaS, anyone including script kiddies can lay their hands on ransomware codes and reinvent their own, according to MacAfee Lab, the writers of 'Cerber' (one of the most dangerous ransomware family) releases a new variant of ransomware every 8 days on average, selling with bonuses and offers of 20% discount (Ashford, 2015; Singh, 2017).

Having said all that, the objective of this research is to identify the popular suggestions given by experts on how ransomware attacks can be dealt with professionally by using the simplest, cost-effective and most effective techniques for mitigating ransomware attacks. Also, to create awareness among individuals and corporate users so that they become aware of what is expected of them during ransomware attacks. However, the data collated for this research were sourced from contemporary works of research including webinars, online interviews, online tutorials, security reports, bulletins, interagency intelligence sharing, news articles, magazine articles, researches done on ransomware and a lot of commentaries from ransomware experts and professionals from different domains and disciplines.

9.2 BACKGROUND AND SURVEY

A recap of the previous years' data has shown a disturbing concern about how ransomware attacks have significantly increased from the last five years. If we trace the impact and growth from 1990 to 2006, ransomware cases were, in fact, insignificant to be recorded. Soon, however, the scale turns the other way round. In a report by Symantec, they highlight that ransomware landscape increased these years dramatically with the appearance of the two variants of self-propagating ransomware threats, namely WannaCry and Petya. The two attacks have caused what Symantec called a "global panic." As many have seen on the news and other media outlets (O'Brien, 2017), ransomware damage costs were estimated to have exceeded $5 billion in 2017, and according to Microsoft, it had risen from $325 million in the year 2015 (Morgan, 2017). Before these massive attacks, ransomware was not seen as a threat of any serious concern, as it was merely a malicious scam campaign. Until in 2016 when ransomware writers started offering ransomware variants as a service with the advent of exploit kits and its wide availability in the black market. According to a report by Symantec, one in every 131 emails contained a malicious link or attachment that could infect a device with a variant of ransomware. Within a short period, security agencies and private companies like Symantec and Kaspersky had blocked and intercepted over 22 million attempted ransomware attacks worldwide (Mort, 2017), and 57% of the victims were individuals while 43% were organizations. The report adds that 59% of the ransomware infections were delivered via email phishing with malicious attachments or compromised URLs embedded within the email (Cyber Intelligence Team, 2017). We must undoubtedly accept the fact that the year 2017 has seen what many would describe the 'fall of security infrastructure.' Modern technologies have failed woefully and proved insufficient as well as inefficient in the fight against ransomware (Downs, Cook, Wright and Kent, 2017).

9.2.1 RANSOMWARE TIMELINE AND HISTORY

A fraction of computer scientists around the world asserts that ransomware might have originated from Russia and was limited to Russians until in early 2012 when several reports proved its presence in other parts of the world including Asia and North America (Micro, 2017). The sudden growth of ransomware happens due to tons of malware variants that are produced because of the stolen exploit kits from the NSA arsenal which allow reusability; hence reinventing new ones isn't difficult (Figure 9.1).

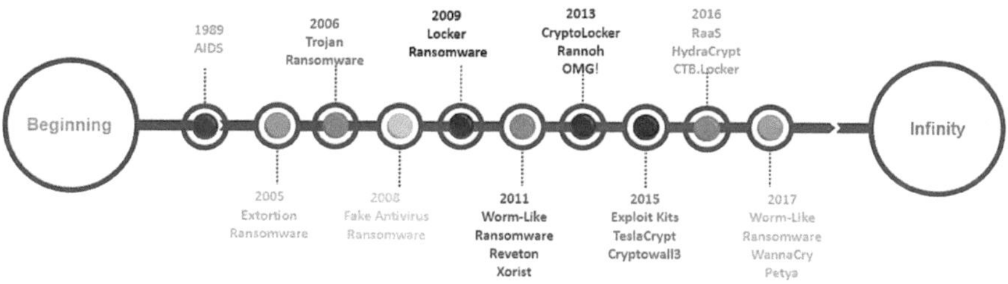

FIGURE 9.1 (See color insert.) Ransomware timeline.

The early found ransomware is known as AIDS. It was first discovered in 1989. It was not spread using the famous phishing attack. Instead, its writer Joseph Popp used ordinary 3.5" floppy disk and sent the malware via postal service to a gathering of World Health Organization (WHO) conference which took place in London. This early ransomware was characterized with encrypting system directory by replacing AUTOEXEC.BAT on the infected machine (Liska and Gallo, 2017) making the system unstable. After infecting the system, a message displayed demanding $189. The ransom was sent via a Post Office Box number in exchange for the decryption key (Trend Labs, 2017).

After a long disappearance of ransomware, it was reported that extortion ransomware emerged in 2005. Most extortion ransomware at that time were mere applications that posed as fake spyware removal tool such as SypSheriff. Others posed as performance enhancement tools and registry cleaners. Their major targeted machines were Windows and IOS Computers. The hackers' trick was an exaggeration of serious issues on the machine. Next, they asked if the user wants their problem fixed for $30 to $90 (Savage, Coogan, and Lau, 2015). Somewhere in May 2005, a new variant of crypto-ransomware that encrypts and demands ransom surfaced known as Trojan. GPcoder. It used a custom-encryption method which was considered weak and easily breakable. Perhaps it was not successful due to the failure of the hackers to make it more effective. Therefore, they kept on modifying the version for a long period (Hampton and Baig, 2015; Savage et al., 2015).

By the quarter of 2006, more experiments had been done, flaws had been identified from previous mistakes, and therefore, more sophisticated ransomware variants had emerged and started to use more complex and rigid RSA encryption algorithm. Reported cases from Moscow revealed that a variant of ransomware called TROJ_CRYZIP. A was discovered. This variant was characterized by certain activities whereby data is copied into specific password-protected archive files (Micro, 2017); therein, the original files are being deleted along the line, leaving a password key saved in a text file for unzipping the files. Other ransomware variants were reported in that period including TROJ.RANSOM.A, Krotten, and Cryzip (No More Ransomware, 2017b).

Furthermore, several ransomware occurrences continued to be the topic of discussion, even in early 2008 to late 2009 when a new threat emerged. Here, the hackers were said to have switched from using malicious software to using fake antivirus programs that appeared legitimate to the users. The fake programs were used to perform a virus scan on the system after they claimed to have detected a large number of security threats that needed immediate action. Thus, users were asked to pay money amounting $40 to $100 to fix their faked threats (Savage et al., 2015). However, this technique did not last much longer; as an ample amount of users decided to ignore the recommendations and chose to address the problem using other available means.

The first locker ransomware surfaced in the last quarter of 2009 and later became prevalent within 2011 to 2012. During that period, attackers no longer use fake antivirus or fake luring-application-software to attack their targets. For the first time, the hackers were able to hijack systems remotely and locked it up. With this new idea, worm-like ransomware equipped with remote-locking-capability started spoofing the Windows Product Activation notice emerged (No More Ransomware, 2017b). A report by Trend Micro suggested that these new variants of ransomware known as TROJ_RANSOM.QOWA and Trojan.Randsom.C have compromised a lot of systems, and both were capable of popping messages that compelled the users into dialing a premium-rate phone number and paid some amount of money "through electronic cash voucher" before the victim's machines are released (Micro, 2017; Savage et al., 2015).

More reports suggested that hackers had continued to be inventive; hence, introduced a new variant of ransomware known as Reveton. This variant began to impersonate law enforcement agencies, including the police and the FBI. The victims were be fooled into thinking that they had broken the law by simply visiting a prohibited site or by downloading a copyrighted item or by visiting a pornographic environment. In that fashion, the hackers used a lot of social engineering techniques in carrying out these attacks as it had to do with convincing and deceiving the victims. It is equally important to note that, Reveton ransomware was the first of its kind to introduce different payment methods for victims. A study by Trend Micro shows that victims were asked to pay the ransom through Ukash, PaySafeCard, and MoneyPak (Micro, 2017).

As technology grew, new types of ransomware known as Crypto-Ransomware and Crypto Locker also emerged in Q4 of 2013. These new variants were considered as the best generation of ransomware since the disappearance of fake antivirus ransomware and were also rated as quite profitable. In the past, ransomware was considered inefficient due to non-standardized use of good infrastructure. The ransomware in question changed the scales of ransom, therefore, collected an unimaginable sum of $300 from each victim, and eventually before the end of the year had earned more than $3 million for its writer. According to Symantec, between 2013 & 2014, there was a massive increase of 250% in the new crypto-ransomware. The newcomer infected more than 250,000 systems globally within two months (Lord, 2017). This time the hackers successfully used more complex encryption algorithms such as RSA 2048, which requires two-way encryption and decryption key (AES + RSA encryptions) (Micro, 2017). In fact, the year 2013 was indeed the beginning of the present-day sophisticated ransomware attacks.

Furthermore, the rise of crypto-ransomware led to the reinvention of even more sophisticated and well-enhanced ransomware in 2015 through 2016. For the first time, hackers started designing exploit kits which were used for creating other series of ransomware variants. Normally, crypto-ransomware does not affect the functionality of the targeted system, it, however, encrypt crucial files to force users into paying the ransom (Kaspersky, 2016; Savage et al., 2015). Thus, the emergence of cryptocurrency in 2009 and its widespread in 2015 has an open door for modern ransomware writers to remain untraceable on the web. Popular ransomware in that period consisted of Crypto wall which earned over $18,000,000 from its victims according to an FBI report (FBI, 2015), Teslacrypt, and CTB-Locker. A new variant called Locky came and overshadowed Teslacrypt and CTB-Locker and became one of the most used and most popular ransomware variants of all times (Fortinet, 2017). It had collected over 1 Billion USD before the end of the year. CryptXXX earned 77 million USD, while its counterpart, Cerber collected over 54 million USD (Crowe, 2016; Korolov, 2017).

It is said that, new breeds of ransomware were born in 2017, ransomware has become first on the list of every organization; in both public and private sector, and it continued to dominate and grow exponentially throughout the year. This spike according to several security experts is due to the emergence of the two new breeds namely WannaCry 2.0 and Petya. The former emerged due to a leak of an NSA exploit kit (known as Eternal Blue) responsible by a group of hackers who identified themselves as Shadow brokers. WannaCry or WannaCrypt has a worm-like feature, self-propagating capability which gives it the ability to spread itself rapidly and widely across the network without any user interaction which makes it invincible and abruptly difficult to stop (Perekalin, 2017; Weatherford et al., n.d.). Petya according to popular assessment is nothing new; it was the handy work of the same exploit kit used by Shadow brokers. Although, the two variants shared certain properties in common viz: Petya also adapted self-propagation technique, it incorporated other SMB network spreading method, and it uses public-key based64 encoded algorithm alongside Salsa20 which makes it extremely difficult to decrypt (O'Brien, 2017). Petya was primarily designed to target organizations in Ukraine, but it later spread to other regions including the US, Russia, France, Germany, and few others.

9.2.2 TYPES OF RANSOMWARE

The types of ransomware vary according to different scholarly opinions. The most prominent and popular types of ransomware can be classified as (Brunau, 2017; Harpur, 2017; Rubens, 2017a; Shinde et al., 2016):

- **Locker or Lock Screen Ransomware:** As the name implies, locker ransomware is the variant of ransomware which overtakes the system or device and denies access to the user lest ransomware is paid. In most of the scenarios, the infected system is left with limited resources to use for interacting with the hacker (s). In a few cases, personal files are not tampered with, while in some cases files may be stolen.
- **Crypto or Encryption Ransomware:** This type of ransomware encrypts and deletes the personal files and folders in the affected machine. Even though not all types of files are encrypted, specific variants of ransomware trace certain types of file formats such as., doc, xsl, xml, zip, pdf, js, and encrypt them only (F-Secure, n.d.). Traditional crypto ransomware, on the other hand, encrypts the entire directory.

Some scholars have gone further to add Master Boot Record (MBR) ransomware, Web server encrypting ransomware, and Mobile device ransomware. Nonetheless, the said types of ransomware if classified according to their function can be sub-categorized under Locker or Crypto Ransomware, respectively (Liska and Gallo, 2017).

9.2.3 RANSOMWARE VARIANTS

Since the emergence of ransomware, new variants of ransomware are being released every day. Perhaps it is important to note that, types of ransomware should not be mistaken with variants of ransomware, the variants are the distinctive types of ransomware family which possess unique features and patterns in which they operate. In tracing the different variants of ransomware family, one might say they are infinite. The earlier variant of ransomware was said to have emerged in 1989 in the form of a Trojan. Since then, hundreds of ransomware variants have been produced annually, coming in different forms, pattern, and structure. Various assessments by security agencies show that a massive increase in ransomware production was seen in the Q4 of 2015 through 2017 (Crowe, 2017; National Cybersecurity and Communications Integration Center, 2016) especially with the advent of RaaS. As mentioned earlier, RaaS gives unskilled hackers the tools to manufacture any variant they desire.

Thus, New Jersey Cybersecurity and Communications Integration Cell (NJCCIC) reports that it has profiled 197 ransomware variants since the second quarter of 2015 (NJ Cybersecurity and Communication Integration Cell, 2018) (Figure 9.2 and Table 9.1).

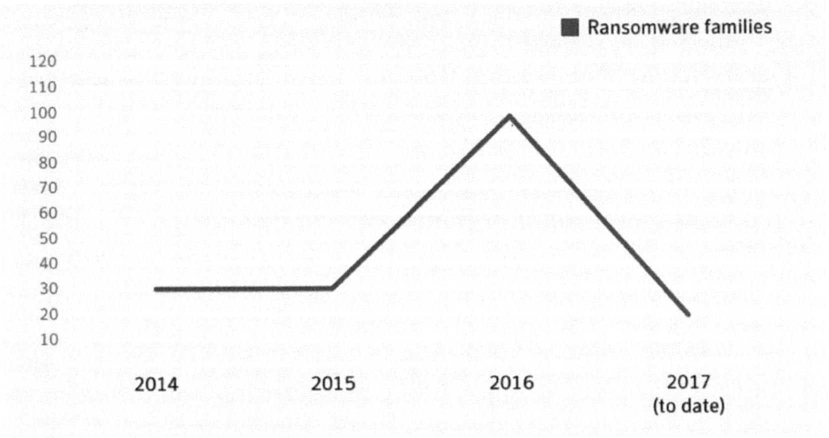

FIGURE 9.2 Ransomware families.
(*Source:* Symantec, (2017). *Fake Adobe Flash Update Installs Ransomware, Performs Click Fraud.* Retrieved from: https://goo.gl/Qn9fbe.)

TABLE 9.1 Popular Ransomware Variants, Attack Vectors and Dates

Alias	Attack Vector
Cerber	Spam campaign, RIG Exploit kit, Magnitude Exploit kit
WannaCry	Phishing campaign, EternalBlue
Jaff	Spam campaign
Sage	Spam Campaign. Botnet, RIG exploit kit
GlobeImposter	Spam campaign
Locky	Spam, Neutrino exploit kit, Nuclear exploit kit
Mamba	Targeted attack, network compromise
CryptoXXX	Angler exploit kit
CryptoWall	Spam campaign, Angler, Nuclear, magnitude exploit kit, Malvertising

(*Source:* New Jersey Cybersecurity and Communications Integration Cell, 2017.)

9.2.4 *HOW RANSOMWARE OPERATES*

Studies indicate that several variants of ransomware have their unique Modus operandi. Several experts describe (Fruhlinger, 2017; Sarah, 2017; Saurbaugh and Liska, 2017) how ransomware works in a lot of works. But, for easy comprehension, we decided to follow an approach which is described by Liska and Gallo (2017) in their book 'Ransomware: Defending against Digital Extortion (Figure 9.3).'

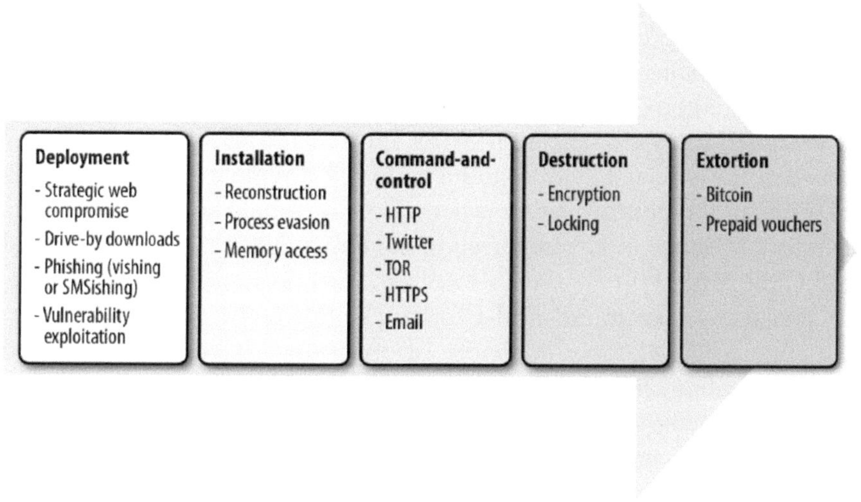

FIGURE 9.3 (See color insert.) Stages of compromise.
(*Source:* Reprinted with permission from Liska, A. and Gallo, T. (2017). *Ransomware: Defending Against Digital Extortion* (First Edit). California: O'Reilly Media, Inc.)

Phase 1 – Deployment: This phase comprises of the distribution of the payload that would later be used to infect the system. This phase cannot be achieved except one or two of the following actors are put in place (Liska and Gallo, 2017):

- **Drive-by Download:** This is a trick whereby a piece of malicious code embedded in a comprised site is downloaded onto the system without the user's authorization.
- **Phishing:** This is one of the widespread methods for ransomware campaign. Through this means, legitimate emails are sent to users

with malicious attachments or compromised links. With a single click, the system becomes infected (Johnson, 2016).

- **Exploiting System's vulnerability:** This method includes conducting a reconnaissance of the target network or system and exploiting the vulnerability found.
- **Advertisement:** Malicious advertisements can also be embedded within legitimate websites to be used for luring victims (Savage et al., 2015).

Phase 2 – Installation: Here, as soon as the malicious malware is delivered to the target, the impacting process also begins. Usually, the infection is being propagated using a method called DDM (Download Dropper Method). In this method, a small piece of file is dropped which is designed to evade detection. After impact, it establishes a connection between the infected system and a server in a remote location known as Command-and-control server where the main malware and instructions are stored. Hence, it is transferred to the target and gets it installed (Liska and Gallo, 2017).

Phase 3 – Command-and-Control: In this phase, the C&C serves as the commanding officer-in-charge of operations. It ensures that smooth communication channel between the infected machine and server is established. Meanwhile, the payload begins to request for instructions on how to carry further action. And these requests include identifying the types of files that are to be encrypted, which encryption algorithm they should used, and whether they should continue to spread at the beginning of the process or not. Furthermore, a synchronization process often takes place in some ransomware variants, where the malware report back significant information such as system information, domain names, IP addresses, operating system banners, information about installed antivirus, and so on (Liska and Gallo, 2017; Sophos, 2017).

Phase 4 – Destruction: At this phase, all malicious files that will carry out the operations in the form of encryption and deletion are put in place; the targeted files are also identified. The encryption keys have also been supplied to the malcode. Now encryption or deletion can begin. In most cases the malcode will encrypt files with extension such as.doc.jpg.gif. pdf.xsl.xml.zip.exe and many more (Liska and Gallo, 2017; Mattias et al., 2016).

Phase 5 – Extortion: At this point, however, it is expected that the hackers have eradicated all backed up files; encryption of sensitive files has also taken place. The next is to generate a notification on the victim's screen informing them that they have been compromised. The different variants of ransomware have their unique ways of displaying their demand messages. However, in this notification, the victim is informed on how to pay the ransom for the release of the system or device which is now paid using bitcoin. If the first allotted time expires, the ransom is doubled (Comtact, 2017; Exabeam, 2016; Klein, 2017; Liska and Gallo, 2017; Mehmood, 2016).

9.2.5 RANSOMWARE INFECTION VECTORS

Attack vectors are typically the transmission channels through which a machine is being infected by a particular attack or another (Zimba et al., 2017). All types of attacks come with their different and unique campaign strategies. Our case study here is ransomware. Hence there are various attack vectors which are peculiar to ransomware attacks. Perhaps the majority of experts believe that the most commonly used attack vector for ransomware attacks is known as phishing attack (Brodsky, 2017; Mehmood, 2016; Zimba et al., 2017), and according to a study by Barracuda networks (Goodall, 2017), 90% of anonymous emails received in the last two years are email phishing. Although, among the newly discovered ransomware variants, exploit kits are now used in the last few months for carrying the attacks. Malvertising has also become very handy these days, as many have fallen victims of this method (Zindell, 2017; Rubens, 2017b). Experts have also identified other types of attack vectors, and they consist of the following -

- **SMSishing:** This is a technique used by hackers to deceive recipient into navigating to a site and providing their personal information. This attack is mostly carried on Android and IOS based mobile devices (Vanderburg, 2016).
- **Vishing:** This is an automated voicemail attack that lures the victim into calling a premium phone number. The caller in the other end usually impersonates a customer representative of a fake company,

who directs the victim into installing the malware on their machine (Vanderburg, 2016).

- **Drive-by Download:** This method allows the victim to download malware on their machine from a legitimate website that is compromised by malicious codes (Lawn, 2016; Vanderburg, 2016).
- **Network Propagation:** This is an instance whereby worm-like ransomware is spread across a given network and affected all the vulnerable machines on the network (Vanderburg, 2016).
- **Freeware Trojans:** Hackers sometimes share bad software with bad codes on the internet with free value and access, to trap unsuspecting individuals. These kinds of software are available for free to download. Use of such could open backdoors and eventually lead to a massive attack.
- **Flash Player:** FP is a small program that is provided by the software giant Adobe Systems. It is intended to support multimedia files on the internet. Several reports came in explaining how some group of hackers used fake FP to coordinate attacks (Adobe Forum, 2017; Collins, 2017). Adobe system confirms that victims were diverted from legitimate websites of Adobe to compromised ones where they get the fake software downloaded on their systems (Symantec, 2017).
- **Messaging Apps:** Hackers embed malicious JavaScript in messages conveyed through Face book Messenger containing images in Scalable Graphics File (SVG) format. With a click, the image directs the victim to a spoofed YouTube site and deceives the victim into installing a codec file (Rubens, 2017b) which eventually compromised the system.

9.2.6 INDICATORS OF COMPROMISE (IOC)

In every instance of a ransomware attack, attackers always leave a trail. Those trails are what experts refer to as indicators of compromise (IOC). Rigorous study and assessment of some selected compromised machines have provided hints on the indicators that prove a machine has been compromised. According to SANS Institute, in one of their periodic white papers indicate that some of the early signs or IOC can be obvious to the user while some cannot be obvious enough. When a machine gets infected by a variant of ransomware, system files extensions begin to change, bulk file renames occurs, explicit ransom notice appears boldly on display screen. While in other cases, possible denial of services also occurs (Majd, 2017). Similarly, CISCO SystemsInc further describes that when machine is compromised certain behavioral indicators begin to show themselves. Some of these indicators include "disappearance of wallpaper background, document file establishing network communication, files being modified in the system directory, VBA Macro uses CallByName, artifact flagged by antivirus engines, a submitted document caused a crash dump file to be created, heavy traffic created by unknown programs, occurrence of DNS (Domain Name System) traffic, unknown processes running unknown activities, system eventually slowing down" and many more (CISCO, 2017). Ericka Chickowski, a columnist at Dark Reading News, while reporting on the IOC states that other critical signs are prevalent in some cybercrime scenes and these indicators comprise of "unusual outbound network traffic, anomalies in privileged user account activity, geographical irregularities, other log-in red flags, swells in Database read Volume, HTML response sizes, large numbers of requests for the same file, mismatched port-application traffic, suspicious registry or system file changes, DNS request anomalies, unexpected patching of systems, bundles of data in the wrong places, web traffic with inhuman behavior, etc. (Chickowski, 2013). Some of the indicators above can easily be traceable on the system, whereas others can only be identified with the help of some tools.

9.2.7 *COMMON RANSOMWARE TARGETS*

In the last few decades, the targets of ransomware attacks were very specific and few. Most of them were organizations, government sectors or individuals. Now, ransomware has widened its attack coverage to all angles. Meaning, it is unpredictable, decisive, and random. Nowadays, ransomware target could be anyone or all of the following entities viz individuals (Mehmood, 2016), Law Enforcement (Krotoski, 2017), Government agencies (Ravindranath, 2016), retailers (Kaspersky, 2016; Starr, 2018), Telecoms companies (Wall, 2018), manufacturers (Perrett, 2018), entertainment, construction, charity organizations (Murray, 2017), transport systems (Williams, 2016), educational institutions (Robbins,

2017), hospitals (Davis, 2016), and financial institutions (Harpur, 2017; Kaspersky, 2017b). Lately, the focus of attack is on IoT devices and other gadgets due to the versatility of their platforms and other factors like zero-day exploits. Furthermore, Kaspersky Security Bulletin: Threat Predictions for 2018 predicts that by 2020, billions of cars will be constructed and there is a chance that 98% of the cars will be connected to the Internet. So with this indication, there is a clear sign that automobiles would be the next ransomware targets (Kaspersky, 2017a) (Figure 9.4).

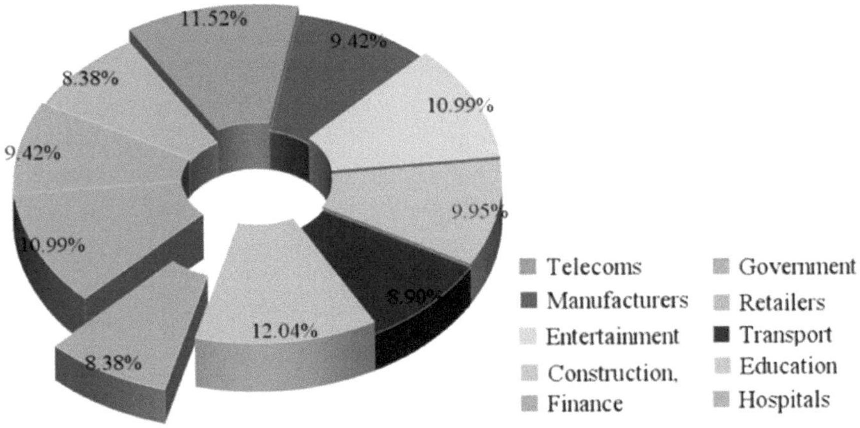

FIGURE 9.4 (See color insert.) Ransomware attack targets.
(*Source:* Kaspersky, (2017a). *Kaspersky Security Bulletin: Kaspersky Lab Threat Predictions for 2018*. Retrieved from: https://goo.gl/nbXMo6.)

9.2.8 MODE OF PAYMENT

The mode of payment for ransom in most cases is categorized based on the hacker's choice. Perhaps, the mode of payment keeps changing dramatically over time. For instance, the earlier ransomware victims were asked to pay the ransom through what was known as Traveler's cheque until later in 1990 when it was completely abolished. Later as technology evolves, a computer scientist and expert in cryptology by the name Satoshi Nakamoto invented a unique digital Cryptocurrency popularly known as Bitcoin in 2009. Hackers use the cryptocurrency because no bank or intermediaries are needed, and it is untraceable, safe, and reliable for anonymous transactions (Kshetri and Voas, 2017). Other recently discovered modes of payments include iTunes and Amazon Gift Cards. Both methods are not popular and in-use (Olenick, 2016). It is important to know that from the time of compromise some ransomware give the ultimatum of 72 hours to pay the ransom. If the deadline is over, the ransom gets multiplied.

9.2.9 CAUSES OF RANSOMWARE ATTACKS

Cybersecurity experts have analyzed hundreds of ransomware cases and concluded that among the likely causes of ransomware attacks are a failure to patch up operating systems regularly and added that lack of proper security policy management is a major cause, too (Figure 9.5).

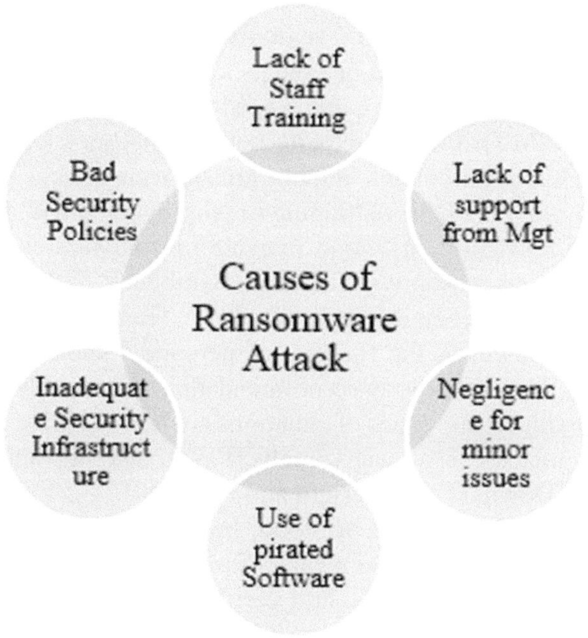

FIGURE 9.5 Causes of ransomware attacks.

9.3 DEFENSE-IN-DEPTH

Defense-in-depth is a systematic strategy of using a layered approach to tackling security issues within a system or organization in case one method or two are compromised (Cartwright, 2017). Bella (2016) in his article "Defense in depth simplified" suggests that when using the defense-in-depth approach in cybersecurity, four main pillars are often put in place namely protection, detection, response, and awareness (Bella, 2016). Similarly, FEMA and NIST also propose what they called "Framework core" for tackling security risks which include identification, prevention, protection, mitigation, response, and recovery (FEMA, n.d.; NIST, 2014). Trend Micro also agrees that "there is no silver bullet when it comes to stopping ransomware attack." But a multi-layered approach could prevent it from penetrating the network and systems (Fimin, 2017; Micro, 2017).

9.3.1 PREVENTIVE MEASURES

- **Conduct Training and Create Awareness:** It is worth noting that end users are usually targeted or used as a gateway into the infrastructure when it comes to ransomware attacks. Meaning, people are the weakest link (Westin, 2017). Lots of experts believe that lack of knowledge of ransomware has contributed to a lot of ransomware attacks in these recent years. Therefore, they believe training of employees and making sure they are aware of how to identify and manage ransomware situations would help in preventing ransomware attacks (Bambenek, 2017; Dawes, 2017; Kalember, 2017; Saurbaugh and Liska, 2017; Trend Micro, 2017; Yaqoob et al., 2017). However, the training of personnel should be focused on certain areas. Hence, experts recommend that trainees should be taught how to deal with certain types of situations such as phishing attacks, social engineering attacks, espionage, malvertising, and advanced persistent threat attack (Downs, Cook, Wright and Kent, 2017).
- **Backup Data:** So far so good, backup has proven to be one of the most effective and reliable means of preventing ransomware damage. Hence, perform regular backup and verify the backup to ensure integrity and validity of the backed-up files before moving on. Also, ensure that backup storage are not connected to network (Bambenek, 2017; Brodsky, 2017; Dawes, 2017; Haley and Sherman, 2017; Januszkiewicz, 2017; No More Ransomware, 2017a; Perekalin, 2017; Phil, Chris and Amber, 2017; Singh et al., 2017; Trend Labs, 2017). According to Unitrends, an effective backup strategy called 3-2-1 rule is one of the best approaches nowadays for ensuring data safety during ransomware outbreak. Many

including Trend Micro considers this strategy as one of the best methods for backing up data. The backup can be done systematically as follows: 3 copies of your data, store in 2 different types of media storages and store 1 copy of data in an off-site location. Unitrends adds that user must test back up several times to ensure that backed up data is valid and restorable (Fimin, 2017; Goodall, 2017; Harpur, 2017; Jordan, 2017; Kalember, 2017; Micro, 2017; Saurbaugh and Liska, 2017; Singh, 2017; Uni Trend, 2017). Haley and Sherman, on the other hand, add incremental backup as a solution to backup issue especially on endpoints (Haley and Sherman, 2017). Similarly, Watson believes users should back up their data using what he refers to as Volume Shadow Copy Service-based backup. VSS is an application based backup system in windows machines which uses two methods namely clone and copy-on-write to create shadow copies of backup from snapshots (Watson, 2017).

- **Apply System Updates, Security Patches, & System Upgrades:** We have seen recently how failure to patch systems had contributed considerably to the WannaCry and Petya ransomware pandemic in 2017. Henceforth, ensure that all systems' operating systems are patched up and up to date. To achieve a successful patch up, use software assets management solution like Symantec Endpoint software manager, TripWire, Corvil which can detect which version of software needs to be patched and on which computer. Also, make sure that firmware and third-party software including Adobe Flash, Java, Web browsers are patched too (Beek and Furtak, n.d.; Dawes, 2017; Goodall, 2017; Kalember, 2017; Saurbaugh and Liska, 2017; Trend Labs, 2017). Likewise, ensure that any application on the system that connects to the internet is always up to date and fully patched (Liska and Gallo, 2017). On top of that, experts warned that users should beware of fake security patches and updates which are downloadable via torrents and popular sites which masquerades the genuine patches. Such files' extensions are labeled as "Patcher," so beware of those (Singh et al., 2017; Trend Micro, 2017).
- **Use Robust Anti-Malware and Anti-Virus Solutions:** Security giants like Trend Micro, Kaspersky, and McAfee claim that using a robust anti-malware and or antivirus could help detect and block ransomware from attacking. According to the giants, they boasted that their labs had successfully blocked over 4 million variants of ransomware from 2015 to date (Perekalin, 2017; Abiodun et al., 2018; Dawes, 2017; Januszkiewicz, 2017; Kalember, 2017; Trend Micro, 2017; Westin, 2017).

- **Proper Security Policy Implementation:** Poor security policy implementation could open direct access to the organization's infrastructure. To start with, make sure administrative privileges are being assigned properly to the right users. Also, make sure that the policy of least privilege assignment is handled efficiently (FBI, 2016a). It is reported that variants of ransomware take advantage of this privileges and exploit the systems. Henceforth, users should use administrative privileges and accounts only when it is necessary (Fimin, 2017; Harpur, 2017; Phil et al., 2017; Trend Micro, 2017; Westin, 2017). Furthermore, it is advised that Access Control is configured on files, directory, and network share permissions; make it read-only on those directory locations (FBI, 2016b; Leong et al., 2016; Mehmood, 2016). In addition to that, set up software control policies to prevent programs from executing themselves automatically without user's authorization especially%AppData% and%LocalAppData% folder (Harpur, 2017). Also enforce UAC (User Account Control) feature found in windows OS; it helps prevent malware from being executed by itself. If it manages to execute itself, UAC will pop up an authentication dialogue box for the user to sanction it (Downs, Cook, Wright and Kent, 2017; Harpur, 2017).
- **Disable Unwanted Functions, Features, & Services:** It is undoubtedly true that operating systems do come with many features and services that are not always used by the operating system or the user, for instance, communication ports, some active directory services, etc. These features and services can be exploited by hackers when they are not secured. Assessment of malware cases shows that different variants of ransomware have used Microsoft office Macros as a point of entry to exploit the system. Hence, disable macros so that files sent through email cannot have an impact on the macros. IC3 also calls that it is advisable to open attached emailed documents using a simple office viewer rather than using the whole Office suite (Downs, Taylor and Whiting, 2017; Harpur, 2017; Kalember, 2017; Saurbaugh and Liska, 2017; Singh, 2017; Westin, 2017). Also, disable WMI, by doing so it will be impossible for ransomware to spread over the networks.
- **Decentralize Resource & Isolate Network Components:** Decentralize and separate critical files and other resources including network resources and devices throughout the organization. That could be achieved by implementing logical/physical separation of resources and

network items based on organizational units and or departments within the corporate organizations (Dawes, 2017; Downs, Taylor and Whiting, 2017; Fimin, 2017; Harpur, 2017; Saurbaugh and Liska, 2017; Stenhouse, 2016). In addition, at the physical layer of the network, implement VLAN as another effective method for segmentation of critical infrastructure. This method if implemented properly could prevent ransomware from spreading over the network (Stenhouse, 2016).

- **Provide Security Support Team:** It is recommended that every organization should have at least a team of IT support personnel who can analyze and deal with threat before it escalates. Also, IT personnel are advised to join the dark web so as to get updated on new threats; that would make them aware of how to address certain threats (Downs, Taylor and Whiting, 2017; Singh et al., 2017). However, it is also a good practice to keep your IT support team on speed dial. That sounds funny, but it can save you time and reduce the impact of attack if help comes quickly (Downs, Taylor and Whiting, 2017; Mehmood, 2016; Yaqoob et al., 2017).
- **Avoid Using Freeware:** Visiting compromised sites for freeware could endanger your entire organizations. Visit only sites that are trusted. To ensure trustworthiness of a site, the user can tighten security by configuring security zone in the internet option from the system's control panel. Raise the zone security level from the default (medium) to high. Users can also use third-party software solutions for monitoring compromised sites (Dawes, 2017; FBI, 2016a).
- **System Hardening and Hygiene:** Security experts believe poor security hygiene is in many cases amongst the causes of cybersecurity risk on systems. Therefore, maintaining proper security hygiene is vital (Westin, 2017). Some of the good practices for sanitizing and hardening system's security are explained as follows. Part of system hygiene is blocking ads and unnecessary web contents from accessing your browser; this is because criminals use legitimate sites for malicious intent (Harpur, 2017). It is a good practice also to run system checks and diagnosis periodically; this could help uncover premature risks. Perform periodic penetration testing on your organization's network. That could help identify vulnerabilities that could endanger the infrastructure (Haley and Sherman, 2017).

9.3.2 PROTECTIVE MEASURES

- **Implement End-Point Security:** One of the major roles played by End Point Security (EPS) is reducing or eliminating chances of getting exposed to non-trusted sources. Implementing Endpoint solutions could help control access to critical and sensitive information within the organizations (Anon, n.d.; Beek and Furtak, n.d.; Brodsky, 2017; Downs, Taylor and Whiting, 2017). End-Point solution could be any form of solution, such as ransomware behavior monitor, vulnerability shielding system, malware profiler, web server protector, email, and gateway protector, spear-phishing protector, etc. (Brodsky, 2017; Dawes, 2017; Goodall, 2017; Saurbaugh and Liska, 2017; Trend Labs, 2017; Westin, 2017). Another strategy for endpoint protection is the implementation of DNS sinkhole. This is a trap designed on the DNS server to prevent resolving hostnames of some selected URLs. Some of its common functions include blocking access to drive-by download websites, controlling, and blocking access to C&C channels and other malicious traffics (Mazerik, 2014).
- **Filter and Monitor Incoming Emails:** Provide a solution that can filter and monitor all incoming emails and make sure links and attachments are not malicious. And do not attempt to open any suspected links or attachments unless you are sure of their authenticity or simply forward them to the IT department for further analysis (Downs, Taylor and Whiting, 2017; FBI, 2016a; Goodall, 2017).
- **Use Virtually Controlled Environments:** It is a good practice to execute suspicious items in a controlled environment such as virtual machines and sandbox to limit and control the impact of the damage in case there is one. (FBI, 2016a) Another smart move is to make Notepad the default program for opening script files such as JavaScript, PS, WSH, etc. (Harpur, 2017; Leong et al., 2016). You can also use a type of endpoint solutions that have sandbox embedded in them which provides a micro virtualized platform for testing email attachments, word docs and so on (Liska and Gallo, 2017).
- **Implement Program Whitelisting:** Implementing application control using AppLocker for white listing programs, software, and applications could help a lot in restricting unauthorized and unknown programs from executing. WannaCry, Petya, NotPetya have benefitted from this vulnerability. And a lot of experts view it among the most effective ways of preventing ransomware from launching itself on the system (FBI, 2016a;

Harpur, 2017; Leong et al., 2016; Singh, 2017; Stenhouse, 2016; Westin, 2017).

- **Migrate Data TO Cloud-Based Service:** Migrating your data to cloud-based data protection services is the smartest move for protecting data in the era of ransomware. It has certain ramifications including cost and espionage. The benefits are boundless as it enhances recoverability and reliability of backups as well as security and compatibility (Reavis and Nielsen, 2017). Alternatively, you can implement DRaaS. According to Margaret Rouse (Rouse, 2017), DRaaS is a physical or virtual server machine that provides third-party failover services in the event that the main systems fail. DRaaS has the ability to backup data with less user interaction and also in the event that the system fails, it quickly restores the data back to its original form.

9.3.3 DETECTION TECHNIQUES

Nieuwenhuizen (2017) in his whitepaper published on April 2017 recommends a behavioral-based approach for detecting ransomware. The approach comprises two techniques namely static-based detection and dynamic or behavioral-based detection technique:

1. **Static-**based analysis detection analyses a suspected application's code before it is executed on the machine to determine whether it can cause harm to the system. Hence, if the analysis turns out to be positive, the executable will be automatically halted from launching.
2. **Behavioral-**based detection, this process monitors the application, assesses its behavioral patterns carefully to determine whether it has malicious intent, if so, the detector will quickly flag it and classify it as dangerous.

- **Deploy Firewalls with IDS/IPS:** By deploying an effective firewall equipped with intrusion detection and prevention capabilities, and then feeding it with updated signatures. The user has a very high chance of detecting and stopping ransomware from establishing a connection with the command & control server (Brodsky, 2017; Dawes, 2017; Harpur, 2017; Saurbaugh and Liska, 2017). As a supplement, the user could impose spam filtering as well as web gateway filtering to tighten the security.

- **Deploy Heuristic Detection Solutions:** There are a variety of solutions out there that help you detect ransomware. Perhaps the choice of heuristic detection solutions would give an added advantage to the user. Heuristic solutions have the ability to learn and adapt to any situation. Their learn ability is what makes them unique. Thus, budget for tools and solutions that detect known and unknown variants of ransomware family and ensure they can automatically feed their repository with latest signatures of new variants of ransomware (Harpur, 2017; Reavis and Nielsen, 2017).
- **Monitor Events and System Logs:** Event logs have been so valuable and resourceful when it comes to uncovering cybersecurity mysteries. However, security experts believe that lots of security failure comes due to reluctance to analyze event logs. Hence, setting log filters and monitors to track the activities of your system (Ambre and Shekokar, 2015; Brodsky, 2017; Grimes, 2013).
- **Traffic Analytics:** DNS traffic analytics have been very handy these days when it comes to detecting anomalies especially when the attack is in motion and communication has been established between the malcode and the C&C server. So by blocking the communication chain, you have automatically stopped the attack (Lemos, n.d.).
- **Deploy Honeypot:** Honey pots are decoys set in the network to help proactively detect suspicious activities in the network before they could make any serious impact (Rouine, 2017). Ideally, honey pots are not deployed to prevent ransomware from attacking; instead, they are a line of defense that gives the administrator a signal that something is about to happen. Hence, deploying honey pot could give the administrator time and opportunity to quickly shutdown computers and network devices before they get infected (Delaney, 2016; Moore, 2016; Stenhouse, 2016; Surati and Prajapati, 2017).

9.3.4 RESPONDING TO RANSOMWARE INCIDENT

9.3.4.1 RESPONDING AT PRE-ATTACK PHASE

- **Design Incident Response Plan:** The response plan is basically a piece of the document containing contingency plans about how an organization can easily respond to a disaster in case it occurs. So, usually, it contains some routine exercises which are to be carried out during the response period. In talking about incident response plan, the National Institute of Standards and Technology (NIST) have published a 79-page response guideline. In one section, it describes the major stages of incident response process which comprises of preparation, detection, and analysis, containment, eradication, and recovery, and post-incident activity (NIST, 2014) (Figure 9.6).

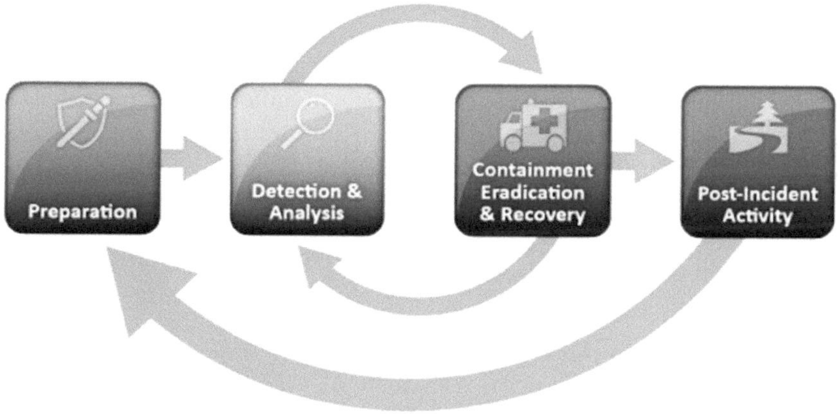

FIGURE 9.6 (See color insert.) Incident response plan.
(*Source:* NIST, (2014). *Framework for Improving Critical Infrastructure Cybersecurity*. US: National Institute of Standards and Technology.)

Similarly, Harpur (2017) says a response plan should have the following checklist of 10 exercises:

- Respond quickly;
- Isolate infected device to contain infection;
- Preserve the encrypted data;
- Determine if you have backups;
- Identify the ransomware;
- Determine if a decryption tool is available;
- Restore or decrypt;
- Last resort is to pay ransom;
- Review and strengthen the infection point, and lastly; and
- Fully wipe and re-image infected device.

- **Get Cybersecurity Insurance:** Some organizations when attacked cannot respond quickly due to shortage of resources to facilitate an effective and quick response. So in order to stay put, it is recommended that an organization avails for an insurance policy to enable the organization to put in place all the resources required for full response and recovery (Hamlin and Rutledge, 2017).

9.3.4.2 RESPONDING AT CAMPAIGN PHASE

- **Situation Analysis Through Tabletop Exercise:** Hold a special meeting among high ranking organization's members to discuss the nature of the emergency. Hence, set up a team to analyze the situation at hand by focusing on the following (Hamlin and Rutledge, 2017; Saurbaugh and Liska, 2017):
    - What exactly you are dealing with;
    - How did the attack happen;
    - What part of the system is affected;
    - How to deal with the incident;
    - Determine whether a Legal action will be taken; and
    - How to improve the situation and get back to work.

- **Report Suspicious Activities:** One of the good etiquettes that are often associated with good users is the ability to report suspicious activities upon discovery. Suspicious activities could be heavy traffic, idle ports becoming actives suddenly, system misbehaving on its own, etc. (Downs, Taylor and Whiting, 2017; Singh, 2017). Thus, establishing an efficient process of reporting incidents as soon as they are detected on the systems is important (Hamlin and Rutledge, 2017).
- **Impose Human Firewall:** Put all hands on deck. Get your best IT support team to try and do everything at their disposal to stop the ransomware from propagating itself at its early stage. That includes the use of endpoint and by imposing filtering technology. Do not let it pass endpoint (Hamlin and Rutledge, 2017).

9.3.4.3 RESPONDING AT INFECTION PHASE

- **Use Sandbox:** Sandboxing allows the user to execute the suspected ransomware in a controlled environment using a special virtual system and observe the behavior of the malware. However, one of the best practices for responding to ransomware attack is to sandbox the threat, reverse engineer it and analyze it (Maass, 2016; Trend Labs, 2017; Westin, 2017).
- **Report Incidents to Law Enforcement:** The Internet Crime Complaint Center in their public service announcement dated September 2016 had urged the public to report any cyber incidents to www.IC3.gov quickly and requests victims to submit the following information while reporting the threat: Date of infection, ransomware family or variant, victim company details, how the infection happened, ransom amount, attacker's bitcoin wallet address, overall losses, and impact statement (FBI, 2016a).
- **Reset Passwords and Entry Codes:** Upon realizing that your entire system has been compromised, it will be smart to quickly change login passwords, entry codes, authentication keys, and what have you. A study reveals that some ransomware not only deploys a ransom malware but also use that opportunity to steal other information including login credentials, financial information and so on (Dargin, n.d.).
- **Exit User Accounts with Admin Privilege:** As soon as you detect that you are being attacked, sign out all user admin with Administrative privileges. And if there are IoT appliances that are managed by computers, like the elevators, room conditioning system, temperature controllers, gas controllers, etc., stay away and shut them down too. If one or two files or script is responsible for controlling them, they could jam off as the system goes down.

9.3.4.4 RESPONDING AT COMMUNICATION (C&C) PHASE

- **Disconnect Machines From Network:** Upon discovering any strange activities or process on your system, simply disconnect from the internet or WI-FI, as some ransomware variants have to retrieve the encryption key from the C&C server. Hence, wait a moment and understand the situation or call IT support for help (Harpur, 2017; Januszkiewicz, 2017; No More Ransomware, 2017a; Harnedy, 2016).

- **Block Non Trusted Domains From Firewall:** Firewall is your first line of defense; hence use the firewall rules to block incoming and outgoing communications that might be established with the C&C server. That could help terminate the exchange between the malcode and the C&C server from taking place (McAfee, n.d.).
- **Block Tor Browser:** When you realize that you are under attack, block access to tor. Experts learned that Tor Onion browser could be used by ransomware to obfuscate C&C. Which means Tor is clearing the passage for the crypto to be delivered (McAfee, n.d.).

9.3.4.5 RESPONDING AT ENCRYPTION PHASE

- **Isolate Infected Source and Control Spread:** If the response team can quickly identify the infected machine or source of the attack, then it is recommended that the infected machine is isolated from the rest of the machines to avoid rapid spread (Stenhouse, 2016). To control the spread, you can start by detaching network connectors from the machine, disabling all shared drive and disconnect external attachments in the form of devices if the infection has not spread yet (Dargin, n.d.; Reavis and Nielsen, 2017; Harnedy, 2016).
- **Use Threshold Alert for Crypto API Calls:** Some variants of ransomware usually use one of the windows feature known as Windows Crypto API to manage encryption. In the process of file encryption, a number of unusual calls are being made to the Crypto API. So by creating a threshold alert to trigger an alarm whenever suspected calls are detected, the user can link the alert to an endpoint solution which will be able to stop the encryption or simply inform the admin about an ongoing attack (Liska and Gallo, 2017).
- **Using Bait for Honey Files and Honey Directories:** A Honey file is also an intrusion detection technique which uses a bait file that is deliberately kept for ransomware to access. A group of files with different formats like doc, pdf, Jpeg are stored in a decoy file server, and the files are given names that can be attractive to anyone with malicious intent like "company salary list" or "Top Secret." As soon as the ransomware hit and begin to encrypt them, an alert notification is sent to the IT team for further action (Liska and Gallo, 2017; Yuill et al., 2004).

- **Stop All Shared Folders and Drives:** When you notice that the encryption of certain files and folders has begun, quickly stop shared folders and mapped driver from having further access. Also, detach all external storage systems from infection zone. Additionally, some ransomware might want to attack file server, turn the server down until the problem is over (McAfee, n.d.; Wahdain and Mohamad, 2014).

9.3.4.6 RESPONDING AT PAYOFF DEMAND PHASE

There are controversies on whether victims of ransomware attack should pay the ransom or not. Lots of experts and even law enforcement agencies have differed in this regard. Liska and Gallo assert that ransom should not be paid unless the system plays a significant role in the organization's network. A commonwealth article (Common Wealth, 2017) discloses that paying the ransom has both pros and cons. The popular European site known as "No more ransom" also discourages paying the ransom as there is no guarantee for restore of data (No-More-Ransom, 2017). Notwithstanding, paying the ransom should be an option when the situation is unavoidable. Hence, below are tips for paying the ransom:

- Before making the payment explore all open options, endeavor to find out the exact family or variant of ransomware you are dealing with by searching online or consulting cybersecurity experts.
- Check whether you have backups, if you do, simply wipe the system clean and install new OS again. After that, discard the ransom demand.
- Visit websites like No More Ransomware for free ransomware decryption tools as they update their reservoir frequently.
- As part of your contingency plan, learn how cryptocurrency works, how to buy a cryptocurrency especially Bitcoin.
- If paying has become the last option, reach out to an expert who knows how Bit coin works, that is, if victims cannot pay for themselves.
- Yaqoob et al., recommend that as part of the payment procedure, the victims should try to negotiate minimum ransom payable with the hackers (Bambenek, 2017; Yaqoob et al., 2017).

- Arrange the purchase and go ahead and make the payment. At the very time we are writing this paper 1 bitcoin is 11,182.23 USD.

9.3.5 RECOVERING FROM RANSOMWARE INCIDENT

Recovery after an incident is a sensitive stage. Experts have given recommendations on strategies for recovering from a ransomware incident. Also, there are quite a number of ransomware recovery solutions that can help you recover from a ransomware attack (Supasatit and Clement, 2017). The following recommendations are considered as the best strategies for recovering from a ransomware attack incident:

- **Design a Recovery Plan:** The first thing we need when recovering from a ransomware attack is come up with a tentative and well-defined recovery plan (Hamlin and Rutledge, 2017). Recovery experts say a comprehensive recovery plan must comprise of the following routines: Risk identification and mitigation, measuring impact of damage, building guidelines, designing recovery strategies, developing plans, and lastly maintain and develop testing routines. It is important to note that the routines above can be altered according to situations (Martin, 2017).
- **Traditional Recovery Method:** Steve Ragan (Ragan, 2017) recommends that some systems can be recovered using the traditional approach; by simply implementing the following steps: Reboot the system (s) in safe mode, then go ahead and install a good anti-malware solution. Then after installation, scan the system and see whether the malware could be detected. If yes, remove the malware. Here, removal of malware will require the system to reboot. After rebooting, go ahead and restore the system to its previous state using system restore and then recover your data (Ragan, 2017).
- **Find Decryption Key Online:** You can sometimes visit websites of some Law enforcement agencies and private security companies to find a decryption key for free online. Some ransomware variants have already been analyzed, and decryption keys are made available online. You can visit the following sites for free decryptors: www.nomoreransom.org, and www.noransom.kaspersky.com (Stenhouse, 2016).
- **Perform Cause Analysis:** It is a good practice to perform a cause analysis after every incident. That will give you a clear picture of

what went wrong and how to prevent or limit its future occurrence. If you cannot perform it due to lack of tools and personnel, you can always hire experts to perform the analysis on your behalf (Stenhouse, 2016).

9.4 CONCLUSION

So far, we have seen in the above discourse how detrimental ransomware could be if unleashed on our businesses, lives, and our security. It is undeniably true that what we have witnessed so far is just the tip of the iceberg. Surely ransomware is growing bigger and smarter every day (Singh, 2017). It is complex enough to evade pursuit; it is becoming more difficult to tackle than ever. With the disturbing WikiLeaks news about how a million-dollar hacking project arsenal comprising of viruses, Trojans, malware, weaponized "zero-day" exploit kits, as well as malware remote control frameworks and related documentation got stolen from the NSA (National Security Agency) servers (WikiLeaks, n.d.). Since then, many have a fear that the next generation ransomware will not just hit small targets; instead, it will target critical infrastructure like nuclear power plants, cruise missile controllers, submarines, drone controllers and so on. When that time comes, we will not just be dealing with crypto-ransomware anymore. Instead, we will be dealing with weaponized ransomware and AI-based ransomware that will be capable of mass destruction. In the future, we might not even know what type of ransomware we will be dealing with.

KEYWORDS

- **defense-in-depth**
- **detection**
- **exploit kit**
- **hacker**
- **malware**
- **phishing**
- **ransom**

REFERENCES

Adobe Forum, (2017). *A New Strain of Ransomware Nicknamed "Bad Rabbit" Asks to Update Adobe Flash Player*. Retrieved from: https://goo.gl/oSq6zr (Accessed on 28 August 2019).

Ambre, A., & Shekokar, N., (2015). Insider threat detection using log analysis and event correlation. *Procedia Computer Science, 45*, 436–445. https://doi.org/10.1016/j.procs.2015.03.175 (Accessed on 28 August 2019).

Anon, (n.d.). *End Point Control*. Retrieved from: https://goo.gl/uVRCsR (Accessed on 28 August 2019).

Ashford, W., (2015). *Ransomware Growing Rapidly, Warns Intel's McAfee Labs*. Retrieved from: https://goo.gl/NjFjmE (Accessed on 28 August 2019).

Bambenek, J., (2017). *Ransomware in the Age of Wannacry: What Fintech Needs to Know?* Retrieved from: https://goo.gl/A5HwbU (Accessed on 28 August 2019).

Barracuda, (2017). *Understanding Ransomware*. Retrieved from: https://goo.gl/qUisCG.

Bella, R. L., (2016). *Defense in-Depth Simplified*. Retrieved from: https://goo.gl/czf78C (Accessed on 28 August 2019).

Brodsky, J., (2017). *Detection of Ransomware and Prevention Strategies*. Retrieved from: https://goo.gl/mpExdZ (Accessed on 28 August 2019).

Brunau, C., (2017). *Common Types of Ransomware*. Retrieved from: https://goo.gl/Kczkyz (Accessed on 28 August 2019).

Cabaj, K., Gregorczyk, M., & Mazurczyk, W., (2017). Software-defined networking-based crypto ransomware detection using HTTP traffic characteristics. *Computers & Electrical Engineering,* 1–16. https://doi.org/10.1016/j.compeleceng.2017.10.012 (Accessed on 28 August 2019).

Cartwright, D., (2017). *Defense in Depth: A "Layered" Strategy can Repel Cold Attackers*. Retrieved from: https://goo.gl/Xsm2st (Accessed on 28 August 2019).

Chickowski, E., (2013). *Top 15 Indicators of Compromise: Dark Reading News*. Retrieved from: https://goo.gl/aE2TqQ (Accessed on 28 August 2019).

CISCO, (2017). *Indicators of Compromise and Where to Find Them*. Cisco BLOGS. Retrieved from: https://goo.gl/VULXTo (Accessed on 28 August 2019).

Collins, K., (2017). *The Latest Ransomware Presents Itself as an Adobe Flash Player Download*. Retrieved from: https://goo.gl/MeLQmQ (Accessed on 28 August 2019).

Common Wealth, (2017). To pay or not to pay: How to survive a ransomware attack. *Common Wealth Financial Network*. Retrieved from: https://goo.gl/VzXLfk (Accessed on 28 August 2019).

Comtact, (2017). *How Ransomware Works*. Retrieved from: compact.co.uk.

Crowe, J., (2016). *Ransomware by the Numbers: Must-Know Ransomware Statistics 2016*. Retrieved from: https://goo.gl/Nn3CRk (Accessed on 28 August 2019).

Crowe, J., (2017). *Must-Know Ransomware Statistics 2017*. Retrieved from: https://goo.gl/FBVFns (Accessed on 28 August 2019).

Davis, J., (2016). *Ransomware: See the 14 Hospitals Attacked so Far in 2016. Healthcare IT News*. Retrieved from: https://goo.gl/TJFie5 (Accessed on 28 August 2019).

Dawes, S., (2017). *WannaCry Ransomware: How to Detect the Vulnerability and Exploits*. Retrieved from: https://goo.gl/A1ukWN (Accessed on 28 August 2019).

Delaney, D., (2016). *5 Methods for Detecting Ransomware Activity*. Retrieved from: https://goo.gl/qcxpUE (Accessed on 28 August 2019).

Downs, J., Cook, D., Wright, J., & Kent, J., (2017). *Protecting Data in the Age of Ransomware*. Retrieved from: https://goo.gl/YJ2f94 (Accessed on 28 August 2019).

Downs, J., Taylor, A., & Whiting, I., (2017). *Fighting Ransomware & Responding if the Worst Happens*. Retrieved from: https://goo.gl/WSrVM3 (Accessed on 28 August 2019).

Exabeam, (2016). *Threat Research Report the Anatomy of A Ransomware Attack*. Retrieved from: https://goo.gl/gpwAKX (Accessed on 28 August 2019).

FBI, (2015). *Criminals Continue to Defraud and Extort Funds from Victims Using Crypto Wall Ransomware Schemes*. Retrieved from: https://goo.gl/b7GSog (Accessed on 28 August 2019).

FBI, (2016a). *Ransomware Victims Urged to Report Infections to Federal Law Enforcement*. Retrieved from: https://goo.gl/8vNpVo (Accessed on 28 August 2019).

FBI, (2016b). *Incidents of Ransomware on the Rise. Federal Bureau of Investigation*. Retrieved from: https://goo.gl/z8Z1Bf (Accessed on 28 August 2019).

FEMA, (n.d.). *Five Mission Areas: US*. Retrieved from: https://www.fema.gov/mission-areas (Accessed on 28 August 2019).

Fimin, M., (2017). Are employees' parts of the ransomware problem? *Computer Fraud & Security, 8*, 15–17. https://doi.org/10.1016/S1361–3723(17)30072–6 (Accessed on 28 August 2019).

Fortinet (2017). *Fortinet Threat Landscape Report*. Fortinet. New York: Fortinet . Retrieved 06 2017, from https://www.fortinet.com/corporate/about-us/newsroom/press-releases/2017/fortinet-threat-landscape-report.html

Fruhlinger, J., (2017). *What is Ransomware? How it Works and How to Remove it*. Retrieved from: https://goo.gl/UJJ1C8 (Accessed on 28 August 2019).

F-Secure, (n.d.). *Crypto-Ransomware*. Retrieved from: https://goo.gl/amkNiU (Accessed on 28 August 2019).

Goodall, D., (2017). *Ransomware: The Best Defense*. Retrieved from: https://goo.gl/z8qyso (Accessed on 28 August 2019).

Grimes, R. A., (2013). *Detect the Undetectable: Start with Event Logs*. Retrieved from: https://goo.gl/eeNyPq (Accessed on 28 August 2019).

Haley, K., & Sherman, M., (2017). *Defense Against a Ransomware Attack: Latest Research and Best Practices*. Retrieved from: https://goo.gl/4fHm8h (Accessed on 28 August 2019).

Hamlin, M., & Rutledge, B., (2017). *How to Recover from a Ransomware Disaster*. Retrieved from: https://goo.gl/n98EAj (Accessed on 28 August 2019).

Hampton, N., & Baig, Z. A., (2015). Ransomware: Emergence of the cyber-extortion menace. *Australian Information Security Management Conference, 13*, 47–56. https://doi.org/10.4225/75/57b69aa9d938b (Accessed on 28 August 2019).

Harnedy, R., (2016). *How to Recover from Ransomware: The First 5 Things You Should Do*. Retrieved from: https://goo.gl/Xr3VeF (Accessed on 28 August 2019).

Harpur, R., (2017). *Cybersecurity Threats: Ransomware*. Retrieved from: https://goo.gl/xFTKKN (Accessed on 28 August 2019).

Januszkiewicz, P., (2017). *Ransomware Protection-Top 3 Prevention Techniques to Use*. Retrieved from: https://goo.gl/PWMQWc (Accessed on 28 August 2019).

Johnson, B., (2016). *How Ransomware Works*. Retrieved from: https://goo.gl/rLMNRU (Accessed on 28 August 2019).

Jordan, M., (2017). *5 Ways Backup Kills Ransomware Threats*. Retrieved from: https://goo.gl/XdxKay (Accessed on 28 August 2019).

Kalember, R., (2017). *Ransomware-The Billion Dollar Thief*. Retrieved from: https://goo.gl/a39FDh (Accessed on 28 August 2019).

Kaspersky, (2016). *Attacks on Business Now Equal One Every 40 Seconds: Ransomware is Kaspersky Lab's Story of the Year 2016*. Kaspersky Lab. Retrieved from: https://goo.gl/NKHKrn (Accessed on 28 August 2019).

Kaspersky, (2017a). *Kaspersky Security Bulletin: Kaspersky Lab Threat Predictions for 2018*. Retrieved from: https://goo.gl/nbXMo6 (Accessed on 28 August 2019).

Kaspersky, (2017b). *Kaspersky Security Bulletin: Review of the Year 2017*. Retrieved from: https://goo.gl/z6yg8W (Accessed on 28 August 2019).

Kashinath, T. (2017). *Command And Control Servers : Things You Should Know*. Retrieved from Secpod Blog: http://www.secpod.com/blog/command-and-control-servers-things-you-should-know/.

Klein, T., (2017). *5 Phases of Ransomware Attacks*. Retrieved from: https://goo.gl/bGbwkm (Accessed on 28 August 2019).

Korolov, M., (2017). *Ransomware Took in $1 Billion in 2016-Improved Defenses May Not be Enough to Stem the Tide, CSO*. Retrieved from: https://goo.gl/26DA32 (Accessed on 28 August 2019).

Krotoski, M. L., (2017). WannaCry ransomware cyberattack raises legal issues. *The National Law Review*. Retrieved from: https://goo.gl/gdYnLp (Accessed on 28 August 2019).

Kshetri, N., & Voas, J., (2017). Do crypto-currencies fuel ransomware? *IEEE, 19*(5), 11–15. https://doi.org/10.1109/MITP.2017.3680961 (Accessed on 28 August 2019).

Lawn, S., (2016). *Ransomware: Current Strains, Attack Vectors and Protection*. Retrieved from: https://goo.gl/m2tVh7 (Accessed on 28 August 2019).

Lemos, R., (n.d.). *Ransomware: 5 Strong Tactics for Defense and Response*. Retrieved from: https://goo.gl/qEfN9w (Accessed on 28 August 2019).

Liska, A., & Gallo, T., (2017). *Ransomware: Defending Against Digital Extortion (1st edn.)*. California: O'Reilly Media, Inc.

Lord, N., (2017). *A History of Ransomware Attacks: The Biggest and Worst Ransomware Attacks of All Time*. Retrieved from: https://goo.gl/A8o978 (Accessed on 28 August 2019).

Maass, M., (2016). *A Theory and Tools for Applying Sandboxes Effectively*. Carnegie Mellon University. Retrieved from: https://goo.gl/Hqcahd (Accessed on 28 August 2019).

Majd, (2017). *Kaspersky Security Bulletin: Predictions for 2017-"Indicators of Compromise" are Dead*. Retrieved from: https://goo.gl/CoLVkB (Accessed on 28 August 2019).

Martin, J. A., (2017). *7 Things Your IT Disaster Recovery Plan Should Cover | CSO Online*. Retrieved from: https://goo.gl/kt5YHc (Accessed on 28 August 2019).

Mattias, W., Frick, J., Sjostrom, A., & Jarpe, E., (2016). A novel method for recovery from crypto ransomware infections. In: 2nd *IEEE International Conference on Computer*

and Communications (pp. 1354–1358). IEEE. https://doi.org/978–1–4673–9026–2/16/$31.00 (Accessed on 28 August 2019).
Mazerik, R., (2014). *Understanding DNS Sinkholes-A Weapon Against Malware*. Retrieved from: https://goo.gl/JwKDoi (Accessed on 28 August 2019).
McAfee, (n.d.). *Understanding Ransomware and Strategies to Defeat It White Paper*. Retrieved from: https://goo.gl/7vdx3d (Accessed on 28 August 2019).
Micro, T., (2017). *Ransomware*. Retrieved from: https://goo.gl/nZaoAa (Accessed on 28 August 2019).
Moore, C., (2016). Detecting ransomware with honeypot techniques. *Proceedings-2016 Cybersecurity and Cyber Forensics Conference, CCC*, 77–81. https://doi.org/10.1109/CCC.2016.14 (Accessed on 28 August 2019).
Morgan, S., (2017). *Ransomware Damage Report*. Retrieved from: https://goo.gl/um3tBe (Accessed on 28 August 2019).
Mort, M., (2017). *Symantec Blocks 22 Million Attempted WannaCry Ransomware Attacks Globally*. Business wire. Retrieved from: https://goo.gl/BG56TR (Accessed on 28 August 2019).
Murray, S., (2017). Charities unprepared for cyberattack risk. *Financial Time*. Retrieved from: https://goo.gl/kJmZcm (Accessed on 28 August 2019).
National Cybersecurity and Communications Integration Center, (2016). *Ransomware and Recent Variants*. Retrieved from: https://goo.gl/UBXGBi (Accessed on 28 August 2019).
Nieuwenhuizen, D., (n.d.). *A Behavioral-Based Approach to Ransomware Detection*. Retrieved from: https://goo.gl/WfDtDG (Accessed on 28 August 2019).
NIST, (2014). *Framework for Improving Critical Infrastructure Cybersecurity*. US: National Institute of Standards and Technology.
NJ Cybersecurity & Communication Integration Cell, (2018). *Ransomware*. Retrieved from: https://goo.gl/EPUW9G (Accessed on 28 August 2019).
No More Ransomware, (2017a). *Prevention Advice*. Retrieved from: https://goo.gl/f16kF3 (Accessed on 28 August 2019).
No More Ransomware, (2017b). *The History of Ransomware*. Retrieved from: https://goo.gl/1Aqxzg (Accessed on 28 August 2019).
No More Ransomware, (n.d.). *Types of Ransomware*. Retrieved from: https://goo.gl/wmhwZh (Accessed on 28 August 2019).
No-More-Ransom, (2017). *Prevention Advice*. Retrieved from: https://goo.gl/GK8LzU (Accessed on 28 August 2019).
O'Brien, D., (2017). *An Internet Security Threat Report Special Report*. Retrieved from: https://www.symantec.com/content/dam/symantec/docs/security-center/white-papers/istr-ransomware-2017-en.pdf (Accessed on 28 August 2019).
Olenick, D., (2016). *New Ransomware Demands Payment in iTunes, Targets Older Android Software*. Retrieved from: https://goo.gl/CnomUi (Accessed on 28 August 2019).
Perekalin, A., (2017). *WannaCry: Are You Safe?* Retrieved from: https://goo.gl/hguon9 (Accessed on 28 August 2019).
Perrett, M., (2018). Food manufacturers warned over "cyber hurricane" events. *Food Manufacture*, Retrieved from: https://goo.gl/2zBPVh (Accessed on 28 August 2019).
Phil, R., Chris, G., & Amber, B., (2017). *How to Recover from the WanaCrypt Ransomware Attack*. Retrieved from: https://goo.gl/iRRxzm (Accessed on 28 August 2019).

Ragan, S., (2017). *How to Recover from a Ransomware Attack | CSO Online*. Retrieved from: https://goo.gl/BhRvjS (Accessed on 28 August 2019).

Ravindranath, M., (2016). *Ransomware Attacks on Government Agencies Tripled in Past Year*. Nextgov. Retrieved from: https://goo.gl/6HC5tF (Accessed on 28 August 2019).

Reavis, J., & Nielsen, A., (2017). *Backup & Recovery: Your Get Out of Ransomware Free Card*. Retrieved from: https://goo.gl/CjkLe2 (Accessed on 28 August 2019).

Robbins, G., (2017). *Los Angeles College Pays $28,000 in Ransomware*. Sandiego Union Tribune. Retrieved from: https://goo.gl/HLj3R3 (Accessed on 28 August 2019).

Rouine, S., (2017). *A Guide on How to Prevent Ransomware*. Retrieved from: https://goo.gl/PKgJ53 (Accessed on 28 August 2019).

Rouse, M., (2017). *What is Disaster Recovery as a Service (DRaaS)?-Definition from WhatIs.com*. Retrieved from: https://goo.gl/85dRV2 (Accessed on 28 August 2019).

Rubens, P., (2017a). *Common Types of Ransomware*. Retrieved from: https://goo.gl/EiQjE3 (Accessed on 28 August 2019).

Rubens, P., (2017b). *Understanding Ransomware Vectors Key to Preventing Attack*. Retrieved from: https://goo.gl/j6Gxjp (Accessed on 28 August 2019).

Sarah, (2017). *Spotlight on Ransomware: How Ransomware Works*. Retrieved from: https://goo.gl/mQX8dJ (Accessed on 28 August 2019).

Saurbaugh, M., & Liska, A., (2017). *Defending Against Ransomware with Intelligence, People, and Automation*. Retrieved from: https://goo.gl/6wRDwz (Accessed on 28 August 2019).

Savage, K., Coogan, P., & Lau, H., (2015). The evolution of ransomware. *Security Response, 57*. https://doi.org/10.5437/08953608X5403011 (Accessed on 28 August 2019).

Shafqat Mehmoon. (2016). *Enterprise Survival Guide for Ransomware Attacks. 80*. North Bethesda, United States: SANS Institute. Retrieved 07 21, 2017, from https://www.sans.org/reading-room/whitepapers/awareness/paper/36962

Shinde, R., Veeken, P., Van Der, Schooten, S., & Berg, J. V. D., (2016). Ransomware : Studying transfer and mitigation. In: *International Conference on Computing, Analytics and Security Trends (CAST)* (pp. 90–95). Pune, India: IEEE.

Singh, A., (2017). *The Second Coming of Ransomware-Insights into New Developments*. Retrieved from: https://goo.gl/p5GMvd (Accessed on 28 August 2019).

Singh, A., Grantz, M., Payne, C., Laing, B., & Wolf, R., (2017). *How the US Secret Service Combats Advanced Ransomware*. Retrieved from: https://goo.gl/hLqj5z (Accessed on 28 August 2019).

Sophos, (2017). *Ransomware: How an Attack Works*. Retrieved from: https://goo.gl/KLDJ4r (Accessed on 28 August 2019).

Starr, R., (2018). *Hackers Will Target Small Business Through the Internet of Things in 2018, New Report Says*. Small Business Trends News. Retrieved from: https://goo.gl/n87Qg7 (Accessed on 28 August 2019).

Stenhouse, J., (2016). *Master of Disaster Webinar-Recovering from Ransomware in Minutes*. Retrieved from: https://goo.gl/1hBKdF (Accessed on 28 August 2019).

Supasatit, T., & Clement, K., (2017). *Backup on the Wire: Ransomware Recovery Based on Precision Packet Capture*. Retrieved from: https://goo.gl/kosr7t (Accessed on 28 August 2019).

Surati, S. B., & Prajapati, G. I., (2017). A review on ransomware detection & prevention. *International Journal of Research and Scientific Innovation Issue IX, IV (Ix)*, 2321–2705. Retrieved from: https://goo.gl/JwDUyV (Accessed on 28 August 2019).

Symantec, (2017). *Fake Adobe Flash Update Installs Ransomware, Performs Click Fraud*. Retrieved from: https://goo.gl/Qn9fbe (Accessed on 28 August 2019).

Trend Labs, (2017). *Ransomware: Past, Present, and Future*. Trend Labs.

Trend Micro, (2017). *Ransomware Recap: Patcher Ransomware Targets MacOS*. Retrieved from: https://goo.gl/L1dk85 (Accessed on 28 August 2019).

Ubale, M. K., & Isyaku, S. M., (2017). A situation analysis on cybercrime and its economic impact in Nigeria. *International Journal of Computer Applications, 169*(7), 975–8887.

UniTrend. (2017). Beat Ransomware in 5 Easy Steps. Burlington, United States: UniTrend. Retrieved 03 23, 2017, from https://www.unitrends.com/wp-content/uploads/Ransomware-White-Paper.pdf

Vanderburg, E., (2016). *The Top 10 Ransomware Attack Vectors*. Retrieved from: https://goo.gl/JDSntN (Accessed on 28 August 2019).

Wahdain, E. A., & Mohamad, N. A., (2014). User acceptance of information technology: Factors, theories and applications. *Journal of Information Systems Research and Innovation, 31*(2017), 17–25.

Wall, M., (2018). *Firms Buy Insurance "in Mad Panic" as Cyberattacks Soars*. UK: BBC London. Retrieved from: https://goo.gl/6oqEiG (Accessed on 28 August 2019).

Watson, G., (2017). *A Foolproof Ransomware Recovery Strategy*. Retrieved from: https://goo.gl/ifiMUt (Accessed on 28 August 2019).

Weatherford, M., Hayslip, G., & Crosby, S., (n.d.). *Wannacry Ransomware & the Future of Cyberattacks*. Retrieved from: https://goo.gl/Q49TZb (Accessed on 28 August 2019).

Westin, K., (2017). *How to Stay Ahead of Today's Ransomware Realities*. Retrieved from: https://goo.gl/DoUg29 (Accessed on 28 August 2019).

WikiLeaks, (n.d.). *Vault 7 : CIA Hacking Tools Revealed, WikiLeaks*. Retrieved from: https://wikileaks.org/ciav7p1 (Accessed on 28 August 2019).

Williams, C., (2016). *Passengers Ride Free on SF Muni Subway After Ransomware Infects Network, Demands $73k*. The Register UK. Retrieved from: https://goo.gl/bg6UKF (Accessed on 28 August 2019).

Yaqoob, I., Ahmed, E., Rehman, M. H., Ahmed, A. I. A., Al-garadi, M. A., Imran, M., & Guizani, M., (2017). The rise of ransomware and emerging security challenges in the internet of things. *Computer Networks*. https://doi.org/10.1016/j.comnet.2017.09.003 (Accessed on 28 August 2019).

Yuill, J., Zappe, M., Denning, D., & Feer, F., (2004). Honey files: Deceptive files for intrusion detection. *Proceedings from the 5th Annual IEEE SMC Information Assurance Workshop* (pp. 116–122). https://doi.org/10.1109/IAW.2004.1437806.

Zimba, A., Wang, Z., & Chen, H., (2017). Reasoning crypto ransomware infection vectors with bayesian networks. *IEEE*, 149–151.

Zindell, P., (2017). *Stopping Ransomware at the Door (and Every Other Threat)*. Retrieved from: https://goo.gl/S8FSD7 (Accessed on 28 August 2019).

CHAPTER 10

Global Challenges and Role of ICT in Empowerment of Teachers

SHWETA TRIPATHI

Devi Ahilya Vishwavidyalaya, Indore, Madhya Pradesh, India,
E-mail: shwetatrip02@gmail.com

ABSTRACT

Teaching is an entangled and reclaimed process, the main goal of which is the conquest of the pupils in every sector of life. Relevance of an educator in the life of a pupil is long-lasting. Technology is the pragmatic understanding of techniques, processes, and other indistinguishable items, embedded in machines and may be dealt with without detailed information of their operation. We can feel technology everywhere in the area of teaching-learning process. But some researchers have made clear that most teachers are slow in adapting technological inventions in their educational process. There remain contradictory evidences that technology and online learning are improving the result of teaching-learning process for most students. This chapter attends to the role of technology in teaching and learning, role of teachers in making aware their students towards cybercrime and cybersecurity and empowerment of teachers.

10.1 INTRODUCTION

Teaching is an entangled and reclaimed process, the main goal of which is the triumph of the students in every field of life: skill development, ability enhancement, knowledge, social ability, etc. "The destiny of India is being shaped in her classroom" has been indicted by the Education Commission (1964–65), and furthermore, the fate of these classrooms is being molded in the organization of education with submitted educators (Kothari Commission, n.d.). Since ages, teachers are holding prominent place in the academic field of our Indian society. Teaching is a complex and challenging process which is directed towards training the students in every field of life: skill development, ability enhancement, knowledge, social ability, etc. Relevance of an educator in the life of a pupil is indispensable and long-lasting. School is like a 'home beyond home' for a pupil, and an educator is like 'second parent' to them. The Relevance of an educator in students' life is very broad and not limited to the teaching of subject matters. Hence the role of teachers becomes very important.

10.2 MAIN FOCUS OF THE CHAPTER

10.2.1 THE ROLE OF 21ST CENTURY TEACHERS

The role of a teacher has undergone drastic changes with time. The role of 21st-century teachers is totally different from the traditional teachers -

- **The In-Charge:** The teacher is the manager of the class, what pupils do, what they act upon, and how they act upon it. The educator accepts this part when novel items/idea are being put forth and exact propagation and penetrating methods are required In this teaching-learning process, the educator is for the most part the main person, the educator may have the enrichment of direction, and can motivate by their learning and skill, at the same time, does this bit honestly allow for enough student talk time? Is it really enjoyable for the students? There is likewise an interpretation that this role could have a dearth of variety in its activities. To improve this role an educator requires being expert in the use of digital media and playing his role dynamically.
- **The Instigator:** The teacher urges student to partake and initiate suggestions about how they may continue in an action. The educator needs to help them just when needed. At the moment, when students are truly 'lost for words,' the guide can empower by circumspectly prodding students. Students may now and again babble or end up being doubtful about how to proceed; the teacher in such manner can motivate and initiate but always in a supportive way.
- **The Resource Person:** An educator is a kind of flexible resource, arranged to offer help if vital, or give students whatever thing they need

while performing exercises. The teacher must be accessible to the pupils so that they can have guidance during their need. As strength the educator can manage students to make the maximum use of available possessions, for example, the web, for themselves, it surely isn't expected to spoon-feed students, as this may have the detriment of making students parasite.

- **The Analyzer:** The educator expects the role to recognize how well students are performing or how well they performed. Here an educator acts as an evaluator and make complete use of digital media to play his role to perfection. Input and revision is composed and completed. There are assortments of ways in which we can review students; the part of an assessor gives educators a chance to improve students. In any case, in the event that it isn't spoken with affectability and bolster it could demonstrate counter-gain to a student's confidence and trust in taking in the targeted information.
- **The Organizer:** Maybe the most exasperating and challenging part the teacher needs to play. Accomplishment of various activities depends upon extraordinary affiliation and on the students knowing exactly what they are to do straightaway. Giving headings is central in this part and furthermore setting-up exercises. The facilitator can moreover fill in as a demonstrator; this part furthermore empowers a teacher to get included and attract with students. The instructor moreover serves to open the impeccably shut practices and moreover give content criticism. Here it is imperative that the instructors playing this job realize how to utilize ICT. As while playing out this job utilization of ICT winds up irreplaceable to spare time, cash, and energy.
- **The Partaker:** This part upgrades the air in the class when the teacher takes part in an action. In any case, the teacher puts it all out there for overpowering the development when performing it. Here the teacher can revive a class; in the event that a teacher can stay back and not transform into the central purpose of thought; it very well may be a marvelous strategy to speak with students without being exorbitantly overwhelming.
- **The Tutor:** The educator goes about as a mentor when pupils are incorporated with venture work or in self-thinking. Here an educator should motivate the pupils to utilize computers to escalate their creativity, knowledge, and divergent thinking ability. The teacher gives counsel and enlightens the way and enables students to illuminate ideas and farthest point assignments. This role can be an incredible way to give mindful recompense to a student. It can similarly enable a teacher to tailor a course

to fit the pupil's requirement. In any case, it can likewise prompt a student to become excessively needy or even increasingly comfortable with one teacher and one strategy or method of teaching.

- **Confident Coach:** An educator should be Empathetic Mentor. One thinks back with gratefulness to the splendid teachers, with admiration towards the pupils who touched the human sentiments. Gone is the time when teachers dole out assignments and send pupils on their cheerful way. A cutting edge teacher understands the plenty of components affecting students in our present world, and endeavors to tailor, realizing where conceivable, toward individualized requirement and absorption. Here he understands the requirement of ICT for students and also realizes the significance of making the students aware of cybercrimes. He understands the significance of teaching the pupils the ways of cybersecurity. This student-focused concentration additionally makes learning open doors for the educators to gain along with students, building up their educating, and shared aptitudes.
- **The Facilitator:** Brown (2007) noticed that educators might assume numerous roles throughout their teaching career and in turn, encourage learning. Their potential to play these roles successfully will depend to a substantial degree on the affinity they build up with their pupils, and obviously, the standard of teaching-learning and their aptitudes. As per Harmer (2007), the term 'facilitator' is utilized by numerous thinkers to depict a specific sort of teacher, one who is fair and democratic in nature (where the educator involves the pupils in leadership and management activities) instead of despotic (where the teacher is liable for everything that goes ahead in the teaching-learning), and the person who encourages student self-governance (where students learn on their own, and assume liability for that learning) using group and pair work and by going about to a larger extent as an asset than as a transmitter of information. This way it strengthens both the pupil and the educator and frees the educator from many loads that having to act as an 'expert' might entail (Nola, n.d.). Here too the responsibility of digital media cannot be avoided.

So here's hoping the next time a teacher is in a class he/she considers how their role might affect students' learning. Is the classrooms teacher-centered, with teachers always exercising control over the things? Or are the teachers able to 'let go,' and allow students to take center stage? The roles are often interchangeable. The educator's task is never static. One activity could see an experienced teacher smooth transformation from one role to another. As said, the present century classroom is developed on the thought that students encounter what they require to face the present century working environment and live in the worldwide condition. The features of the present century classroom therefore set it apart from the last century classroom.

To meet up the above role of 21st-century teachers, an educator should be very much aware of the ongoing improvements, one of which is the advancement of ICT in the field of education. As of late, a few investigations and reports have featured the openings and the capable advantages of digital world for enhancing the nature of educational process. ICT is viewed as a noteworthy instrument for building learning social orders (Sangrà and González-Sanmamed, 2010) and, especially, as a component at the school level that could give a way to reconsider and update the teaching-learning frameworks and procedures, in this manner providing quality education to all. ICT plays significant role in upgrading the quality and commitment of educators and the educational process. As per Wikipedia, technology is the accumulation of systems, abilities, techniques, and procedures utilized in the generation of products or administrations or in the achievement of targets, for example, logical investigation. Innovation can be the data of techniques, shapes, et cetera, or it tends to be embedded in machines which can be worked without distinct learning of their workings (Wikipedia, 2017). Teaching, utilizing the ICT, does not, without anyone else's input prompt accomplishing educational programs goals, since part from aiding class planning, a great information of the ICT enables us to help our students in their class exercises involving the digital media. Moreover, Professional growth is a vital component to refreshing abilities and for professional success for teachers. Of other things, we may utilize the ICT to participate in a group discussion, subscribe to a news gathering, take classes, and stay in contact with colleagues. To effectively utilize the ICT for teaching, a teacher should be aware of how to get to the different administrations available through the ICT. Besides, it has been accounted for that the greater part of educators who utilize the ICT in teaching are the individuals who trust that the ICT is new path for getting things done. It isn't simply the PC now, how they are utilized makes the effect. They stresses on teacher training as a key to powerful use of ICT innovation in the classroom (Richardson, 2007). Each new innovation has its pros and cons. No one has set aside opportunity to break down the negative effect of presenting students to the ICT, may have on their social improvement. A teacher should be well versed and aware of the cyber crimes that are increasing day by day. They should be up to date with the new ways of committing these crimes so that they make their students aware too in this direction and can help them as and when needed. We can sense technology everywhere in education: classroom, playground, library, laboratory, etc., meaning, all over the school climate.

10.2.2 SCOPE OF ICT TOOLS IN EDUCATION

ICT as an apparatus to develop teaching-learning practice by means of Internet (i.e., digital content, multimedia, teaching-learning strategies, learning condition) – Utilizing PCs and the Internet makes the learning procedure unique and different and prompts increment in intellectual activity. The assortments of web assets improve the educator and student's activities. Nowadays, data advancements have infiltrated into all circles of instructive fields. This reality is associated with a changeless enlarging of capacities of the World Wide Web from one perspective, and makes it conceivable to put any critical data concerning instruction on its larger part of servers. ICT has transformed into a virtual space where people connect with their accomplices, cherished ones. Web is a technique for strengthening correspondence among teachers and students, it is a strong instrument used for socialization.

1. **ICT as an Administrative Tool, i.e., Education Management Information Systems (EMIS):** Administration and management uses of ICT are as of now well known in schools because of its abilities in encouraging organization exercises from information stockpiling to learning administration and basic leadership. There are three fundamental organization bunches that apply ICT in their daily organization and administration in schools: Administrative Heads, Administration Teachers and Administration Staffs.
2. **ICT As An Expanding Learning Opportunity, i.e., Distance Learning, E-Learning:** The utilization of ICT in E-Learning and distance education can open entryways for adapting all the more broadly and fairly over the teaching power. It can moreover upgrade the quality and combination of the benefits and bolster accessible to teachers, opening up new procedure for capable changes. If the social value is to be expert regardless, with respect to estimation of informative opportunity and organizations,

the game plan ought to be made in ways that make it available, open, palatable, and flexible to all teachers and head-educators, empowering them to settle on choices in what and how they understand. It also needs engaging game plans in help of these focuses. Receiving a school-based system maintained by ICT, it gave more far-reaching access than the past courses of action while meanwhile extending learning openings past the points of confinement of the school. It moved the emphasis from a supply-driven game plan to a request-driven one, giving teachers and head-educators more proprietorship and choice in their master progression. It was moreover flexible, satisfying different kinds of learning and getting ready activities, going about as an open learning center, changing its course of action and degree as demonstrated by neighborhood setting while in the meantime holding a common focus mission and set of activities.

3. ICT as a Facilitator of Higher-Order Thinking Skills, i.e., Learner-Centered, Self-Directed Learning, Tailored Learning: Technology has assumed a greater part in our lives, our abilities in basic reasoning and investigation, visual aptitudes have moved forward. By utilizing more visual media, students will process data better, innovation has rolled out numerous improvements in all parts of life, and in addition, changing our desires for what students should figure out how to assume a part in the new worldwide economy. Students must figure out how proportional through tremendous data skylines and pick up the capacity to investigate and settle on choices and adjust to the subjective areas. There is more probability that logical and innovative advance is liable to scholarly advance and not scholarly learning alone. Scholarly advance is the after-effect of crafted by reason and illumination, prompting innovativeness, development, and great administration. The part of man in this time is not restricted to adjusting to the real world. Surely, it is now equivalent with the desires of the unending. This leads us to the need to supplant our customary techniques for influence, which tries to sanctify reality and proportionate with human goals, and it is occupant upon instructors to perceive the colossal part of educational program in the improvement of reasoning among understudies (Gangapathy, 2017). Thinking does not develop without presentations. We, as teachers, must support the student and help him in gaining learning, aptitudes, and data that make him the important logical foundation such that he looks, confirms, and checks for the task of his psychological capacities. Rather than being aloof and latent, he should be dynamic and positive all the while. This is absolutely one of the destinations of the instructive procedure in us to build up the reasoning capacities of our students, particularly inventive reasoning, by making an air of school exercises animating understudies, and in addition making a learning situation discovering that the student dynamic part in it. Using the PC to execute particular objectives to think inside the goals of training with the recognizable proof of instructive exercises to actualize and in this manner conditions for the advancement of reasoning.

For the Educators, it is an inception into:

- **Exploring Educational Possibilities of Technology:** Utilization of PCs in conveyance of education is a standout amongst the most rapidly developing patterns in educational uses of technology. Passing on substance and direction on the web can similarly allow students at schools to fit additional courses into a clamoring logbook, besides, it can allow students who can't go to schools, for whatever reasons to just carry on their dreams. In any case, while providers of online guideline assume that it is ground-breaking in coming to and serving a broad assortment of students, little has been studied to review its reasonability contrasted with up close and personal direction.
- **Learning to Make Right Choices of Hardware, Software, and ICT Interactions:** There are assortments of decisions before an educator regarding the utilization of ICT for the improvement of his students. The ability and intelligence of an educator lies in the right decisions at right time for ideal usage of assets for the ideal advancement of the students.
- **Growing To Become a Critical User of ICT:** Before presenting ICT in classrooms, an educator needs to end up as a capable client of it himself/herself. For smooth running of keen classes an educator needs to wind up as brilliant first. To do as such, training must be given to the teachers toward this path. ICT is a consistently developing field, in this way, an educator should be constantly refreshed toward this path, thus, workshops conferences and seminars should be maintained every once in a while to keep the teachers refreshed with the present advancements and improvement in this field.

For the STUDENT, it is an initiation into:

- **Creativity and Problem Solving:** Creativity is as often as possible saw as a capacity or as a normal for people. Specific character ascribes have been perceived to embody an innovative identity. Meanwhile, various researches found out that imagination can be improved and developed. Very few researches appear to have conveyed to manage upon the development of inventive reasoning by the utilization of online resources. During a time of omnipresent change as result of the digital media and other correspondence advancements, student-focused guideline gives chances to construct higher-order thinking, teamwork, and apprenticeship. Critical thinking requires innovative aptitudes, divergent thinking and also the capacity to actualize thoughts and speculations in down to practical ways (Wheeler et al., 2002). In addition, intelligent, and self-handled issue-comprehending encounters advances the inspiration of the students as initiated and developed through the formative movement of students' Meta intellectual abilities, for example, self-checking and self-augmentation. Viable learning dependent on constructivist didactics, including self-sorted out learning in mix with dynamic and innovative issue-explaining in synergistic settings, progresses students' attendant cognitive and meta-cognitive processes. Consequently, students' co-development of information entrusted in social unique learning conditions, for example, school-based errands grasp the semantic connections arising from exercising, confirming, and testing of learning through data sharing and discussion. Future investigations ought to center around planning intelligent, versatile, sick-characterized, genuine-world learning situations to inspire students' psychological and meta-intellectual procedures as a key factor for the viable preparing of issue-fathoming skills.

- **An Initiation into the World of ICT:** By using technology students are able to independently organize their learning process; hence, instead of being a passive recipient of information, they become the active users of information. By introducing the students to the world of information and technology, we in fact open new venues for them to enhance their capabilities and actively use their inner strength and be aware of all the possibilities available to them to shine like a star in this world full of opportunities.
- **An Opportunity to Shape Career Pursuits:** Information and Communication Technologies (ICT) ought to be completely coordinated into the conveyance of profession benefits as the world is currently in the global era. Such coordination would possibly accomplish more as well as increment the adaptability of services and upgrade their quality. Professional decision is an imperative choice a man needs to make in his life. Shockingly, numerous youthful people think that it is not easy to take a suitable professional decision. To solve this problem, help of ICT ought to be taken. Vocation guides ought to incorporate utilization of ICT, using

talk rooms, email, and long-range informal communication. At the end of the day, a few decision of vocation could be shown to the understudies on DVD and CD Video. They ought to likewise be presented to Television and radio Chat on a Career decision.

10.2.3 USEFULNESS OF ICT

- **Economical in Terms of Time and Money:** Given proper engagement from the part of the learner, reduced learning retention time has been recorded but the difficulty of transferring such learning to real-life situations has also been observed, therefore care should be taken on the practical implementation aspect of the knowledge gained through ICT.
- **Easy to Implement:** There are various elements which have been recognized which identify with the apparent convenience of ICT, which for this situation is for knowledgeable practicing ICT/IT users. By using ICT it is easy to think of new lesson ideas, and easy to control the class. Regular utilization and encounter with ICT beyond the classroom and owing of personal computers nowadays makes it user-friendly.
- **Allows Flexible, Self-Paced Learning:** By utilizing ICT, students are, to a calculable degree, ready to pick what they'd jumped at and the chance to center around and invest variable measures of energy in it in light of the impression of their adapting needs and positions.
- **Breaks the Monotony:** ICT in schools and classrooms tend to attract school learners' interest and motivation. For those students who do have a reach to the computer and other ICT equipments, the classroom becomes more attractive when networked.
- **Anytime, Anywhere:** One characterizing highlight of ICTs is their capacity to rise above time and space. ICTs make conceivable no concurrent learning, or learning portrayed by a period slack between the conveyance of guideline and its gathering by students. Online course materials, for instance, might be gotten to 24 hours per day, 7 days a week. ICT-based instructive conveyance (e.g., instructive programming communicate over radio or TV) likewise gets rid of the requirement for all students and the teacher to be in one physical area. Also, certain sorts of ICTs, for example, remotely coordinating advances, empower direction to be gotten at the same time by different, topographically scattered students (i.e., synchronous learning).

- **ICTs can Enhance the Quality of Education in Several Ways:** By expanding student inspiration and commitment, by encouraging the acquisition of essential aptitudes and by improving teacher preparing; ICT can upgrade learning to a large degree. ICTs are likewise transformational apparatuses which, when utilized properly, can elevate the move to a student-focused environment.
- **Curriculum Development:** In recent years ICT has been used in curriculum development in areas like curriculum design, needs assessment, forecasting, selection of objectives, generating test items, monitoring student's progress, evaluation, material selection, analysis of scope and sequence, development of objectives, grouping students and reporting of student's success (Wilson, 1997).
- **Digitalized Instruction Material:** As schools step by step move far from print items to creative, innovation-based technology format, a new digital evolution is happening in the classroom, the utilization of applets and applications and online supplementary assets are presently on increment among students. ICT in classroom can be used in many forms.

10.2.4 ROLE OF ICT IN ENHANCING TEACHING-LEARNING PROCESS

We can utilize ICT to add esteem, deal with classroom or enhance management and planning. We need to utilize ICT since we need to change, enhance, add another measurement to our educating, or to fluctuate the sorts and increment the nature of exercises appointed to students. In taking in, the students require some inspiration. To enable them to utilize the ICT in their learning is a motivational push to those who are exhausted by the conventional methods for data conveyance, and in this way speed up the exchange of data from the fleeting memory to the long term memory. The ICT enables us to persuade a portion of the lost students in the class, and in this manner quicken the absorption procedure.

As per an African axiom, a person who asks questions understands a new language faster. Utilizing the ICT for instructive exercises gives an alternate road to revelation through addressing, and for basic assessment of data. It is additionally critical that teachers answer students' inquiries appropriately and conveniently. This is the reason that teachers should continually retrain as indicated by the necessities of the time. For this data age, teachers should be vexed with the individual and instructive employments of the ICT. By and large, Communication is a method for utilizing information. It is a hotspot for re-authorization. The ICT advances quick correspondence crosswise over geological hindrances and in this way gives students a chance to communicate with a wide scope of individuals not envisioned conceivable previously. Likewise utilizing information in various settings give a superior comprehension of the distinctive shades of significance of the ideas engaged with a giving learning circumstance.

Other than to deal with the classroom, you can display class exercises, homework issues, and practical exams on the Net for students to access from residence, from the school, or from wherever that approaches the Net. Guardians can get to the Web pages to affirm the homework issues their kids are required to do. Besides, the homework could be organized such that the student can advance from less difficult to all the more difficult exercises by methods for hyperlinks. Notwithstanding posting class exercises and homework, we can have a class bulletin on site. Along these lines, students can get associated with class discourses in a way impractical in the classroom. This medium can likewise be utilized to determine issues between class periods. Taking everything into account, the ICT enables a teacher to add substance to his/her exercises and to distribute helpful data to students and guardians without misuse of important class time and those give numerous focal points for students in learning.

10.3 MAIN APPROACHES TO ADOPT ICT IN CLASSROOM

10.3.1 INTEGRATED APPROACH

Integrated Approach to Teaching was prepared by Beverly Allen. Integrated approach is one in which learners learn by doing in a learner-centered environment in learner-interest contexts (Webb, 2011).

10.3.1.1 THE VALUE OF THE INTEGRATED APPROACH

It enables students to participate in intentional, pertinent learning under the guidance of their teachers which leads to their all-round development. It urges students to see the interconnectedness and interrelationships between the educational curriculum zones and this leads to the clearance in their concept and they come to know about the importance and the utilization of the things included in their curriculum in their daily life. Students can utilize their past learning to present situations. Integrated units are child-focused; they give chances to culturally diverse sharing. This approach gives opportunities to the students to show competence in various fields, as opposed to depending on a composed or oral test. Proficient advancement of educators is done by utilizing Professional Development of Teachers. ICTs improve student's learning with just in time learning. Curriculum integration crosswise over branches of knowledge is done. Expanded joint effort and correspondence utilizing, ICTs conduct universal learning all through school. Expansion of accessibility of value through ICT instructive assets is done for both, the teachers and the students. This approach enhances students' commitment and leads to improvement of higher-level thinking aptitudes. It cultivates the role and the part of the teachers in ICT Integration, a facilitator of learning. It leads to setting student-focused educational programs related errands. Setting task that are open-ended keeping in mind the end goal to help critical thinking and basic reasoning, creating contextualized exercises. Setting assignment that creates synergistic gathering abilities and utilizing ICT to upgrade and fortify reasoning around the issues introduced in exercises of coordinated effort and sharing learning that is significant for 21st-century students.

10.3.1.2 LEARNING APPROACHES FOR EFFECTIVE ICT INTEGRATION PROJECT-BASED LEARNING

- **Task-Based Learning (TBL):** TBL is an approach in which learning centers around the successful accomplishment of relevant tasks. In the TBL approach, the main focus is the correct use of language for genuine communication.
- **Active Learning:** It is for the most part characterized as any teaching strategy that draws in students in the learning activity. To put it plainly, active learning expects students to do significant learning exercises and consider what they are doing.
- **Collaborative Learning:** It is an educational outlook to teaching-learning that involves students working together to solve a problem, complete a task, or create a product.

- **Inquiry-Based Learning:** Inquiry-based learning (also inquiry-based learning in British English) is a type of dynamic discovering that begins by suggesting questions, issues or situations as opposed to just introducing built up facts or depicting a smooth way to information. The procedure is regularly helped by a facilitator.
- **Cooperative Learning:** It is a fruitful showing technique in which little groups, each with students of diverse levels of capacity, utilize an assortment of learning exercises to enhance their comprehension of a subject.
- **Problem-Based Learning (PBL):** PBL may be a student-centered instructional method in which students learn almost a subject through the encounter of fathoming an open-ended issue found in trigger material. The PBL procedure does not center on critical thinking with a characterized arrangement, yet it takes into consideration the advancement of other attractive aptitudes and traits. This incorporates information obtaining, upgraded assemble joint effort, and correspondence. The PBL procedure was produced for medical education and has since been widened in applications for different projects of learning. The procedure takes into consideration students to create abilities utilized for their future practice. It improves basic examination, writing recovery, and supports continuous learning inside a group situation.
- **Peer-Led Team Learning (PLTL):** PLTL is a demonstrated technique for little gathering peer learning utilized at schools and colleges everywhere throughout the world. PLTL bunches ordinarily comprise of 6–8 students who cooperate to tackle issues, and are encouraged by a Peer Leader.
- **Inquiry Guided Learning (IGL):** IGL alludes to a variety of classroom rehearses that advances student learning through guided and, progressively, the free examination of inquiries and issues for which there are no single answer. As opposed to educating the consequences of others' examinations, which students learn inactively, educators help students in acing and learning through the procedure of dynamic examination itself. This procedure includes the capacity to define great inquiries, recognize, and gather suitable confirmation; present results systematically, examine, and decipher results, plan conclusions, and assess the value and significance of those conclusions. It might like-wise include the capacity to recognize issues, analyze issues, produce

conceivable arrangements, and select the best arrangement with proper legitimization.

- **Just-In-Time Teaching (JiTT):** JiTT is an instructive strategy that uses feedback between classroom activities and work that students do at home, in arrangement of the classroom meeting.
- **Small Group Learning:** It is an instructive approach. The gathering work must be deliberately arranged and every now and again requires a facilitator to guarantee aggregate advance. What's more the gathering capacity and the discovering that happens should be surveyed and assessed. The material learned is similarly as imperative as the gathering's capacity to accomplish a shared objective. Facilitators' aptitudes are essential and require the teacher to guarantee that both the undertaking is accomplished and the gathering working is kept up. Small group teaching enables students to create critical thinking, relational, presentational, and relational abilities, all helpful to life outside the classroom. These nonexclusive aptitudes are hard to create in detachment and require input and collaboration with different people.
- **Project-Based Learning (PBL):** PBL is a creative, deliberate instructing technique that advances student commitment through significant examinations of complex inquiries. Put basically: It's learning by doing. To be put more precisely, the PBL center around bestowing particular learning and aptitudes while rousing students to address effectively, think fundamentally and draw associations between their investigations and this the real world.
- **Direct Instruction (DI):** DI is a general term for the unequivocal educating of a range of abilities utilizing or showings of the material to students. A specific subset of DI, indicated by capitalization as DI, alludes to a particular illustration of the approach created by Siegfried Engelmann and Wesley C. Becker. DI educates by expressive teaching, rather than exploratory models, for example, demand-based learning. DI incorporates instructional exercises, participatory research center classes, dialog, recitation, courses, workshops, perception, and dynamic learning, practice, or entry-level positions. DI depends on an orderly educational programs configuration, conveyed by usage of recommended conduct content. On the preface that all students can learn and all educators effectively show if given viable preparing in particular systems, teachers might be assessed in light of student learning.

10.3.1.3 ROLE OF THE TEACHER IN ICT INTEGRATION

- A supporter of learning;
- Constructing learner-centered curriculum-related tasks;
- Setting assignments that are open-ended keeping in mind the end goal to help critical thinking and basic reasoning;
- Creating content related task;
- Creating activities that develop collective group skills;
- Utilizing ICT to upgrade and animate reasoning around the issues displayed in lessons.

Hence, we lead to the conclusion that ICT Integration approach aims to Student involvement, Development of critical thinking, cultivating of collective and distributive Learning that is meaningful for the present century learners; therefore it is highly beneficial but only when used by experienced and trained teachers.

10.3.2 ENHANCEMENT APPROACH

It focuses on enhancing the subject matter. ICT enhanced learning is student-directed and diagnostic. Not at all like static, content, or print-based instructive advancements, ICT improved learning perceives that there is a wide range of learning pathways and various enunciations of information. ICTs allow learners to explore and discover rather than merely listen and remember.

- **The Value of the Enhancement Approach:** Enhancement approach is highly interactive and diagnostic. It leads to applied empathy and establishes discrepancy. Its advantage lies in understanding resistance, encouraging self-efficacy, and development of tolerance among teachers and students. There is participative learning. The curriculum integration is done crosswise over branches of knowledge. There is a use of wide varieties of methodology of teaching and learning in this approach. It develops self-dependence and confidence among the students, solves the past issues of students and creates a free and fearless environment, emphasis on the enhancement of self-management and social skills.

- **Learning Approaches for Effective Enhancement Approach:** Of three common approaches to learning-surface, deep, and achieving, the last two are more effective in enhancement approach. Application of these two approaches of learning leads to a significant increase in certain indices of deep and achieving attitude among the students, with evidence for long term improvements in academic performance.
- **Role of the Teacher in ICT Enhancement Approach:** Here, a teacher is to act as a knowledgeable and experienced guide and fellow explorer in the journey of education. Here a scope of points of view is investigated, there is no 'right' assortment of information to be educated and the desire is that the teacher will learn alongside the students. A teacher is relied upon here to urge understudies to consider an assortment of elective origination is a critical component in realizing enduring applied change in the learner. The teacher's job has changed from being a dependable master, in charge of a last item, to being a guide who is receptive to the setting in which the learning is occurring. Henceforth, ICT Upgrade Approach can turn out to be profoundly advantageous and benevolent for the improvement of both, the educators and students. It makes an imaginative personality which inquiries as it lays weight on profound and accomplishing adapting in this manner, it ought to be connected well by the instructors in showing learning process for greatest all-round improvement of the students.

10.3.3 COMPLEMENTARY APPROACH

In this approach, ICT acts as a commendatory in improving teaching-learning in the classroom and outside. Teachers take maximum help of ICT to enhance the teaching-learning process (Reitzes, 2011). Use of ICT is mainly done in the following areas by the teacher -

- **In Classwork:** Students display positive approach towards the utilization of ICT in reading, writing, and communicating while doing classwork as it raises the feeling of personal empowerment and reduces the control of the teachers.
- **Homework:** ICT can be used as effective, stimulating, and motivating homework tool. It increases the creative learning by students while doing homework. Use of ICT increases interest and motivation level among the students while doing homework.

- **Communication:** ICT helps to communicate beyond face to face. The limitation to be physically present is removed by the ICT. 21st century teachers make the use of this phenomenon to the maximum.
- **Use of E-Mail:** The trend of using E-mail by the teachers to relate with the students beyond classroom is increasing. E-Mail is acting as an effective tool of giving home works, clearing doubts, handing projects and sharing materials. The students are highly benefited through this.
- **Other Social Media:** The use of other social media like Facebook, twitter in educational field is increasing day by day. Present day teachers are motivating the students to use these Medias in their project and research like for data collection and for advertising their new innovations and thoughts.
- **The Importance of Complementary Approach:** Makes the class lively, students are active and are motivated. Students become critical minded and question things, not just welcome things as they are. Use of modern techniques is done. Students become self-dependent, and the teaching becomes interesting when aided by modern techniques.
- **Learning Approaches for Complementary Approach:** It stresses on audiovisual teaching. Traditional method will not do, innovative new method is welcomed and is adopted which will allow and motivate the students to use all their sensory organs actively. Teaching did through these methods have a long-lasting effect.
- **Role of Teachers:** Teachers acts as a supervisor and facilitator. Teachers' main role as facilitator is to offer support and advice when needed, and to provide the necessary scaffolding and teaching of skills when necessary. It is indispensable that educators make sure to educate and develop in their students any specific ability or bit of learning that is required for thc assignment. In the supervisory role the teachers keep their students under supervision, they move away from the main role but never lose control. They let the students to come on the front and be more self-directed, taking the advantages of modern techniques and equipments. Peer interaction, communication, sharing, and at times expert guidance is given to extend student thinking. Students-not teachers-should drive discussions as they show, protect, expand upon, and react to each other. In complementary approach both the teacher and the ICT play important role in student's development. They both (teacher and the ICT) complement each other, not replace each other. They both have an indispensable role to play in the teaching learning scenario. Nowadays, most of the schools

adopt this approach for maximum development of their students through maximum utilization of resources at their disposal.

10.3.4 GUIDANCE AND COUNSELING APPROACH

Guidance and counseling as an academic support system provided in colleges is meant to produce scientific data to assist students to manage their psychosocial and learning issues. Scientific data acquisition and dissemination is the major thrust of any significant guidance and counseling practice. A counselor is professionally prepared to disseminate such information based on counseling to the clients. Counselors are duty-bound to hitch the globalization venture by applying the data and communication technology ways in counseling practice. The web are often employed in a faculty setting for electronic discussion forum, delivering counseling sessions, gathering data and depositing analysis knowledge for general access. The personal computer-driven age has expedited the utilization of internet sites for job search, delivering career data and other guidance services including interpersonal counseling. The advantages are endless-speed, accuracy, storage, security of information and above all turning the citizens into science destined people. No wonder, a nation with scientifically uneducated citizenry cannot expect to make any reasonable headway in socio-economic and political decisions.

- **Teacher as a Counselor:** As students spend half their years in class and faculty, academics have a big role to play in shaping the temperament of the scholars. Whereas college counselors are trained to deal with a variety of problems like adore broken relationships, strained relationships with oldsters, vanity, and body-image problems, addiction, and thoughts of suicide or potential career methods (Blanco and Ray, 2011). An educator who is in constant contact with the scholars also can initiate conversations and encourage students to open up. This makes the role of a counselor imperative in an educational originated, and academics will become the first supply of on-campus support for the scholars. In spite of the fact that the Central Board for Secondary Education (CBSE) India suggests that schools utilize a fulltime on-campus counselor; in general, this isn't executed. Sometimes it is an aftereffect of lack of engagement by the administration, and regularly because of the inaccessibility of trained counselors. The other challenge is to motivate students to meet school counselors when they end up in disturbed circumstances. Teachers play an important role in bridging this gap. A teacher, who can be relied upon and is sympathetic, can provide support and take students to the school counselor when necessary. However, not all teachers have what it takes to become one. A teacher as counselor needs to:

- **Build Rapport:** Characterized as the capacity to keep up agreeable connections in view of partiality, is all the more conversationally thought of as what happens when two individuals get well together, they interface, associate well, and react to each other positively. Keeping in mind the end goal to wind up a legitimate guide a teacher needs to get in great terms with their students so that they opens up before them and effortlessly examine their issues with them (Rogers, 1943).
- **Enable Students to Express Themselves:** A teacher as a decent guide enables the students to express their more profound emotions decisively. For this the essential prerequisite is to wind up a companion of the students. It takes time to become friend of introvert students; here a teacher needs to act patiently.
- **Be Non-Judgmental:** A teacher as a counselor should get in non-judgmental listening, have the attitude of acceptance, empathy, and genuineness.
- **Consistent in Emotions and Maintain Complete Confidentiality:** When a student approaches his teacher with his problem expects his/her teacher to keep it confidential (Rogers, 1943). A student is very sensitive towards such issues. Therefore, a teacher needs to keep up this faith intact for future correspondences. A teacher in order to become a good counselor should have control on his/her emotions. He/she should not be fickle minded.

Guidance and counseling by teachers is not a new approach, we can trace it back since gurukul age but the utilization of ICT in this task is a recent development which has enhanced its effect. Care should be taken and the teachers involved in this task should be trained and experienced; otherwise, the result will not be positive, and the time, money, and energy invested would go waste.

- **Use of ICT in Guidance and Counseling by Teachers:** Most teachers utilize ICT for composing letters and reports, calling guardians, and keeping records. They additionally utilize ICT for guidance related data through web. In any case, costs, secrecy, security are essential issues that the teachers need to address in this issue.

10.3.5 MANAGEMENT OF LEARNING APPROACH

Administration and management applications of ICT are at present, well known in schools because of its capacities in encouraging organization exercises from information storage to learning administration and basic leadership. This refers to the application software used for organizing and management of institutions broadly refers as Educational Management Information System (EMIS). Use of ICT for record-keeping and database, evaluation, and other managerial activities are some few applications of this area. Change has been going on at an uneven pace in any development-oriented sector, and the education sector is no exception. Speedy growth in the educational arena has made governance in academic sector a very complicated errand. The present century has seen huge progressions in innovation which has prompted far-achieving improvements in the managerial framework. Cost-viable innovation joined with the adaptability in learning and regulatory exercises is fundamental to upgrade proficiency. ICT assumes a crucial part in supporting intense, proficient administration and organization in instruction segment. It is determined that innovation can be utilized ideal from student administration to different resource organization in an educational establishment. There are three main administration groups that apply ICT in their daily administration and management in schools: Administrative Heads, Administration Teachers and Administration Staffs (Ghavifekr, 2013).

- **Administrative Work Performed by Administrative Teachers besides Teaching:** The administrative work performed by Administrative Teachers besides Teaching is that of curriculum head, students' affair head or co-curriculum head. In order to meet their additional responsibilities they have to maintain many records for which they take the help of ICT. However, the most common use of ICT by teachers is in the area of preparing notes, teaching-learning resources or for examination purpose.
- **Administrative Heads:** This should be proficient in the use of ICT as it makes the tedious, lengthy, and time taking work, easy, and quickly completed for them. As an Administrative Head they have to maintain

large amount of data, records, and have to do so many file works. ICT helps them to meet their responsibilities quickly and smartly. It also helps them to safely save their records.

- **Administration Staffs:** This uses ICT in order to meet up their daily work and responsibilities easily and quickly. They use various types of equipments to handle their financial work or to communicate with any concerned authority, maintain records or to collect relevant data. In maintaining financial data like balance sheet, payslips, audit reports or stock keeping ICT proves very effective and makes their work easy and lively. Administration Teachers that apply ICT in their daily administration and management in schools should see to the fact that they have enough time for this task as in no case the student's learning should suffer. Therefore, the management should see to the fact that Administration Teachers should not be overburdened and are having enough time and energy to invest in the classroom with students as it is the pupils for whose development this whole castle is being constructed.

10.3.6 ASSESSMENT OF LEARNING APPROACH

Assessment assumes a gigantic job in the teaching-learning process. It helps teachers and learners to improve teaching and learning. It causes educators and students to enhance instructing and learning. Assessment is an endless method and an irregular exercise. It makes a difference in confining the estimations of judgment, informative status, or achievement of the students. ICT-enhanced learning is understudy facilitated and demonstrative. Not in any way like static, content or print-based educational progressions, ICT-enhanced learning sees that there are an extensive variety of learning pathways and different clarifications of information. ICTs empower students to research and investigate as restricted to just tune in and recollect.

- **Teacher as an Evaluator:** Evaluation is the after-impact of the instructor's worry with the targets of education. In his part as an evaluator, the instructor postures such request as the going with: "are students increasing any ground toward the goals fitting to his learning situation? What proof shows paying little mind to whether they are advancing? What measure of development expecting any, would they say they are making? To what degree can their prosperity (or disappointment) be credited to the encounter that they had in the classroom? "As educators it is the commitment to collect information on students learning experiences and their capacities and advances to ensure that they have demonstrated what they know also, can do. Students learn and develop their work in different unmistakable ways so it is essential as an educator that students are empowered the opportunity to demonstrate their work to ensure that they have met the outcomes in prospectus reports. As an educator, one should achieve these through using unmistakable sorts of examination counting formative midyear, analytic, and input which will have the ability to grasp what the students know, what they don't understand and how a teacher can upgrade the activity transport to students.

- **Use of ICT by Teachers During Evaluation:** Technology has a major part to play in the compelling and proficient appraisal of learning. Present-day development offers instructors a variety of new gadgets that can be used as a piece of the classroom. Advancement can empower instructors to overview their students' learning and furthermore it's execution in the classroom. Usage of ICT in assessment incorporates the usage of automated contraptions to help the advancement, movement, stockpiling or announcing of student's examination assignments, responses, assessments, or its criticism. Instructors can use PCs to assemble their evaluation undertakings, pass on these errands to the concerned students, and to record and give feedback and evaluations to these students. PCs can in like manner be used to examine students' reactions, both to give criticism to the understudy on the quality and significance of their reaction, and additionally to give input to the instructor on whether the task can recognize students with particular abilities. Development enhanced learning conditions, offer a promising street for assessment of the more personality boggling type and lead estimations of Key Competencies, in perspective of Learning Analytics. A substantial number of the present open advancements have the capacity to enhance learning conditions, instruments, and structures and recreate learning conditions which require complex thinking, basic reasoning, and composed exertion philosophies and along these lines think about the progression of non-particular aptitudes. A segment of these conditions empowers students and educators to assess execution, appreciate messes up, and gain from them (Wilson, 1997).
- **Nature of Education:** Evaluation has two types:

- **Assessment of Training or "Evaluation of Learning Accomplishments"** which attempts to give insights in regards to the educative strategy discoveries and to help educational choices gaining in light of real ground; and
- **Summative Evaluation or "Learning Evaluation"** tries to exhibit a definitive outcome and impact of the teaching-learning procedure diverged from expected standards of results.

To date, the two sorts have been associated for different settings and times. Developmental assessment is typical of school classrooms and is associated with little get-togethers of understudies where it is possible to practice separation, while institutionalized appraisals are associated with huge number of students, conventionally picked indiscriminately working with gatherings of control.

10.4 BENEFITS OF UTILIZING ICT FOR EDUCATIONAL PURPOSES

As was called attention to already, the Internet furnishes students with the instruments they require to find and claim information. What's more, give students the snares and formats they have to affix data to the long-term memory. There are a few focal points of utilizing ICT for learning -

- **Motivating Factor:** ICT can go about as a propelling instrument for some students. Youngsters are exceptionally enthralled with innovation. Teachers must gain by this premium, fervor, and energy about the Internet to enhance learning. For officially excited students, the Internet enables to furnish them with extra learning exercises not promptly accessible in the classroom.
- **Quick Communication:** ICT advances quick correspondence crosswise over geological hindrances. Students can join community-oriented activities that include students from various states, nations, or landmasses. This sort of learning background was unrealistic before the Internet. This is an exceptional learning knowledge extremely basic for every one of our students, as the world is getting to be one major network.
- **Helpful Learning:** ICT encourages helpful learning, energizes discourse, and makes an all the more captivating classroom. For instance, an attendance registers of all the sections of class will enable students to

get associated with class discourses through messages in a way unrealistic within the four wall dividers of the classroom.

- **Finding Research Materials:** Aside from correspondence, 'investigate' is the thing that takes various people to the Internet. There is huge learning on the Internet than the school library can give. We can ask students to misuse this bounty of advantages on the Internet for their examination.
- **Obtaining Varied Writing Skills:** In case students are required to distribute their take a shot at the Internet, they have to make hypertext aptitudes. These capacities empower students to get association in non-progressive creations. Moreover, and since the Internet is accessible to all with get to, students distributing their work on the Internet are constrained to know about their tongue and to write to non-ace social occasion of people.

10.5 THE BARRIERS TO ICT TRAINING FOR TEACHERS

- One supporter of the aptitudes hole for instructors is by all accounts age-with augmentation in age the try and capacity to increase some new helpful information decreases when appeared differently in relation to that of energetic age. Instructors are ordinarily a man who has crossed the age of 35, as such, they vacillate to learn and bring into preparing, a totally outsider thought before them. This issue can be overpowered by engineering workshops, course, and gatherings every so often to make them familiar with ICT. Preparing additionally should be masterminded them from time to time.
- Reluctance to acknowledge can be an aftereffect of an absence of a confidence or weakness of utilizing ICT for learning. Instructors, as often as possible stress that their data level does not arrange with those of their 'carefully local' students. They falter on working with ICT in the savvy classes. This dread and wavering can be evacuated by giving appropriate preparing to them (Crawford, 2000).
- Some educators are not persuaded of the advantages it can convey to their instructing-customary instructors are not effectively persuaded about the utilization and advantages of ICTs. To defeat this legitimate exhibit on the employments of ICT in training should be possible before them. Orchestrating workshops, courses, and gatherings every now and then on the utilizations and advantages of ICT can be useful in this case.

10.6 DRAWBACKS OF THE UTILIZATION OF TECHNOLOGY FOR EDUCATIONAL PURPOSES

Each coin is two-faced; same is the situation with utilization of innovation in instructing and learning. Alongside points of interest there are a few weaknesses also. Some of them are setting up the gadgets can be exceptionally troublesome particularly where the instructors are not master in the utilization of technologies. Excessively costly, making it impossible to manage the cost of particularly in Government Schools of creating nations which generally keeps running in shortage where money related issues are concerned. It is hard for educators to use with an absence of experience of utilizing ICT tools. The usage of Internet for preparing isn't without issues. Along these lines, one should foresee that the issues will be knowledgeable about using the Internet in instructing to progress too.

- **Plagiarism:** Aside from Web goals that guarantee to empower students to make explore ventures, there are different examples of students downloading information from the Net and giving them over for evaluations. We can manage the issue by requesting that students allude to state about sources. There is an online organization at www.plagiarism.org, which can help us in restricting cases of composed adulteration in the class. This organization neutralizes composed distortion by choosing whether an exploration venture has been copied from the Internet or not.
- **Student's Privacy:** Crooks, publicists, and diverse individuals can without quite a bit of stretch get information from students when they are on the web. These could display chance on students' lives or may even incite prosecution against the school. To evade this issue, students' ought to be told on the threats of offering data to individuals on the web. Guardians and educators need to manage students' online activities.
- **Low Income Groups:** As illustrated, most government schools have a lower rate of Internet access than private schools. The comparable was authentic for instructional rooms in those schools. In like way, students' from low-wage families presumably won't have PCs at home or may have PCs at home with no passageway to the Internet. Along these lines, students' in low-pay frameworks might be disturbed. To diminish the impact that social or monetary status may have, we should give Internet assignments that students can without a lot of a stretch finish while in school. On the off chance that basic, schools may need to keep PC labs

open for more timeframes or at possibly odd hours. The use of PCs at open libraries should in like manner be engaged.

- **Planning Time:** It requires an impressive proportion of arranging venture to enough use the Net for preparing. Despite sketching out Internet-based exercise structures, we may need to surf the Internet to download practice plans and change them to help the instructive modules targets or visit goals to pick those appropriate for classes. We should pick between restricted choices in any case, prepare with a particular ultimate objective to empower students to wind up as tried and true customer of the Internet.
- **New Administrative Responsibilities:** Utilization of the Internet to pass on or to hold up under another course of action of administrative demands on the teacher and the school association. These fuse enhancement and utilization of sufficient use game plan, getting ready, developing new appraisal criteria as required, and keeping an eye on the guardian's stresses.
- **Cyber Crime:** Crime is expanding step by step. The programmers are concocting better approaches to hack information and use them in unlawful ways which may hurt the notoriety of the blameless individuals. Students are of youthful age and energetic towards the utilization of web-based social networking. They are generally thoughtless while sharing their data and picture via web-based networking media and thus can be simple prey for these programmers. These students should be cautioned towards such issues by their gatekeepers and instructors. They should be educated about the approaches to stay secure while utilizing web.
- **Unwanted Contents:** Now day's sexual substance are effectively accessible for review on web. Tragic part is that they are even accessible on some instructive locales. This is unsafe for students who are utilizing web for their insight upgrade. Watchmen and educators should be careful about this and see to the way that the PCs utilized by students are anchored by Quick Heal or other such measures.

The above inconveniences or blocks don't show that ICT can be sworn off within the instructive field. The genuine truth is that ICT has turned into an indispensable piece of our training framework. For the upliftment of any instruction framework legitimate utilization of ICT must be presented.

10.7 ROLE OF TEACHERS MAKING STUDENTS AWARE TOWARD CYBER CRIME & CYBERSECURITY

The Internet is a colossal bit of everyone's lives today, including children's. Indeed, even created, cunning, now, and again actually educated developed-ups can likewise much of the time capitulate to traps on the web. It tends not out of the ordinary effortlessly that an adolescent overflowing with intrigue, anyway obliged being developed and care, can surrender to the dangers of the web. It's fundamental for watchmen and instructors to open up to their youngsters about online security, however then again it's more valuable if teachers get occupied with exchange with their students in neighborly way (Ayer, 2017).

Youngsters take development classes, surf the web for school errand, and unmistakably use the Internet in their accessible time for an extensive variety of things, so it's a keen idea for instructors to consolidate some advanced security talk in their activities. All things considered cybercrime has ended up being to a more noteworthy degree a peril than customary sorts of bad behavior. Infringement like robbery, theft thievery, and even unpleasant bad behavior have on a very basic level declined over the latest a long time since wrongdoers have found another way to deal with break into our homes through our screens, giving them access to fundamentally more than what is physically in our homes or our pockets. Over robbery, we in like manner have predators to worry over.

- **Digital Natives:** Consider kids nowadays-most have been imagined specifically into this propelled age and have never known a world without the Internet. They see the Internet as this great immense thing with enormous measures of information and answers to everything. So an instructor should make them careful that the Internet has flaws; that people with horrendous points can spill through the parts and get to them and their own information on the off chance that they are not vigilant.
- **Digital Security in the Classroom**: A couple of schools may not start educating about computerized security at introductory stage yet the truth of the matter is that in present-day period even children are using tablets as routinely as young people thus these schools need to change their attitude and give careful consideration toward this path. Students should be made mindful of digital wrongdoing and digital security. This may not be something that is straightforward for youth to understand, yet fortunately there are ways to deal with make it more tyke pleasing like

with the comic, The Cynja, a web security delight that allows children to learn by doing, that watchmen or instructors can benefit. Indicating kids about computerized security is by and large same as examining the standard and threats of present reality; however, as far as advanced life. You should moreover prepare them to make sound passwords and not use a comparable one for everything.

- **Youngsters and Technology:** Teenagers, it may be a brilliant idea to show to them some real instances of 'Web sharing turned out severely.' Youths get a kick out of the opportunity to share what they are doing, their personality with, what they're wearing, and various more parts of their lives by means of web-based systems administration media. In any case, they generally don't comprehend that what they post can be seen by anyone. Moreover, there are courses for pariahs to hack straightforwardly into their records and see those posts they thought were private. Despite deleting a post isn't as sound as it may show up, in light of the way that once something goes up on the Internet; there are ways to deal with reveal it back paying little mind to whether it's been eradicated. Every one of these actualities should be managed and displayed before the students by their instructors. Students need to understand this can remarkably impact their picture now. It can change their shining resume and capabilities into a massive "no" in light of something they posted 5 years back.

10.8 CONCLUSION

Schools today are better place for teachers. This growing awareness may be both cause and effect of rising standards of all-round development of students. Pupils, in standard in the fundamental, better qualified and more versatile to the utilization of devices and aptitudes as per circumstance. In developed nations, the schools and classrooms that are better resourced, with littler class sizes and latest mechanical help consequently more chances to apply ability and gifts by the pupils. Pupils appreciate better open doors for long-lasting learning and expert advancement. Appraisal procedures accessible are more advanced and, as "broadened experts," they practice a more extensive, more mind-boggling and professionally requesting dispatch than in the experts before them. The underdeveloped and the developing countries are seriously lacking such facilities and advancements in the education sector. In order to relish the full-fledged advantages of current developments in the educational arena, all the hurdles in the path of the student's development. There is no doubt that finding the time to integrate technology and the impacts of globalization with teaching is an overwhelming task for a teacher. Throughout the course of a day, teachers find themselves pulled in many directions, especially in the developing and the underdeveloped countries. The positive side of the story is that slowly, but steadily, teachers are able to seek out, identify, and use technology to enhance student learning. The teachers are also making the students aware of the dangers of using digital media carelessly. The teachers are now increasingly making their students aware of cybercrimes and the safety measures to be adopted by them. In the time to come it can be hoped that ICT along with trained, lively, experienced, and motivated teacher will play an indispensable and marked role in the development of students on the every front, cognitive, affective, and conative.

KEYWORDS

- **cybercrime**
- **cybersecurity**
- **information communication and technology (ICT)**
- **student counseling**
- **teaching-learning process**

REFERENCES

Ayer, R., (2017). *The Importance of Teaching Students About Cybersecurity*. Retrieved from: https://www.emergingedtech.com/2017/05/teaching-students-about-cyber-security (Accessed on 28 August 2019).

Blanco, P. J., & Ray, D. C., (2011). Play therapy in elementary schools: A best practice for improving academic achievement. *Journal of Counseling & Development, 89*, 235–243.

Brown, D. H., (2007). First language acquisition. *Principles of Language Learning and Teaching* (5th edn., pp. 24–51). Pearson ESL.

Crawford, R., (2000). Information technology in secondary schools and its impact on training information technology teachers. *Journal of Information Technology for Teacher Education*, 183–198.

Gangapathy, M., (2017). *How can Technology Help in the Teaching of Higher Order Thinking Skills?* Retrieved from: https://www.researchgate.net/./How_can_technology_help_in_the_teaching_of_higher 24–51.

Ghavifekr, S., (2013). *ICT Application for Administration and Management: A Conceptual*. Retrieved from: https://www.sciencedirect.com/science/article/pii/S1877042813042419 24–51.
Harmer, J., (2007). *The Practice of English Language Teaching*. Harlow, England: Pearson Longman.
Kothari Commission, (n.d.). Retrieved from: https://en.wikipedia.org/wiki/Kothari_Commission (Accessed on 28 August 2019).
Nola, A., (n.d.). *7 Roles of a Teacher in the 21st Century | Eton Institute*. Retrieved from: https://etoninstitute.com/blog/the-7-roles-of-a-teacher-in-the-21st-century (Accessed on 28 August 2019).
Reitzes, B. M., (2011). *Integrating Technology with Student Learning*. Quincy: Nellie Mae Education Foundation. http://www.nmefdn.org" www.nmefdn.org A Report to the Nellie MAE education Foundation prepared by Babette Moeller & tim Reitzes | July 2011
Richardson, J. W., (2007). *The Adoption of Technology Training by Teachers in Cambodia: A Study of the Diffusion of an ICT Innovation*. ProQuest Dissertation, University of Minnesota, United States, Awarded 2007. https://www.learntechlib.org/author/Jayson+W+Richardson"
Rogers, C. R. (1943). *Counseling and Psychotherapy*. Houghton Mifflin, Boston,The Ribertsite Press, Cambridge, Massachusetts, USA.
Sangrà, A., & González-Sanmamed, M., (2010). The role of information and communication technologies in improving teaching and learning processes in primary and secondary schools. *ALT-J, 18*(3), 207–220. doi: 10.1080/09687769.2010.529108.
Webb, L., (2011). Supporting technology integration: The schools' administrators role. *National Forum of Educational Administration and Supervision Journal, 28*(4), pp. 1–7.
Wheeler, S., Waite, S. J., & Bromfield, C., (2002). *Promoting Creative Thinking Through the Use of ICT*. Retrieved from: https://www.researchgate.net/./227619540_Promoting_creative_thinking_through (Accessed on 28 August 2019).
Wikipedia, (2017). *Technology*. Retrieved from: https://en.Wikipedia.org/wiki/Technology (Accessed on 28 August 2019).
Wilson, B. G., (1997). The postmodern paradigm. In: Dills, C. R., & Romiszowski, A. A., (eds.), *Instructional Development Paradigms*. Englewood Cliffs: Educational technology publications, p. 297–309. Retrieved from: http://www.cudenver.edu/~bwilson (Accessed on 28 August 2019).

CHAPTER 11

Cybersecurity and the Internet of Things: Anticipating the Leverage of Cyber Crimes with Interconnected Devices

A. REYES-MENENDEZ,[1] J. R. SAURA,[1] and P. PALOS-SANCHEZ[2]

[1]*Rey Juan Carlos University, Department of Business Economics, Pº Artilleros s/n, Madrid-28032, Spain, E-mails: ana.reyes@urjc.es (A. Reyes-Menendez); joseramon.saura@urjc.es (J. R. Saura)*

[2]*University of Seville, Department of Financial Economy and Operations Management, Avda. Ramón y Cajal 1, Seville- 41018, Spain, E-mail: ppalos@us.es*

ABSTRACT

These days, technology is reaching a critical point in its evolution. The increase of technologies – as cloud computing, data analysis, and mobile communications – has encouraged other technologies as, for example, the Internet of Things to reach their full potential. Resulting, the Digital Business Environment has been rapidly changing, leading to an exponential growth of the number cyber crimes. The present study aims to identify the challenges the Digital Business Environment has to face and to propose viable solutions that would allow avoiding cyberattacks that result in a reinforcement of cybersecurity in the Internet of Things. So that, with the aim to deepen the impact that cybersecurity has on the Internet of Things for individuals, society, nations, and global humanity, the authors undertake a review of the literature and perform an exploratory analysis of the results on this ubiquitous topic.

11.1 INTRODUCTION

Throughout the last decade, the effects of cybersecurity on both public and private sectors and data have gained a wide scholarly attention. In 2017, NASDAQ cybersecurity corporations, such as Finjan Holdings, KeyW Holding, and MobileIron, provided important revenue for investors. At the same time, cyber attacks, such as ransomware WannaCry, made it to the headlines worldwide, making the society sensitive to these crimes and emphasizing the importance of leveraging technology (Harrington, 2016). By 2012, business recovery and other negative consequences of cybercrimes are expected to cause trillions in damages.

In parallel, technology is reaching a critical point in its evolution. The increase of technological advancements – such as cloud computing, data analysis, and mobile communications – has led to the rapid development of a trend that has previously been latent for decades, waiting for the technological advance to allow for it demonstrating its maximum potential. This trend is the Internet of Things (IoT) which is expected to have a dramatic economic impact comparable to the one caused, in their time, by the entry of PCs. The Internet of Things, based on the fact that objects or devices are connected to the Internet, opens up possibilities that were previously unthinkable (Burkitt, 2014). However, along with boosting economic development and multiple business opportunities, IoT also presents challenges that need to be addressed.

This chapter has the aim to expand the knowledge of the Internet of Things with the focus on its advantages and disadvantages, as well as its implications for individuals, communities, business, societies, and nations. The authors seek to provide meaningful insights about the scope of interconnected devices to support executives concerned with the strategic and practical management of cyberspace, cybercrimes, and cybersecurity in the pervasive field of Internet of Things. To this end, the authors identify the technology behind the Internet of Things and its application, highlighting the main challenges that arise in terms of ethics, government legislation, data privacy, and cybersecurity.

11.2 BACKGROUND

11.2.1 INTERNET OF THINGS

The widespread adoption of technology and digitalization will in all probability increase the opportunities for digital crimes. New interconnected technologies appear to be the most probably victims of the next generation of ransom-based attacks. At the center of this target are IoT devices, i.e., connected objects in which security considerations remain largely absent. Although many consider Internet of Things to be still in its infancy (Deloitte, 2014), IoT is a pervasive and fast-growing technology. As acknowledged by many authors (Yan et al., 2008), Internet of Things it is on its way to transform the current rigid Internet into a fully integrated Internet known as Future Internet (Townsend et al., 2011; Schaffers et al., 2011; Calabrese et al., 2009; Könnöllä, 2007).

Despite the fact only a few decades have passed since the introduction of the Internet, during the passage of these years. Internet has been changing at the pace of technology advances, shaping the Digital Business Environment. In particular, Future Internet will be determined by the ability to relate multiple devices in a connected network (Saura et al., 2017; Hernández-Muñoz et al., 2011; Lemke, 2010; Townsend et al., 2009). At the same time, communication between devices has given rise to new challenges in cyberspace. Above all, these concerns relate to cybersecurity and cybercrime.

The roots of the concept of The Internet of Things go back to the beginning in the mid-1990s. Kevin Ashton, father of the term "Internet of Things," had the idea to attach tiny radio frequency chips to products to provide information for the sales department. In 1999, the term "Internet of Things" was used as the title of one of the presentations for Procter and Gamble (Maney, 2015). Kevin Ashton imagined a world where the Internet would be connected to the physical world through sensors and a platform based on real-time feedback, which would offer a great potential to improve comfort, security, and control of our lives.

Several years afterwards, members of Kevin Ashton's group at MIT used this concept again, defining IoT as "an intelligent infrastructure linking objects, information, and people through the computer networks, and where the RFID technology found the basis for its realization" (Ashton, 1999). Despite technological advances, it was not until 2005 when the International Telecommunication Union (ITU), 2005 made and published a report on the subject, defining the Internet of things as "a new dimension has been added to the world of Information and Communication Technologies (ICTs): from anytime, anyplace connectivity for anyone, the authors will now have connectivity for anything. Connections will multiply and create an entirely new dynamic network of networks – an Internet of Things."

Furthermore, combining the various precedent concepts, Borgia (2014) provided the following description of IoT: "The future Internet of Things links uniquely identifiable things to their virtual representations in the Internet containing or linking to additional information on their identity, status, location or any other business, social or privately relevant information at a financial or non-financial pay-off that exceeds the efforts of information provisioning and offers information access to non-predefined participants." The provided detailed and convenient information may be accessed in the right amount and condition, at the appropriate time and place, and at the right price. While The Internet of Things is not synonymous to man similar phenomena – such as ubiquitous/pervasive computing, the Internet Protocol (IP), communication technology, embedded devices, their applications, the Internet of People, or, for that matter, the Intranet/ Extranet of Things – yet it combines aspects and technologies of all of these approaches.

Since then, this network of inter-connected devices (Vermesan et al., 2014) has grown and expanded. By 2025, this network is expected to generate over 20 trillion gigabytes of data (Bonetto, 2015). According to a different estimate, by 2025, the IoT will have a huge potential economic impact and its potential annual economic value will reach as high as US$ 11.1 trillion (McKinsey, 2015). According to the Global IoT Forecasting published in Forbes (2017), by 2020, the expected growth rate of IoT will be 38.3%, increasing each year and ending by a market share size of US$ 457.29 billion (Figure 11.1).

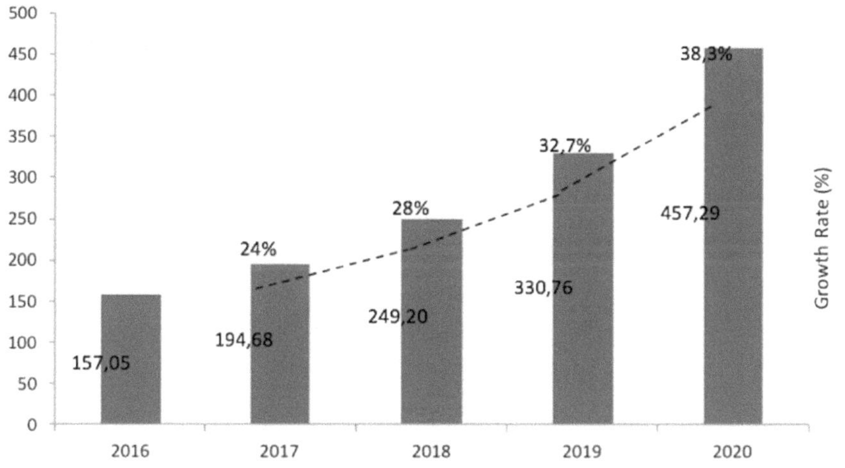

FIGURE 11.1 Global IoT market share.

11.2.1.1 IOT PRECURSORS: M2M COMMUNICATION

The background and roots of the Internet of Things can be found in Machine-to-Machine communication (M2M). This type of communication is relevant in this context, because it involves two or more entities that do not need human interaction in the communication process, although they can count on such an interaction. Although M2M is often limited to serving as a transport network, it is supported by mobile networks. According to Ministry of Communication of India (2015), M2M is a different kind of communication, because it has a series of specific characteristics. These characteristics include the following:

- M2M arises in view of new market scenarios or different scenarios;
- M2M requires lower costs and less effort;
- M2M has a number of potentially very large communication terminals;
- In M2M, there is generally little traffic for each terminal.

The Machine to Machine (M2M) communication can also be called Machine Type Communication (MTC) in 3GPP. The applications of

Machine to Machine (M2M) are very varied and range from payment management to security systems (Table 11.1).

TABLE 11.1 M2M Applications

M2M Category	Applications
Security	Surveillances, Alarm systems, Access control, Car/driver security
Tracking & Tracing	Fleet Management, Order management, Pay as you Drive, Asset tracking, Navigation, Traffic information, Road tolling, Traffic optimization/steering
Payment	Point of sales, Vending machines, Gaming machines
Health	Monitoring vital signs, Supporting the aged or handicapped, Web access to telemedicine points, Remote diagnostics
Remote Maintenance/Control	Sensors, Lighting, Pumps, Valves, Elevator control, Vending machine control, Vehicle diagnostics
Metering	Power, Gas, Water, Heating, Grid control, Industrial metering
Manufacturing	Production chain monitoring and automation, Facility management, Home/building/campus automation

(*Source:* Ministry of Communication of India, (2015). *M2M/IoT Technical Reports.* Retrieved from: http://tec.gov.in/pdf/Studypaper/White%20Paper%20on%20 MachinetoMachine%20 (M2M) Communication.pdf.)

11.2.1.2 THE INTERNET OF THINGS TECHNOLOGY

Several others technologies, viewed in their entirety, can also be seen as precursors of the Internet of Things. All of them are innovative technologies that allow the development of capacities to put the online world in contact with the offline world. The different capacities resultant from this technological development and that shape the Digital Business Environment of the Internet of Things are listed in Table 11.2. Taking into account that connected devices are usually designed to connect in a network, develop activities, and interact with their environment, it is interesting to identify how they obtain information, transform signals, recognize their location, as well as communicate with each other and the outside world.

TABLE 11.2 Capacities Supported by Technological Development

Capacities	Description
Capacity of response	The devices can be configured remotely to develop the activities for which they have been designed.
Communication and cooperation	Several devices connect in a single network that can be connected to the Internet, enabling data sharing and update performance. The technological development of wireless networks, such as GSM, UMTS, Wi-Fi, Bluetooth, ZigBee or WPAN, is essential for this functionality.
Identification	Once the devices are connected to the network, each of the connected objects has an identifier that provides the specific information and the current status of each of those connected objects. The technology that supports these objects is RFID, NFC, or barcodes. This technology also requires a reading device that identifies the object.
Detection	Each of the connected objects obtains information from its environment with the help of sensors. Based on that information, they can respond or only register or share their status.
Actuation	The connected objects contain actuators that allow the object to transform the signals it receives and thus respond to the stimuli from the environment.
Embedded Information Processing	The information received through the sensors can be interpreted through processors or microprocessors with storage capacity allowing the object to remember precedent use of this information.
Location	The connected devices have their location identified. The technologies supporting this capacity are GPS, mobile telephony network, ultrasound, UWB, radio beacons (neighboring WLAN or RFID readers), and optical technologies.
User interfaces	Given the connected objects communicate with people in some way; there are remarkable advances in flexible screens, as well as voice, image, and gesture recognition.

Source: Mattern and Floerkemeier, (2010)

As mentioned above, Ashton (1999) argues that RFID has laid the foundations for the ubiquitous development of the Internet of Things. According to Yunjie (2010), the development of the Internet of Things was promoted by both RFID and the WSN technology and communication protocols. However, Tripathi (2017) postulates that the Internet of Things has also been promoted by several other highly innovative technologies, such as embedded technology or nanotechnology.

In what follows, all these technologies will be outlined in further detail:

- **Radio-Frequency Identification (RFID):** RFID is an automatic scanning system that allows for managing large amounts of data. For that reason, it plays a critical role in tracking, identification, and security check-process automation (Asghar, 2015). Owing to the use of automatically detected radio signals, RFID is also used to obtain and store data remotely. On the other hand, the increase in the use of RFID gives rise to cybersecurity problems. In this context, it is essential to thoroughly understand these threats so that to be able to propose solutions to the issues of cybersecurity, cyberspace, and cybercrimes of the new Digital Business Environment. Owing to the use of radio waves, data are transferred from the data transmitter to the data receiving device (Yan et al., 2008). Both have an antenna that favors the transmission of data. In addition, data can be stored in the cloud (Figure 11.2).

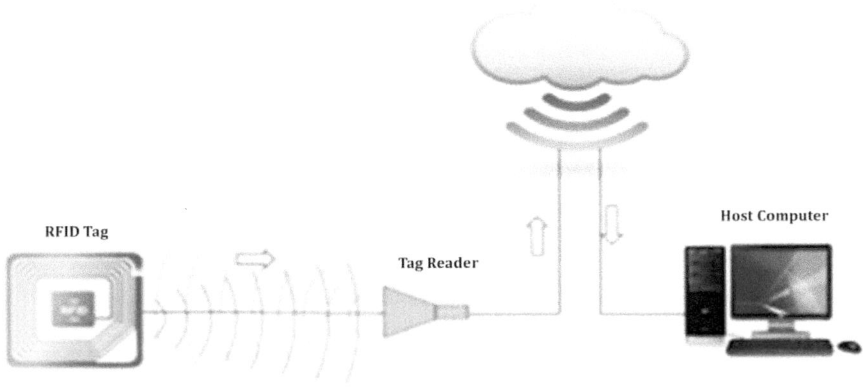

FIGURE 11.2 (See color insert.) RFID block reader diagram.
(*Source:* Asghar, (2015))

The components of RFID are tags, reader tags, antenna, information management software, and database (Table 11.3).

TABLE 11.3 RFID Components

RFID Component	Description
Tags	The piece that sends the data is usually called a tag, while the receiver of the information is referred to as reader or tag reader. Tags are usually located in the objects. Depending on the energy source used, three categories of tags can be distinguished. The first category embraces active tags that use energy from a mobile battery. The second category includes passive tags that use the energy of the electromagnetic radiation emerging from the tag reader. Tags of this type are smaller in size, but also have a smaller scope. The third category consists of semi-active tags that use both energy sources.
Antenna	Radio signals are transmitted between the reader tags and tags because both use the antenna.
Information management software	It is in charge of collecting and organizing data on a local server. The data can be retrieved at a later point in time.

- **RFID, Wireless Sensor Network (WSN), and Communication Protocols:** According to Yunjie (2010), RFID together with WSN Technology and communication protocols of objects constitute the most important parts of IoT. RFID technology is a wireless Automatic Identification and Data Capture (AIDC) technology (Wamba et al., 2006) that enables accurate and automatic identification, as well as tracking and tracing, of every product from the factory, through shipping and warehousing, to the retail location without human involvement (Lai and Hutchinson, 2005). WSN technology can be generally described as a network of nodes that cooperatively sense and control the environment, enabling thus an interaction between persons or computers and the surrounding environment (Verdone et al., 2008). A network of devices, referred to as nodes, perceive sensorial information about the environment and communicate the information gathered from the monitored field (e.g., an area or volume) through wireless links (Akyildiz et al., 2002).

At the connectivity layer, communication protocols facilitate the communication between an individual thing and a clod (Wortmann and Flüchter, 2015):

- **IEE 802.15.4:** This protocol is a standard compatible with many Internet applications. However, it has limitations in making communications between different devices separated from each other by large distances.
- **Bluetooth:** It has always been so present that it seems strange that it could have evolved during the last several years. Overall, Bluetooth is a wireless connection standard whose purpose is to connect devices while always sharing data at short distances. The main evolution of Bluetooth is found in Bluetooth Low Energy (BLE). Therefore, the two main types of Bluetooth are classic Bluetooth and BLE. Although both types of Bluetooth share the same name and some common characteristics, they have important differences. While classic Bluetooth consumes much more power, BLE is smaller, lighter, and connects devices without requiring security (Table 11.4).

TABLE 11.4 Technology Differences Between Bluetooth Classic and BLE

Technology	Topology	Data Rate	Range	Peak Power
Bluetooth Classic	Direct connect	Up to 3 Mbit/s	10m typical 100m in theory	< 15mW
BLE	Direct connect	Up to 1 Mbit/s	10m typical 100m in theory	< 75mW

(*Source:* Mindtribe, (2015). *Wi-Fi, Bluetooth, and Internet of Things*. Retrieved from: https://mindtribe.com/wp content/uploads/2015/04/WiFi Bluetooth and the Internet of Things 0.5 edited.pdf.)

The EPC (electronic product code) networks are intended to identify objects and facilitate processing and use of obtained data. The development of this technology has been very useful for the Internet of Things.

- **Other Innovative Technologies:** In recent years, along with the technologies overviewed above, several other technologies have also boosted the rapid growth of the Internet of Things market (Tripathi, 2017). Among these new technologies are the nanotechnology and the embedded technology. The former technology facilitates connection and interaction of smaller-volume things, which can lead to a decrease of the consumption of the system. The latter technology can provide a certain degree of intelligence to object in IoT.

11.2.1.3 ARCHITECTURE SYSTEMS FOR THE INTERNET OF THINGS

Basically, Internet of Things-Architecture (IoT-A) is based on the idea of "a globally interconnected continuum of devices, objects, and things in general emerged with the RFID technology, and this concept has considerably been extended to the current vision that envisages a plethora of heterogeneous objects interacting with the physical environment" (Bassi, 2013). Several methodologies are available to develop the architecture of systems (Rozanski and Woods, 2012). As specified earlier, IoT has its background in the architecture of M2M. In fact, M2M, or Machine to Machine communication, has laid the foundations of what the researchers know today as the Internet of Things. In what follows, the authors discuss the architecture and components of M2M, as compared to that of IoT, in further detail.

11.2.1.3.1 Architecture and Components of M2M

Deriving from the similarity between M2M and IoT, the architecture and components of the two systems are similar. Therefore, to better understand the architecture and components of IoT, it is necessary to first discuss the architecture and components of M2M. The architecture of an M2M system is composed of the following elements:

- **M2M Devices:** Their mission is to respond when they receive a data request; they can also transmit said data independently. The devices are usually connected to an operator's network and are usually connected to each other using different technologies, such as Bluetooth, Wi-Fi, or WPN.
- **M2M Area Network (Device Domain):** Its function is to connect M2M devices with M2M Gateways.
- **M2M Gateway:** It ensures that the M2M devices work and are connected to the Communication Network. The Gateway ensures that the devices on the network have adequate communication with the outside. It also constitutes a distributed system in which generic and abstract capacities are implemented.

- **M2M Communication Networks:** It serves as a connection between the M2M Gateway and the M2M application. Relevant examples include xDSL, LTE, WiMAX, and WLAN.
- **M2M Applications:** It is a layer of middleware that obtains data for later use by other specific engines.

11.2.1.3.2 Internet of Things-Architecture (IoT-A)

The architecture of the Internet of Things is based on the components outlined earlier. However, owing to technology advances and various improvements, the architecture of IoT is more advanced as compared to that of M2M. Conforming to the Commendation of the ITU, the network and architecture of Internet of Things have five categories. These are the sensing layer, the access layer, the network layer, the middleware layer, and application layer (Teixeira et al., 2011) (Table 11.5).

TABLE 11.5 Layers of the Internet of Things

Layer	Description
Sensing layer	The main function of this layer is to transfer data from the sensing layer to the network layer through the existing mobile network, wireless network, wireless LAN, satellite, and other network infrastructure.
Access layer	The target and aim of the access layer is to capture the data and information of interest by different kinds of sensors and Intelligent Identification devices, and share the gained information with the relevant units in the network.
Network layer	This layer aims to integrate information resources of network into a large intelligence network through an Internet platform so as to establish an efficient and reliable infrastructure platform by upper-class servicing provided by management and for large-scale industrial applications.
Middleware layer	The function of this layer function is to manage and control network information real-time so as to provide a good user connection for upper-layer applications, including different business support platforms, information processing platforms, and intelligent computing platforms.
Application layer	This layer aims to integrate the function of the bottom system for the purpose of the making the practical application for various industries, such as smart environment and logistics, smart grids, smart transportation, smart agriculture, and smart medical care.

11.2.2 STAGES OF THE INTERNET OF THINGS

As described by Borgia (2014), the ubiquitous technology of IoT is characterized by different and interacting technologies and protocols with different functions and processes. Among others, these include collection, transmission, and processing, managing, and utilization phases. During the data collection phase, the sensors capture information from the physical environment. These data obtained in real time allow a reconstruction of the surrounding world. While some technologies overviewed above, such as RFID, allow for an identification of the objects located in the environment, other technologies, such as IEE 802.15.4 or Bluetooth, collect the data. In the data transmission phase, different mechanisms are involved that allow the data obtained to be downloaded to external applications or servers. At this point, the gateways also come into play.

In the phase of data processing, management, and use the main objective is that these data can be analyzed, processed, and used by the applications for decision making. Each of these phases will correspond to different layers of the architecture of the Internet of Things System. As for the cybersecurity, each layer-and, by extension, each stage-will need specific attention to avoid cyber attacks that may cause cybercrime.

11.3 MAIN FOCUS OF THE CHAPTER

On the description and analysis of the technical scope of the Internet of Things, the authors will now proceed to the discussion of the different applications of this technology. These applications shape the Digital Business Environment and favor the need to promote cybersecurity to avoid cyber attacks and cybercrime. To this end, a review of the literature has been undertaken, which has allowed the identification of the main sources in the scientific literature on the topic and enabled us to conduct an exploratory analysis of the results. Since IoT is a ubiquitous and ever-changing technology, a discussion of its applications will allow us to more deeply discuss its impact on individuals, societies, nations, and global humanity (Vermesan et al., 2014; Schuermans and Vakulenko, 2014; Porter and Heppelmann, 2014; Yoo et al., 2010).

11.3.1 APPLICATIONS OF THE INTERNET OF THINGS

The huge potentialities of Internet of Things create numerous development opportunities in almost every field. The various applications are distributed among the following three major categories -

- **Health and Wellbeing Domain:** As shown in Table 11.6, the applications related to health and well-being can vary from those aiming at patient care to the development of techniques that facilitate the care of the elderly.

TABLE 11.6 Examples of Health and Wellbeing Applications

Application Categories	Examples
Medical and healthcare	- (Remote) monitoring medical parameters, diagnostics
	- Medical equipment tracking, secure/access indoor environment management
	- Smart hospital services, entertainment services
Independent living	- Elderly assistance, disabled assistance
	- Personal home/mobile assistance, social inclusion
	- Individual well-being, personal behavior impact on the society

(*Source:* Borgia, (2014).)

- **Industrial Domain:** As can be seen in Table 11.7, the industrial applications of the Internet of Things cover a broad spectrum that ranges from logistics management to management of agriculture and livestock feeding.

TABLE 11.7 Examples of Industrial Applications

Application Categories	Examples
Logistic and product lifetime management	- Identification of materials/goods/ product deterioration
	- Warehouse management, retail, inventory
	- Shopping operation, fast payment

TABLE 11.7 *(Continued)*

Application Categories	Examples
Industrial processing	- Real-time vehicle diagnostics, assemblage process, assisted driving
	- Luggage management, boarding operation, mobile tickets
	- Monitoring industrial plants
Agriculture and breeding	- Animal tracking, animal certification, animal trade control
	- Irrigation, monitoring agricultural production and feed
	- Farm registration management

(*Source:* Borgia, (2014)).

- **Smart Cities Domain:** With regard to the applications in the domain of smart cities, the Internet of Things can play a key role in its conception and development. Owing to this technology, important aspects of the management of a city-such as mobility, public safety, transport, or energy management-can be covered (Table 11.8).

TABLE 11.8 Examples of Smart Cities Applications

Application Categories	Examples
Smart mobility and Smart tourism	- Traffic management, bike/car/van sharing, multi-modal transport
	- Road condition monitoring, parking system, waste collection
	- Payment systems, entertainment/Tour guide services
Smart grid	- Power generation/distribution/storage, energy management
	- Sustainable mobility, booking charging slot, customer recognition
	- Load management, storage service, entertainment services
Smart home/building	- Plant maintenance HVAC/lighting/irrigating, energy management
	- Video surveillance, access management, children protection
	- Entertainment, comfortable living

TABLE 11.8 *(Continued)*

Application Categories	Examples
Public safety and environment monitoring	- Environmental and territorial management - Video/radar/satellite surveillance - Emergency site/rescue personnel tracking, emergency plan

(*Source:* Borgia, (2014)).

11.3.2 CHALLENGES FOR THE INTERNET OF THINGS

The development of Internet of Things is opening up huge opportunities for individuals, communities, business, societies, and nations. However, it also represents an enormous social and technical challenges and it involves risks mainly related to the following (Accenture, 2014):

- **Ethics and the Internet of Things:** According to the European Group on Ethics in Science and New Technologies, IoT will radically change "the relationship between humans and the interconnected autonomous objects, giving to the last ones autonomy towards the interaction with human beings" (Weber, 2010). Among others, the kinds of ethical issues that these types of technology raise are related to the following categories:
    - **Autonomy:** Autonomy of things and autonomy of humans
    - **Security:** Understanding security with a dual use, either related to freedom and to liberty.
    - **Equity:** In this regard, equity is equivalent to equality, justice, fairness.

Finally the concept is very similar to the access to goods and services, the treatment received and the possibility of discrimination or the existence of discriminatory interfaces.

- **Governmental Legislation for the Internet of Things:** Another concern regarding the use of IoT technologies in marketing is of legal character. If personal data are leaked or illegally transferred to a third party, consumers may sue a company demanding moral and monetary compensation. Such cases may cause huge damage for a company, depending on the amount, type of data, and number of consumers involved. In this respect, although laws forbidding transfer of personal information and illegal entry in databases are already in place, legislation must correspond to new challenges entailed specifically by implementation of IoT technologies.

The Internet of Things has a great importance for the sustainable development and the global economy (Palos-Sanchez and Saura, 2018). Along with the importance of issues such as resilience, governance, and privacy, one of the main concerns of the institutions responsible for developing the relevant legislation and governments is precisely security. As specified by CERP-IoT (2009), "there is the need and will for international co-operation. China, Japan, Korea, and the USA are on board. Europe has taken the lead and now needs to drive the initiative as a truly global partnership. It has also been shown that governments, industry, and business lacked awareness of the Internet of Things and of what it offered."

- **Data Privacy and the Internet of Things:** According to Gallup (2017), Americans are more worried about cybercrime than conventional crime, as their major concern is having their personal, credit card, or financial information stolen by computer hackers.

With respect to data privacy, relevant regulations are being implemented at different levels and from different organizations. Among others, these regulations highlight the right to be forgotten, approved in the European Court of Justice in 2014. This right is the last manifestation of the need to preserve people's privacy in the face of the threats posed by technological progress (Muñoz, 2016). Of note, the role that this right plays in deleting the information of users on the Internet in a context in which companies use a large amount of data to perform predictive analysis through the Big Data techniques. The right to be forgotten has meant a before and an after in the way companies use their data, and its impact also extends to strategies. Companies can no longer develop their strategies the way they did until this legislation came into force. For instance, the way in which Social Media strategies are planned has been modified.

At the European level, a dramatic change in the Digital Business Environment has occurred with the adoption of General Data Privacy Regulations (GDPR) that force companies to change the approach with which they subscribe their customers in their databases. Instead of making an automatic subscription of users to the company databases and giving the option (in case when those individuals do not want to continue to stay in the aforementioned corporate database) to unsubscribe through an "opt-out," now companies have to obtain the explicit consent of users to be able to include them in their databases. Therefore, the focus has now shifted to "opt-in." This regulation is effective since May 2018 and extends not only to the new users, but also to those users who were already part of the aforementioned databases (Smouter, 2018; Dominici et al., 2016). Likewise, in order to have an adequate understanding of the Digital Business Environment, the economic impact of these measures pursuing the privacy of users must be carefully considered (Miller and Skiera, 2017).

In its turn, the United States has not kept pace with European countries. Its regulation on the privacy of user data is much less strict. Although, in May 2017, the BROWSER Act was filed with the US House of Representatives, it will not be until the next few years when the points included in it regarding the options of "opt-in" on digital platforms will be made effective. In this context, implementation of these new limitations on the use of the data becomes both a vital need for companies and a part of their value proposition. That is, if the privacy of users is respected, it means that those users are located at the center of the corporate strategy, leading to the creation of value. Accordingly, a new trend has emerged where corporations start to differentiate themselves from other companies based on the degree to which their customer privacy is respected (Ehret and Wirtz, 2017; Keskin and Kennedy, 2015; Novales et al., 2016).

- **Cybersecurity and the Internet of Things:** Our focus, mirroring the focus of the Chapter will be on this point. Numerous studies have investigated the impact of the Internet of Things on cybersecurity. Since the IoT is a disruptive trend, a comprehensive understanding of its impact on security requires the awareness that the security of IoT transactions will depend on both the existing business models and those that will emerge in the future (Burkitt, 2014; Ehret and Wirtz, 2017; Keskin and Kennedy, 2015; Novales et al., 2016; Porter and Heppelman, 2014; Rong et al., 2015; Verdouw et al., 2013). As such, IoT is altering previous business models. Therefore, a thorough investigation of business models is necessary in terms of both revealing their weaknesses and finding solutions before the occurrence of cyberattacks and cybercrimes.

With respect to the business models of the Internet of Things and the security threats each of them presents, available literature offers several perspectives on the realities of the Digital Business Environment in the upcoming years. For instance, Burkitt (2014) predicts that companies will be able to understand the different types of players in the Internet of Things market, as well as associated security limitations. Therefore, this study emphasizes that there are three strategic categories of opportunities derived from the Internet of Things. The first is the Enablers, who will develop the technology behind the Internet of Things. The second category is the Participants, who will design, develop, and integrate the services related to the IoT. Finally, the third category embraces the Breeders who will design their own Internet Services of Things.

Similarly, Ehret and Wirtz (2017) argue that the threats posed by the Industrial Internet of Things (IIoT) cannot be fully understood from the current business models. Accordingly, the authors identify three types of commercial Internet Business Models of Things. The first is the model for the supply of manufacturing, maintenance, and repair assets. The second is information and services such as Big Data, analytics, or artificial intelligence. The third and last model consists of all services targeting end-users and seeking to integrate these end-users of the chain into the model.

On the other hand, Keskin and Kennedy (2015) postulate that the business models that are designed must include the externalities generated in the markets. At this point, it is also necessary to be specific concerning which markets one has to attend to. In their comparison of the advantages and disadvantages of the Internet of Things of each model, the authors focus specifically on the models of commercial services. Cybersecurity occupies an important position in the analysis of the business models generated by the Internet of Things (Fleisch et al., 2014; Fleisch, 2010; Atzori et al., 2010). Furthermore, in their in-depth literature review, Novales et al., (2016) identify five conceptual elements that form the basic components of the connected objects. Each of the identified elements presents great challenges in terms of both guaranteeing safety of these connected objects and ensuring adequate management by the companies that develop and market them.

Another relevant study in this respect is the one by Porter and Heppelmann (2014) who explain that, although industries are still at the early stages of the connected products revolution, it is now high time to ask relevant questions and reconsider their business models. This way, they will be able to understand the new Digital Business Environment and avoid the threats posed by cyberattacks, thus reinforcing cybersecurity in the cyberspace. As noted by Verdouw et al., (2013), the Internet of Things can be very useful in supply chains to dynamically virtualized operational processes to optimize them. From a different angle, Rong et al., (2015) contend that the Internet of Things is not merely a supply chain-rather, in the authors' view, it is much more all-encompassing phenomenon that gives rise to a new rapidly-evolving business ecosystem. The authors offer some practical advice that will guide professionals in this new ecosystem, as both its opportunities and threats are new to the companies that enter it.

Once a thorough understanding of the different business models related to the Internet of Things and the challenges they pose is acquired, it is necessary to focus on the implications of cybersecurity in this specific context. To this end, in the present chapter, the researchers have provided detailed descriptions of the threats of potential cybercrimes developed through the connected devices of the Internet of Things. In this case, the issue is not only the access to a multitude of devices that belong to different connected networks, or that all these devices give access to private data about their users; rather, concerns arise due to the nature of their applications that can result in crimes affect numerous aspects of people's lives. This is exactly the case of smart cities, health, and wellbeing, as well as the industrial domain. For example, an insulin manager, a household appliance, or a connected car can not only jeopardize access to the private data about their owners, but also become tools to commit cybercrimes.

From the marketing point of view, hacking a company's database containing personal information of consumers can be harmful not only for consumers as victims of the attack, but also for a company in terms of massive financial loss, damage of reputation, crisis in public relations, and even bankruptcy. One of the first attacks using inter-connected devices was the Mirai botnet. Mirai hijacked millions of DVR devices to create a huge DDoS attack and shut massive segments of the Internet, including, among many others, Netflix, Spotify, Twitter, and Amazon (Harrington, 2016). This Internet-of-Things attack to the Dyn DNS server was followed by other attacks of its kind, such as Hajime and Devil's Ivy, simultaneously imposing sophisticated ransomware on thousands of victims, rather than targeting a single large network.

In addition, according to the target of the attack, cybercrimes can be largely classified into two types. The first type includes the attacks whose objective is to get access to the data stored in the device itself. By contrast, the second type aims to directly attack the device (Tripathi, 2017). In both cases, both the different components that intervene in the management of information in the connected devices and the protocols designed to minimize such attacks ca serve as potential barriers to a cyber attack.

Therefore, in what follows, the authors will discuss both these measures in further detail.

To avoid the aforementioned threats, there are a series of security parameters that must be followed in order to create a safe Internet System of Things. At this point, the different components of the Internet of Things that have been discussed in detail earlier come into play. The importance of these components lies in the fact that all these elements can either reinforce the security of networks of connected devices or allow the entry of cyberattacks. While the key points to be respected so as to ensure security of the transmission of a connection are many and varied, they all are related to the technology of the Internet of Things. In Table 11.9, the authors will proceed to explain each of them, as well as the role they play in a network of Internet of Things devices (Tomic and McCann, 2017).

TABLE 11.9 Description of Security Issues in IoT

Issues	Description
Authentication	The information received by a reader must come from an electronic tag.
Access control	These are the access rights of different connected devices. Some questions related to this access are related to the management of the data or the identification of the devices with access rights.
Policy enforcement	This includes performing rules that come from order, security, and consistency of data to the system.
Trust	It is a concept related to the access points and the criteria measured to identify whether or not to trust a device.
Mobile security	Mobile nodes in IoT are usually transferred from one cluster to another, so the security protocols should be specifically designed contemplating this fact.
Secure Middleware	Given the large amount of data collected in a system of connected devices, security in the middleware is a pivotal concern. The use of ad hoc gateways to ensure such security is common practice.
Confidentiality	Sensitive information must be protected with an RFID electronic tag to prevent it from being read during the transmission process.
Integrity	It ensures that the information received by a device has not been modified or altered during the transmission process.
Privacy	This refers to the protection of user identity.

The IoT architecture security requirements are different at different layers. In an IoT set up several ubiquitous devices are connected simultaneously and exchanging data of different kinds so that security is always a matter of importance. Due to this reason, the development of an IoT security set up requires an approach founded on a layer-by-layer basis.

In what follows, the authors discuss different challenges faced at each layer in further detail and outline solutions proposed thus far in the literature:

- **Security of the Perception Layer:** This is a challenge, since devices in this layer do not usually have sufficient computational capacity or memory to ensure complete security technology. Therefore, attacks usually come from external sources at the nodes where the collected data are transferred to the transport layer. It is usual that the attacks come from the gateways, external networks, or the Internet-connected to wireless sensor networks (WSN). The aim of the attacks in this layer is usually to disrupt object identification through the Mac addresses of devices and jamming of networks that connect sensor nodes. Hence, the data cannot be efficiently collected.
- **Security in Network Layer:** The attacks at this layer can directly target the sensor and actuator nodes and alter their ability to collect and share data. Some of the attacks in this layer include a change in the destination of information, actual information sources or records and the information itself. Other alterations as the block of connections between the perception layer and the application layer may result in a complete breakdown of the service. Since the devices in this layer do not have much memory or computational capabilities, it does not take much bandwidth to achieve flooding in a network of simple nodes. Flooding originates from the network layer, but targets the perception layer, its storage and processing capabilities of the devices. Selective forwarding is where an intermediary node only transmits selective data packets while blocking or dropping other packets.
- **Security in the Application Layer:** Attacks at the network layer can modify source of information and data destination. The exposure of data during the exchange between the different entities is a challenge for security as databases and applications are the departure point and data is exposed to all kinds of attacks before it gets to the user system. As data are being exchanged between different entities like databases and applications, they are exposed to all kinds of attacks before they get to the users of the information. There are several threats in terms of security that will

affect the application layer accessing without permission, stealing data, providing fake data, worms, and viruses.

11.4 SOLUTIONS AND RECOMMENDATIONS

The technical knowledge of the different layers that make up the technology of the Internet of things is presented as a central element to understand the reach of cyber attacks and the ability of the layers to protect themselves from those attacks, as well as to anticipate their occurrence. The development at the protocol level and the cryptographic algorithm level as HMAC may present limits to cybercrimes committed today and those that will occur in the upcoming years (Potlapally et al., 2016). However, the technology advances at high speed and, while these algorithms can temporarily limit cyberattacks, they must continue to advance in their complexity faster than the attacks.

In this respect, the Cluster of European Research Projects on the Internet of Things lays the groundwork for some of the solutions that are outlined below. In the upcoming years, the development of cybersecurity should prioritize the context, as not all situations are equal, and the circumstances surrounding each of the attacks must be deeply studied with all their particularities. There is also an urgent need for an automatic evaluation of the security and privacy needs of the data contained in the devices belonging to Internet of Things systems. Further reinforcement of security in the cyberspace should entail an adequate development of policies pertinent to various fields, such as education, workforce management, investment, entrepreneurship, as well as research and development. The activities needed for adequate response to cyberattacks should include software development, law enforcement, intelligence, incident response, and national defense.

In this regard, policies are of a paramount importance, as the public sector regulates the relationships between the others, both companies, and users. From a policymaking standpoint, cybersecurity should encompass the security of all elements of the cyberspace, including the devices, infrastructure, data, and users. In order to limit cyberattacks, security protocols developed specifically for the Internet of Things can also be used. Among these protocols, the researchers will focus on explaining two of them-Open Trust Protocol (OTrP), and X.509-that are specially indicated for connected device systems (Table 11.10).

TABLE 11.10 Description of Cybersecurity Protocols for the Internet of Things

Cybersecurity Protocol	Description
Open Trust Protocol (OTrP)	It is a protocol that manages security in a Trusted Execution Environment (TEE) through the installation, update, and removal of applications.
X.509	It enables the management of digital certificates and public-key encryption. It is known as a standard for public key infrastructure (PKI) that forms the protocol of the Transport Layer Security used to ensure communications by email and on the web.

In what follows, the authors outline solutions proposed thus far in literature for the different challenges faced at each layer of interconnected devices. For the Security of the Perception Layer, the objective is the protection of data when it is transferred to the transport layer. To this end, intrusion detection and wireless encryption mechanisms can be used to improve the security in this layer. The Security Network Layer aims to protect the ability of the sensor and actuator nodes to collect and share data. In this instance, security can be improved by access security to ensure that data and data routes are accessed by the authorized and authenticated entities. Therefore, intrusion detection systems and encryption of data should be implemented.

The Security in the Application Layer protects data until it gets to the users of the information. In this instance, access control management and privacy protection should be used to improve security in the application layer. Obviously, following the indicated security parameters and implementation of the indicated protocols cannot guarantee that cyberattacks will not occur. However, the more the authors deepen the knowledge of the structures of the connected device system, and the more defined the moments or stages of each of them become, the more capable the researchers are or proposing specific solutions for this ubiquitous trend that is changing the way of understanding the technology.

11.5 FUTURE RESEARCH DIRECTIONS

In the future, one of the key objectives will be to apply Artificial Intelligence to the Internet of Things. In this sense, it should be no longer the need to program the connected devices to perform an automatic evaluation of the needs of cybersecurity; rather, the goal is that machine learning algorithms and methods should be elaborated that will allow the connected devices to learn from their own experiences and failures so that to be able to auto-program their own protocols and security mechanisms to continuously improve their respective cybersecurity systems.

In recent years, along with RFID or the Internet, several other technologies have also boosted the rapid growth of the Internet of Things market. Among these new technologies are the nanotechnology and the embedded technology. The former technology facilitates connection and interaction of smaller-volume things, which can lead to a decrease of the consumption of the system. The latter technology can provide a certain degree of intelligence to object in IoT. The second research direction has objective of identify and work on disruptive technologies that support the current possibilities of the Internet of Things.

Another line of work and research is aimed at improving the security protocols of connected devices. During the previous section two of the security protocols that are most widely used today, OTrP and X.509, were detailed. Both protocols can avoid some cyber threats now however, technology advances not only for the Internet of things but also for the development of sophisticated cyber threats. For this reason, it is necessary to investigate the alternatives and improve protocols that offer security support to connected devices. This fact leads us to the hybrid Encryption algorithms. Encryption algorithms are encryption systems based on algorithms that favor security for the Internet of Things. In the case of hybrid algorithms, symmetric encryption is combined with asymmetric encryption to avoid cyber attacks. Symmetric encryption is one that uses the same key at both the transmitter and receiver side, while asymmetric encryption does not use the same key. Hybrid encryption algorithms combine both types to strengthen cybersecurity in the Internet of things and reduce cyber threats. In addition, the memory consumption of these algorithms is lower, and the security offered is greater.

The future of the Internet of Things is based on technology and simplifications, as long as users might realize that connected devices are simple to use and adopt the Internet of Things will continue to grow as the Digital Business Environment does.

11.6 CONCLUSION

Due to the massive adoption of Inter-connected devices and objects, there is an urgent need to conduct research to fill the gap in the literature and to anticipate the consequences of this trend regarding the management of cyberspace, cybercrimes, and cybersecurity. The very nature of the Internet of Things-i.e., the fact that connected objects and networks might involve not only data, but can have personal and physical consequences-makes IoT particularly sensitive to cyberattacks. A thorough and deep understanding of the landscape, stages, and applications of the Internet of things, attained through with a systematic literature review, might help to this end. While this technology offers numerous valuable opportunities for individuals, communities, business, societies, and nations, it is also imbued with many challenges, such as data privacy, governmental laws, ethics, and cybersecurity. The implications might be addressed by policy-makers to anticipate and avoid potential harm and damage for individuals, society, nation, and global humanity.

KEYWORDS

- **business environment**
- **cyber crimes**
- **cybersecurity**
- **digital trends**
- **future internet**
- **Internet of Things (IoT)**

REFERENCES

Accenture, (2014). *The Internet of Things: The Future of Consumer Adoption.* Accenture Interactive. Retrieved from: https://accntu.re/2m4meW1

Akyildiz, I., Su, W., Sankarasubramaniam, Y., & Cayirci, E., (2002). A survey on sensor networks. *IEEE Commun. Mag., 40*, 102–114.

Asghar, M. H., (2015). RFID and EPC as key technology on internet of things (IoT). *International Journal of Computer Systems and Technology, 6*(1), 121–123.

Ashton, K., (1999). That internet of things thing: In the real world things matter more than ideas. *RFID Journal*, p. 1.

Atzori, L., Iera, A., & Morabito, G., (2010). The Internet of Things: A survey. *Comput. Netw., 54*, 2787–2805.

Bassi, (2013). *Introduction (IoT-A). IoT-A*. Retrieved from: http://www.iota.eu/public (Accessed on 28 August 2019).

Bonetto, L., (2015). *Marketing to Millennials*. US. Retrieved from: https://reports.mintel.com/display/716482/?__cc=1

Borgia, E., (2014). The internet of things vision: Key features, applications and open issues. *Computer Communications, 54*, 1–31.

Burkitt, F., (2014). A strategist's guide to the internet of things. *Strategy+ Business, 77*, 1–12.

Calabrese, F., Ratti, C., & Kloeckl, K., (2009). Wiki city: Real time location sensitive tools for the city. *Handbook of Research on Urban Informatics: The Practice and Promise of the Real Time City*, pp. 390–413. Published by IGI.

CERP-IoT., (2009). *Internet of Things Strategic Research Roadmap*. Retrieved from: http://www.internet-of-things-research.eu/pdf/IoT_Cluster_Strategic_Research_Agenda_2009.pdf

Deloitte, (2014). *How the Technologies Become Harmful for Individual, Society, Nation and Global Humanity*. Deloitte AG.

Dominici, G., Roblek, V., Abbate, T., & Tani, M., (2016). Click and drive. *Business Process Management Journal, 22*(2), pp. 420–434.

Ehret, M., & Wirtz, J., (2017). Unlocking value from machines: Business models and the industrial internet of things. *Journal of Marketing Management, 33*(12), 111–130.

Fleisch, E., (2010). What is the internet of things-an economic perspective? *Auto ID Labs White Paper*. Retrieved from: http://www.im.ethz.ch/education/HS10/AUTOIDLABS WP BIZAPP 53.pdf (Accessed on 28 August 2019).

Fleisch, E., Weinberger, M., & Wortmann, F., (2014). *Business Models for the Internet of Things*. Bosch lab white paper. Retrieved from: http://www.iotlab.ch/wpcontent/uploads/2014/09/EN_BoschLabWhitePaperGMimIOT1_1.pdf (Accessed on 28 August 2019).

Forbes, (2017). *2017 Roundup of Internet of Things Forecasting*. Retrieved from: https://www.forbes.com/sites/louiscolumbus/2017/12/10/2017-roundup-of-internet-of-things-forecasts/#2e5a4a5a1480 (Accessed on 28 August 2019).

Gallup, (2017). *The Crimes Americans Worry the Most*. Retrieved from: https://news.gallup.com/poll/221270/cybercrime-tops-americans-crime-worries.aspx

Harrington, T., (2016). Life after the IoT botnet eleven key takeaways. *Internet of Things Agenda*. Retrieved from: http://internetofthingsagenda.techtarget.com/blog/IoTAgenda/LifeaftertheIoTbotnetelevenkey-takeaways (Accessed on 28 August 2019).

Hernández, M. J. M., Vercher, J. B., Muñoz, L., Galache, J. A., Presser, M., Hernández, G. L. A., & Pettersson, J., (2011). Smart cities at the forefront of the future internet. In: Domingue, J., et al., (eds.), *The Future Internet* (pp. 447 462). Springer.

International Telecommunication Union, (2005). *The Internet of Things*. International Telecommunication Union, Geneva.

Keskin, T., & Kennedy, D. (2015, January). Strategies in smart service systems enabled multi-sided markets: Business models for the internet of things. In 2015 48th Hawaii International Conference on System Sciences (pp. 1443–1452). IEEE.

Könnöllä, T., (2007). *Innovation Roadmap: Exploring Alternative Futures of Industrial Renewal*. Working Paper submitted to the 2007 conference on corporate R&D, IPTS.

Lai, F., & Hutchinson, J., (2005). Radio-frequency identification (RFID) in China: Opportunities and challenges. *International Journal of Retail & Distribution Management, 33*(12), 905–916.

Lemke, M., (2010). *Open Innovation for Future Internet Enables Services in 'Smart' Cities*. CIP ICT-PSP info day, European Commission, DG Information Society and Media.

Mattern, F., & Floerkemeier, C., (2010). From the internet off to the internet of things. In: Sachs, K., Petrov, I., & Guerrero, P., (eds.), *From Active Data Management to Event-Based Systems and More. Lecture Notes in Computer Science,* (Vol. 6462). Springer, Berlin, Heidelberg. Retrieved from: https://doi.org/10.1007/978 3 642 17226 7_15 (Accessed on 28 August 2019).

McKinsey, (2015). Retrieved from: https://www.mckinsey.com/businessfunctions/digitalmckinsey/ourinsights/theinternetofthingsthevalueofdigitizingthephysicalworld (Accessed on 28 August 2019).

Miller, K. M., & Skiera, B., (2017). *Economic Damage of Cookie Lifetime Restrictions*. NET Institute Working Paper No. 17 02. Retrieved from: https://ssrn.com/abstract=3046254 or http://dx.doi.org/10.2139/ssrn.3046254 (Accessed on 28 August 2019).

Ministry of Communication of India, (2015). *M2M/IoT Technical Reports*. Retrieved from: http://tec.gov.in/pdf/Studypaper/White%20Paper%20on%20MachinetoMachine%20(M2M)Communication.pdf (Accessed on 28 August 2019).

Muñoz, A., (2016). Elimination of personal data on the internet: The recognition of the right to be forgotten. *Chilean Journal of Law and Technology, 4*(2). doi: 10.5354/0719 2584.2016.37426.

Novales, A., Mocker, M., & Simonovich, D., (2016). *IT Enriched "Digitized" Products: Building Blocks and Challenges*. In AMCIS 2016 Proceedings-AIS.

Palos, S. P., & Saura, J., (2018). The effect of internet searches on afforestation: The case of a green search engine. *Forests, 9*(2), p. 51.

Porter, M. E., & Heppelmann, J. E., (2014). How smart, connected products are transforming competition. *Harvard Business Review*, pp. 64–88.

Potlapally, N., Ravi, S., Raghunathan, A., & Jha, N., (2006). A study of the energy consumption characteristics of cryptographic algorithms and security protocols. *IEEE Transactions on Mobile Computing, 5*(2), pp. 128–143.

Rong, K., Hu, G., Lin, Y., Shi, Y., & Guo, L., (2015). Understanding business ecosystem using a 6C framework in Internet of Things based sectors. *International Journal of Production Economics, 159*, 41–55.

Rozanski, N., & Woods, E., (2012). Software systems architecture: Working with stakeholders using viewpoints and perspectives. *Choice Reviews Online, 49*(10), pp. 495717.

Saura, J., Palos, S. P., & Cerdá, S. L., (2017). Understanding the digital marketing environment with KPIs and web analytics. *Future Internet, 9*(4), p. 76.

Schaffers, H., Komninos, N., Pallot, M., Trousse, B., Nilsson, M., & Oliveira, A., (2011). Smart cities and the future internet: Towards cooperation frameworks for innovation. In: Domingue, J., et al., (eds.), *The Future Internet* (pp. 431–446). Springer.

Schuermans, S., & Vakulenko, M., (2014). IoT: breaking free from internet and things. *Vision Mobile, London*. Retrieved from: http://www.visionmobile.com/product/iotbreakingfreeinternetthings (Accessed on 28 August 2019).

Smouter, K., (2018). The year of the GDPR. *Research World, 68*, 48–49.

Teixeira, T., Hachem, S., & Georgantas, N., (2011). Service oriented middleware for the internet of things: A perspective. In: Abramowicz, J., Witold, L., Ignacio, M., Surridge, M., Zisman, A., & Vayssière, (eds.), *Towards a Service-Based Internet* (Vol. 257178, No. 257178, pp. 220–229). Springer Berlin Heidelberg.

Tomic, I., & McCann, J., (2017). A survey of potential security issues in existing wireless sensor network protocols. *IEEE Internet of Things Journal, 4*(6), 1910–1923.

Townsend, A., Maguire, R., Liebhold, M., & Crawford, M., (2011). *A Planet of Civic Laboratories, The Future of Cities, Information and Inclusion*. Institute for the Future.

Townsend, A., Soojung-Kim, P. A., & Weddle, R., (2009). *Future Knowledge Ecosystems: The Next Twenty Years of Technology Led Economic Development*. Institute for the Future, IFTF Report Number SR 12361.

Tripathi, V., (2017). Internet of things: Opportunity and security issue. *International Journal of Emerging Technologies in Engineering Research, 5*(10), 82–86.

Verdone, R., Dardari, D., Mazzini, G., & Conti, A., (2008). *Wireless Sensor and Actuator Networks*. Elsevier: London, UK.

Verdouw, C. N., Beulens, A. J. M., & Van Der Vorst, J. G. A. J., (2013). Virtualization of floricultural supply chains: A review from an internet of things perspective. *Computers and Electronics in Agriculture, 99*, 160–175.

Vermesan, O., Friess, P., Guillemin, P., Sundmaeker, H., Eisenhauer, M., Moessner, K., et al., (2014). Internet of things strategic research and innovation agenda. In: Vermesan, O., & Friess, P., (eds.), *Internet of Things-from Research and Innovation to Market Deployment* (pp. 7–142). River Publishers, Aalborg.

Wamba, S. F., Bendavid, Y., Lefebvre, L. A., & Lefebvre, É., (2006). RFID technology and the EPC network as enablers of mobile business: A case study in a retail supply chain. *International Journal of Networking and Virtual Organizations (IJNVO), 3*(4), 450–462.

Weber, R. H., (2010). Internet of things-new security and privacy challenges. *Computer Law & Security Review, 26*, 23–30.

Wortmann, F., & Flüchter, K., (2015). Internet of things: Technology and value added. *Business and Information Systems Engineering, 57*(3), 221–224.

Yan, L., Zhang, Y., Yang, L. T., & Ning, H., (2008). *The Internet of Things: From RFID to the Next Generation Pervasive Networked Systems*. (Hardback)-Routledge.

Yoo, Y., Henfridsson, O., & Lyytinen, K., (2010). Research commentary-the new organizing logic of digital innovation: An agenda for information systems research. *Inf. Syst. Res., 21*(4), 724–735.

Yunjie, L., (2010). *Internet Industry Development Present Situation and Future Prospects*. RFID and Internet of Summit Forum, Hong Kong.

CHAPTER 12

Cloud Computing and Security

DINESH GOYAL and R. S. RAJPUT

Suresh Gyan Vihar University, Center for Cloud Infrastructure and Security, Suresh Gyan Vihar University, Mahal, Jagatpura, Jaipur, Rajasthan – 302017, India,
E-mails: (Dinesh Goyal) dinesh8dg@gmail.com;
(R.S. Rajput) rajpoot.rs@gmail.com

ABSTRACT

Data security has always been a significant matter in information technology. In the cloud computing environment, it becomes mainly severe because the data is placed in different places, even anywhere in the world. Data security and privacy protection are the most critical aspects of a user's concerns about cloud technology. The significant difference in cloud security is that the cloud user does not control the physical system. The application of security, infrastructure, and a platform is under the provider's control. Cloud security depends on some factors that are responsible for forming the cloud more vulnerable to threats and attacks. Thus, security can be implemented into four sections namely: Cloud Security behind the Virtualization, Securing Data over the Cloud, Cloud Channel Security, and Securing Cloud System Management. In this chapter, the discussion is about the factors leading to cloud being vulnerable to threats and attacks and also on the methods and solutions to make secure cloud system.

12.1 INTRODUCTION

Information security, abbreviated as InfoSec, is the manner of restricting unauthorized access, illegal use, personal disclosure, unauthorized modification, personal inspection, illegal recording, or destruction of information. Therefore, security is the prevention and protection against the access to all the resources, viz., data software, hardware, network, etc. by unauthorized recipients; intentional and illegal alteration or destruction of information and protection against all the vulnerabilities the system may exhibit (Figure 12.1).

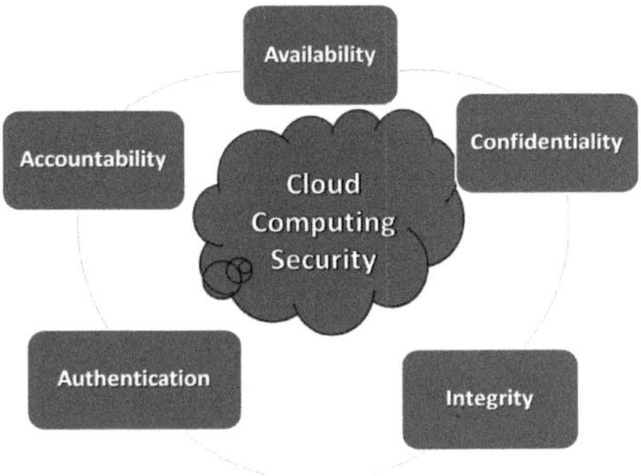

FIGURE 12.1 Cloud computing security.

Infrastructures of cloud computing consist of services. Those are granted and allotted through a data center that is known as a cloud service center. Services can be accessed anywhere in the world through various user interfaces such as the web browser, mobile app, thin client, terminal emulator, remote desktop, etc., A cloud computing system divides it into two primary segments — the front end segment and the back end segment. Both segments joined to each other through a communication network, usually the Internet. The front end segment is the part of the cloud computing model seen by the cloud client, i.e., the cloud user. That includes the client's network and the cloud resources accessed as a service. Backend segment of the cloud computing model is cloud itself; It is a pooling of computing resources like computing units, storage units, memory units, network units in the physical or virtual forms. Availability Zones of Amazon Web Services (AWS) is one of an example of the cloud backend segment.

Cloud computing system has four primary deployment models and three fundamental service models with four key attributes. Deployment models of the cloud computing system are defined by the type of access to the cloud resources, i.e., how the cloud located? Cloud computing systems' access models are public access, private access, hybrid access, and community access. Service model of cloud computing system is categorized into three fundamental service models namely infrastructure as a service (IaaS) model, platform as a service (PaaS) model and software as a service (SaaS) model. The primary attributes of the cloud computing system are broad network access to cloud resources, the rapid elasticity of capacity of resources, measurable of utilization of resources, and on-demand self-customized service respectively.

All items used in the circumstances of cloud computing are cloud resources. Cloud resources include computing, storage, networking, and energy, resources directly or indirectly associated with the set of cloud applications. A cloud resource is any physical or virtual component of the cloud system, which has limited availability within a cloud computing system. Every component connected to a cloud system is a cloud resource. Cloud resources can be classified into two subparts. Physical cloud resources, e.g., CPU, Memory, Storage, Network, Sensors/actuator. Logical cloud resources, e.g., Operating System, Data, Software, Energy, API, Network throughput, Information Security Protocols, Network loads, delays. Cloud computing security deals, securing of its resources both physical and logical.

Data security has always been a significant issue in the study of information technology. In the cloud computing environment, it becomes especially critical because the data, as well as resources, are located in various areas even anywhere in the universe. According to Yunchuan et al., (2014), privacy protection and data security are the two principal factors of cloud customers.' Those are becoming significant for the future growth of cloud environment in business, industry, and government. Data security and privacy protection cases are relevant to both software and hardware in the cloud architecture. In cloud computing, clients can access resources online at any time through the internet without managing the original resources like physical and technical management. As one of the key attribute 'rapid elasticity' of cloud resources, cloud resources are also scalable and dynamic. The dissimilarity in cloud security is that the cloud user has not controlled the physical system. The application of security, infrastructure, and a platform is under the cloud provider's control. The advancement of the cloud computing system is from various latest technologies such as virtualization, grid computing, autonomic computing, and some other techniques. In 2017, Mebrahtu and Srinivasulu published a study; new challenges arise whenever a new technology comes (Mebrahtu and Srinivasulu, 2017).

In general, cloud computing security identifies the following primary objectives:

- **Availability:** The availability of cloud computing systems is to ensure that services and data are always available for cloud users at any time, at any spot as the necessary attribute 'on-demand self-service' of cloud resources, cloud resources are available on-demand and the control of the cloud user. Sabahi (2011), states that Cloud providers guarantee that their servers' uptime matches with cloud users' data centers and users applications. An enterprise must be assured that a cloud provider is always running hosted applications and delivering valid results. The scheduled maintenance and unscheduled maintenance is another availability factor that exists, and it can harm the availability ratio of the cloud provider. Although regularly scheduled maintenance does not consider as downtime, unscheduled maintenance increases downtime and affects availability.
- **Confidentiality:** The purpose of confidentiality is to hold users' data secret in the cloud computing systems and available only for the authorized entity. Gawande et al., (2014) indicate that data confidentiality refers to the ability to share sensitive data among a group of people where the privileges of data access granted by the data owner to each group member. Any user external to the group is assumed to have no privilege at all. Data privacy is a particular status of data confidentiality. Data privacy means that the data held by an individual and no means be disclosed to anybody.
- **Integrity:** The object of data integrity in the cloud environment is to confirm that the data has not been modified in any manner during it has stored or during its transport over the communication network. Al-Saiyd et al., (2013), state that the verification of data integrity at untrusted servers and dealing with the sensitive data are some challenging tasks. These tasks are not easy to maintain sensitive cloud data, which required many software applications for clients. In cloud computing environment the cloud customer does not know which server their data have stored and that server implemented with secure data service or not.
- **Authentication:** The object of authentication is to assure the identity of the entity involved in the use of cloud resources through the communication channel. According to a study (Barhate and Dhore, 2016), the cloud user authentication corresponds to cloud user type, and user of the cloud application with users role in the application. There are two different types of users in cloud computing environments, the first type

is end-users, and another class is cloud resource users. End-users use the cloud applications that are deployed on the cloud environment, and these cloud users register and identify with the use of cloud application only not for infrastructure resources, end users usually are not aware of the cloud application is executing on which cloud resources and which data center. Cloud resource users are typically administrators of the cloud resources and able to set permissions for the cloud resources based roles, access lists, IP addresses, domains. The second kind of users is of more significant concern from an interoperability perspective.

- **Accountability:** The mechanisms for assessing responsibilities for decisions and action were taken. The object of accountability is to confirm that no entity can deny its participation in a data transfer between them. Zou (2016) state in the business circumstances, accountability contains transparency, responsibility, responsiveness, and willingness for assuming liability. Accountability defined as answerability or an obligation to account for how well resources used to meet specified outcomes (Table 12.1).

TABLE 12.1 Security Objectives and its Role in the Cloud Environment

S. No.	Security Objectives	Role in the Cloud Environment
1	Availability	Availability of cloud services, faithfully running services, service delivery time with valid results
2	Confidentiality	Data secret, sharing sensitive data, data privacy, restricting unauthorized access, illegal use
3	Integrity	unauthorized modification or alteration of data and service
4	Authentication	Assure the identity of the entity, personal inspection
5	Accountability	transparency, responsibility, responsiveness, and willingness

12.2 VULNERABILITY IN CLOUD COMPUTING

There are many advantages to cloud computing system like mobility, disaster recovery, flexible costs, always availability, and expenses can be quickly decreased, elastic capacity and limited environmental impact when they are appropriately safeguarded. When an unusual amount of data is stored, advances in digital communications and software have magnified vulnerabilities. Through digital communication networks, information systems in different locations can be interconnected. The potential for fraud, abuse, or unauthorized access is not limited to a single cloud computing data center but occur at any point in the network.

Vulnerabilities are imperfections, errors or bugs in software applications, operating systems, or networking protocols that can be employed by rivals to obtain access or elevate access privilege of the computer systems. Hussain and Gummadi (2013) state that there are several vulnerabilities in virtual machines those propose to a cloud computing environment as vulnerable. Though the cloud is hosted in the data center, all servers are set up similar to physical ones. The only difference is that more software is used to support a virtual machine that makes up the cloud. There has been the number of the vulnerabilities discovered on such software. These vulnerabilities cover hypervisor vulnerabilities, vulnerable hypervisors, virtual machine-based rootkit, and denial of service attacks, VM hopping, and data leakage. According to the study (Ramachandran and Ramachandran, 2012), vulnerability management is a security exercise by which institutions set up methods to identify vulnerabilities and to fix it. Patches are supplementary programs combine to the original software or systems to address the weaknesses. There are a few concepts that are responsible for making the cloud more vulnerable to threats and attacks that are not present in the typical IT system these are (Figure 12.2):

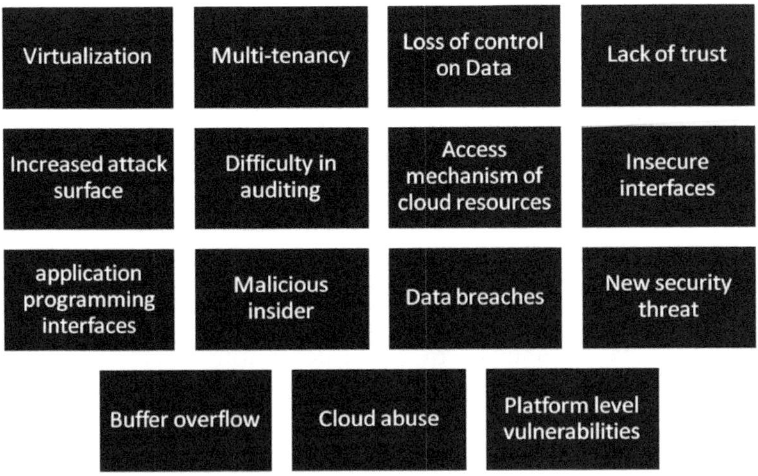

FIGURE 12.2 Vulnerability in cloud computing.

- **Virtualization:** It is perhaps the most significant magic in the cloud that makes cloud efficient regarding shared resources. Cloud providers must make the most in virtualizing and share the resources to maximize the cloud resource utilization, and on the other site to create well-defined isolation across the cloud customers' virtual machines. However, the control is with the hypervisor, and this now becomes the point of weakness in the virtualized environment. The hypervisor layer becomes an easy target compromising which an attacker can access the instances. BLUEPILL rootkit, SubVirt, and DKSM are remarkable attacked examples in the virtualization environment. Through attacks, an attacker can modify the hypervisor and be able to access the host resources. Similarly, memory access to the host can be got by violating the read/write accesses.
- **Multi-Tenancy:** Cloud computing service providers perform their scalable services by sharing virtualized infrastructure, platforms or software applications. That can lead to vulnerabilities that can probably be employed in all cloud computing delivery models. The multi-tenancy specialty is a capability of running multiple instances on top of the same platform. Here, multiple users share the platform concurrently, while being isolated from each other through virtual machine instances. Therefore, all the tenants (virtual machine users) accessing a virtual machine instance are available on the shared platform. As a typical real-life example, we talk of a multi-storied apartment building where the same ground space is shared by multiple tenants living on various floors of the building. In all service levels viz., IaaS, PaaS, and SaaS, multi-tenancy forces different applications share the same underlying platforms at different layers. Moreover, in all the cases, the cloud customer's data may be stored in a shared physical location. It is relatively easy for attackers to obtain control of the data of another application in the form of co-location, co-residence, or co-tenancy attacks through multi-tenancy feature of the cloud. Further, it is easy to achieve a Denial of Service attack due to the underlying shared platform. As in the earlier virtualization discussion, in multi-tenant structure as well, virtualization is used to offer shared on-demand services (Vani and Priya, 2015). Hence vulnerabilities in a hypervisor may increase possibilities of such attacks as well. Possible solutions to this problem lie in implementing appropriate isolation of virtual machine and also service level agreements. Further having stronger authentication and access control mechanisms also are needed.

- **Loss of Control on Data Cloud Encourages Outsourcing:** The industries are growing in this outsource business model. According to this model, the responsibilities of acquiring resources and using these resources be it storage or computing requirements, are delegated to contracted third-party service providers, with the customer concentrating on more delicate parts of the work. This way, a customer saves her capital expenditure on the physical resources and the operational expense of maintaining the equipment, running the latest software or any other operational responsibilities as well. Hence both capital expenditure and operational expenditure of customers are taken care. There is a strategy that has rendered cloud popularity. However, the glitch in this scheme is that data, perhaps the most critical asset of the business, must also be outsourced for this scheme to work. Hence, users, that is the business houses have to part with their data, which will now be maintained by the third party in an off-site location, outside the premises of the organization to whom the data belongs. Therefore, cloud, while providing in numerous facilities, renders customers to lose control over their data. Hence there must be trust between the service providers and consumers for the model to work.
- **Lack of Trust:** In the cloud, the customer infrastructure, data, and compute all located off-premise and are managed by a third-party entity. It is essential for the consumer to be able to trust a provider. However, trust is a concept rather than a measurable element since it is more subjective than objective. Measuring confidence must involve several factors that make the decision-making and evaluating a decision harder. Hence, security management must provide any quantifiable measures that help the consumer gain trust about a provider and the provider, on their end, must enable such factors. From this perspective, trustworthy systems have proposed wherein systems must always perform as expected, even under the influence of hostile disruption. Such trustworthiness combines reliability and security that uses security measures that are visible and measurable at any point in time, thereby allowing consumers to gain trust on a particular provider by testing the provider at the unknown frequency and on different aspects. Since confidence is a human aspect, rather than a machine aspect, a mixture of people to the machine, people to people, and machine-to-machine interactions must be provided in such measures.
- **Increased Attack Surface:** An attack surface is one of vulnerability that gives an opportunity to malicious users may exploit and utilize. Cloud has an increased attack surface since for an organization the data is stored

outside the organization with an unknown entity being in charge. This entity is typically doing even all the computations on the stored data. Also, all the communications during, before, and after storing data in cloud take place over the Internet, thereby making the cloud more vulnerable through the communication links which are public or insecure. The attack surface further increases since cloud forces consumers to share the infrastructure.

- **Difficulty in Auditing:** The cloud customer does not have direct entrance to any cloud resources since everything in the cloud have virtualized. Hence it is difficult to have a monitoring mechanism. Therefore auditing can be done only by the service provider of cloud resources.
- **Access Mechanism of Cloud Resources:** Generally, access to the cloud depends upon only username and password in a public network. Hackers may target the client's computer to know username and password using some attacks like phishing, fraud, etc. The hijacking of users' account or service is not new now, but cloud services increased platform for this type of threats. If attackers obtain access to a cloud user's credentials, and they can monitor the system's activities, transactions, data manipulation, return falsified information and redirect cloud clients to some illegal sites. Social engineering sites such as fake registration websites and counterfeit emails demand the cloud user disclose his password or authentication keys (phishing attack). Another type of attack is known as the masquerade attack in this attack. The attacker acts to be a user and attempts to get the password from the server (man-in-the-middle attack). The attacker represents to be a verifier of the cloud service provider and inquire about authentication keys from the cloud user.
- **The Insecure Interfaces and Application Programming Interfaces (APIs):** Cloud providers provide to cloud customer remote access, terminal emulator, web browser-based application or APIs to communicate and manage with cloud resources. Cloud provisioning, resource management, and monitoring are all performed with the help of these interfaces, and the availability and security of cloud services depend upon provided interfaces.
- **Malicious Insider:** The insider threat is a real rival. A malicious insider in the cloud computing system may be a rolled employee of data center like as a system administrator that is capable of accessing potentially sensitive information. Security of the cloud computing system solely depends upon cloud service providers. The malicious insider has belonged to a higher security risk category.

- **Data Breaches:** A data breach is one of targeted attack objective or simply the result of human error in the data validation, program loading, program execution or software vulnerabilities. The risk of data breach is common to cloud computing, due to primary interest of hackers it consistently ranks as a primary concern for cloud consumers.
- **New Security Threat:** Improvement of computing technology, increasing internet bandwidth, availability of malicious methods, tools or programs on the internet, are some supporting equipment for hackers.
- **Buffer Overflow:** This phenomenon happens when a computer program in a virtual machine tries to store extra data in a buffer memory (temporary data storage area). Buffers of a virtual machine are formulated to contain a prescribed volume of data; the additional data can overflow into adjacent buffers that may be caused to corrupt or overwrite the existing valid data. Generally, it may happen accidentally through a computer program error, and buffer overflow is also considered a particular kind of security attack that produced a data integrity problem. In the buffer overflow attacks, attackers used some specific triggers for sending instructions and data into a targeted virtual machine beyond its buffer limit for damage files, and data, unauthorized modification of data or reveal confidential information.
- **Cloud Abuse:** Some cloud providers grant free limited trial periods for learning or promoting their services. Hackers are the damaging system through the malicious code, spammers, password key cracking and other criminal activities hiding original identity in the restricted free period. A malicious hacker uses cloud server instances to launch a distributed denial of service (DDoS) attack, deliver malware, or share illegally copied of software. The challenge is here is to determine the best processed for identifying trial users.
- **Platform Level Vulnerabilities:** Some Platform level vulnerabilities are cross-site malicious code scripting and SQL injection malicious code. If malicious code inserted into the data fields of a standard SQL query using which an attacker gets illegal access to databases, this type of attack is known as a SQL injection Attack. In the cross-site scripting (XSS) attack, some malicious programming scripts or programming codes are injected into web contents thereby forcing a website to execute the attacker-supplied codes (Table 12.2).

TABLE 12.2 Vulnerabilities in the Cloud Computing

S. No.	Vulnerability	Impact of Vulnerability
1	Virtualization	Loss control of the hypervisor
2	Multi-tenancy	Loss control of the hypervisor, denial of service attack, the effect on cloud resources availability
3	Loss of control on data	Loss of data, the effect on data integrity, the effect on confidentiality
4	Lack of trust	Effect on confidentiality
5	Increased attack surface	Effect on confidentiality and authentication
6	Difficulty in auditing	Monitoring not possible
7	Access mechanism of cloud resources	Effect on confidentiality and authentication
8	Insecure interfaces and application programming interfaces	Effect on data integrity, confidentiality, and authentication
9	Malicious insider	Illegal monitoring by a malicious insider, the effect on data integrity, confidentiality, and authentication
10	Data Breaches	Effect on integrity, confidentiality, and authentication
11	New security threat	Effect on data integrity, confidentiality, and authentication
12	Buffer overflow	Effect on availability and data integrity
13	Cloud abuse	Effect on data integrity
14	Platform level vulnerabilities	Effect on data integrity, confidentiality, and authentication

12.3 SECURITY RISK IN CLOUD COMPUTING

There are numbers of security risks connected with cloud computing, the Cloud Standards Customer Council (2017), indicates some risks those are under (Figure 12.3):

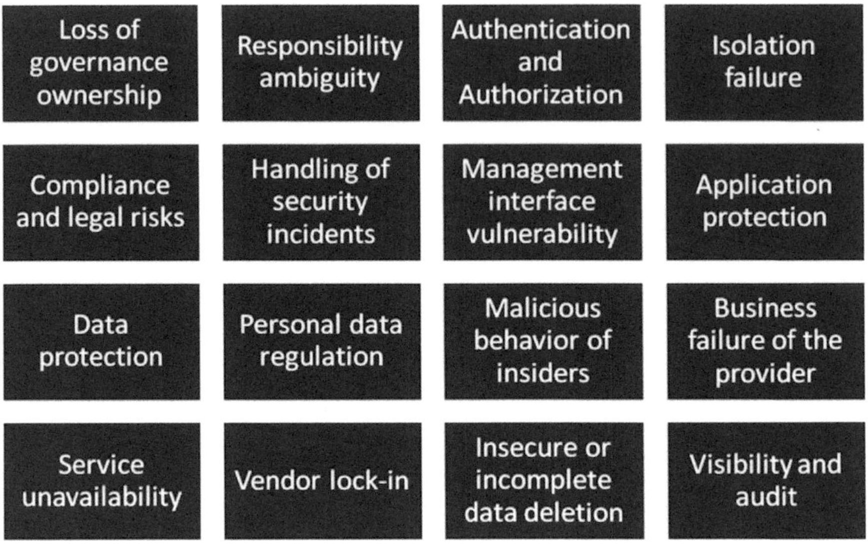

FIGURE 12.3 Security risk in cloud computing.

- **Loss of Governance Ownership:** In the deployment of public cloud, cloud clients surrender control to the cloud service provider covering several points that may affect security and privacy. Cloud service agreements may not offer a commitment to settle such matters on the part of the cloud service provider, thus leaving holes in security defenses.
- **Responsibility Ambiguity:** Responsibilities over appearances of security and privacy are shared among the cloud service provider and the cloud customer. The separation of responsibilities among the cloud provider and cloud customer are required, and depends on the cloud service models (e.g., SaaS, PasS, and IaaS).
- **Authentication and Authorization:** The delicate cloud resources are obtained from anywhere, any time in cyberspace so that there are necessary to establish certainty and identity with the cloud user. If users are employees, service contractors, business partners, and customers, secure authentication and authorization become a critical matter.
- **Isolation Failure:** The multi-tenancy and resources sharing are characteristics of the deployment of the public cloud. The risk covers in the situation the collapse of mechanisms dividing the usage of memory, storage, routing or still reputation among the tenants of the cloud.

- **Compliance and Then Legal Risks:** The cloud client's investment in the managing certification (e.g., regulatory requirement or compliance) may be lost if the cloud service provider cannot provide evidence of their compliance with the relevant provisions. The cloud customer must verify that the service provider of the cloud has proper and appropriate certifications.
- **Handling of Security Incidents:** The threat detection, vulnerability reporting, and consequent management of security occurrences may be assigned to the cloud provider, but these occurrences impact to cloud customers. Advice rules are necessary to be negotiated in the agreement of cloud service, so that cloud customers are informed of the unnecessary delay.
- **Cloud Management Interface Vulnerability:** Interfaces between cloud resources to cloud customer to be managed by public cloud resources (self-provisioning), are generally accessible through the Internet using remote access, web browser, thin client, and API. Self-provisioning is allowing access to larger sets of cloud resources compared to traditional hosting. The self-provisioning method increased risk when the cloud resources access with client interface like the web browser, and that web browser already has some vulnerabilities.
- **Application Protection Workload:** Traditionally, the security solutions developed for application protection imagine a clear separation of physical and virtual resources. Infrastructure security of cloud resources is the responsibility of the cloud service providers. Organizations need to replan boundary of security at the network level, applying more controls at the user level, application level, and data level. That requires designing and managing workload-centric guidelines as well as executing centralized management beyond distributed workload instances.
- **Data Protection:** The primary concerns are showing or making public of personal and sensitive data, the loss of data or unavailability of data, and over maintenance of data. It may be challenging for the cloud service customer (in the role of a data controller) to satisfactorily check the data handling manners of the cloud service provider. This problem arises in cases of multiple transfers of data, (e.g., between multiple cloud services or cloud provider that uses subcontractors and third party service providers), appearing in a reduction of control transparency and unclear purposes for the processing of the data.

- **Personal Data Regulation:** The personal data must be managed according to the terms of standard jurisdictions laws and regulations. That now extends typically beyond the protection of such personal data, also includes rights granted to the data focus to scrutinize, correct, or delete their data, and in some cases, to request that their data is transferred anyplace. Any use of the cloud service to process or hold personal data must satisfy these requirements at the same time securing the data.
- **The Malicious Behavior of Insiders:** The destruction created by the malicious activities of people working inside an organization can be substantial, given the access and authorizations they hold.
- **Service Unavailability:** This could be originated by the communication network, software, or hardware failures.
- **Vendor Lock-In:** Dominion on proprietary services of a demanding cloud service provider for a particular service could lead to binding the customer. The portability of cloud applications and cloud data across the one cloud service providers to another service provider poses a risk of service unavailability in case of new server provider not able to provide that service. Therefore it is an essential but sometimes overlooked perspective of security. Lack of interoperability of interfaces linked with cloud services also ties the cloud customer to a demanding cloud service provider and can make it hard to switch to a different cloud service provider.
- **Incomplete or Insecure Data Deletion:** The ending of an agreement with a cloud provider may not result in the deletion of the customer's data from the provider's and providers' third-party systems. Backup copies of cloud customers' data usually exist and may be combined on the same storage media with other cloud customers' data, making it challenging to erase selectively. The real advantage of multi-tenancy is the sharing of physical resources, thus performs a more significant risk to the cloud customer than dedicated hardware.
- **Visibility and Audit:** Some business users are developing a "shadow IT" by acquiring cloud services to build IT products without specific organizational approval. Significant challenges for the security team are to know about all uses of cloud services within the organization (e.g., what resources are being used, for what purpose, to what extent, and by whom), understand what laws, regulations, and policies may apply to such uses, and regularly assess the security aspects of such uses.

12.4 CLOUD SECURITY AREAS

Some factors are responsible for forming the cloud more vulnerable to threats and attacks that are not present in the typical IT system. These factors categorized into four significant sections –

12.4.1 CLOUD SECURITY BEHIND THE VIRTUALIZATION

Virtualization is the most critical magic in cloud computing that makes cloud efficient concerning shared resources. Cloud providers must make the most of virtualizing and sharing the resources to maximize the resource utilization on the one hand and to create well-defined isolation across user virtual machines on the other. However, the control is with the hypervisor, and this now becomes the point of weakness in a virtualized world. The hypervisor layer becomes target compromising which an attacker can access the instances. BLUEPILL, SubVirt, and DKSM have attacked examples on virtual layer. Some concepts those are responsible for developing vulnerability behind the virtualization -

- **Multi-Tenancy:** It is the capability to run multiple instances on top of the same physical platform. Here, multiple users share the platform concurrently, while being isolated from each other through virtual machine instances. Therefore, all the tenants accessing an instance of virtual machine those are available on the shared platform. In SaaS, multiple cloud users may be sharing the common application stack. Here stack means a logical connected structure of the database, software application, web servers, and networking. That indicates the data from several clients may get stored in the same database software, may get backed up and archived together used on same unencrypted networking channel, and managed by common application processes. That puts a heavyweight on logical security built into the application to separate one tenant's users from others. In PaaS, platform stack is shared between the tenants. Vulnerability in the platform stack can allow flowing between tenants for shared data backups and archives. In IaaS there are changes of cross-virtual machine attacks or cross-network traffic attack.
- **Cloud Misuse:** One of the prime cloud's functionality called "Infrastructure as a Service " which offers the virtualization of physical cloud resources like CPU, storage, and the network does not have a secured

certification process. It means that a person having a proper credit card can log in into the cloud environment and can instantly start using the cloud. Cybercriminals have already exploited and misused cloud computing due to weak registration procedures of obtaining cloud computing services and lack of adequate security controls; also, hackers can abuse cloud services by obtaining unauthorized and illegal access to other cloud customer's confidential data residing on the same cloud platform.

- **Issues Due to Shared Technology:** IaaS functionality of cloud provider provides high-end scalability by allowing the user to access shared virtualized devices. A hypervisor allows a guest operating system to connect to other physical resources. These places the cloud at risk as the guest operating system gains access even to the additional levels which influences other systems on the network. Joanna Rutkowska released a tool name of it is Red Pill. The goal was to detect when a program was running under virtualization instantly. The Blue Pill rootkit is malware that executes as a hypervisor and to obtain control of physical resources. The hypervisor installs without need a restart and the computer functions normally, without degradation of speed or services, which makes detection difficult. The original Blue Pill implementation based on AMD-Virtualization (AMD-V).
- **Resource Attacks:** This type of attack involves manipulating the available resources into mounting a large-scale botnet attack. This kind of attack is targeted either cloud user or cloud provider. This type of attack run several infected computers that spread malware run in the cloud environment. Botnets can borrow adequate computing power to bring down entire networks and businesses. Without defensive detection on networks, organizations continually run the chance of infection. Botnets spread to give the botnet owners massive dark cloud computing power that they can use to conduct highly successful cybercrime. Botnet owners may lease out their army of infected machines to criminal enterprises for a variety of techniques.
- **Denial of Service Attacks:** In the DoS attack, attacker overloads the targeted cloud system with enormous service requests so that it stops reacting to any new requests and hence made resources unavailable to its users. Some cloud security alliance has recognized that the cloud is more vulnerable to DoS attacks because it used by so many users which make it much more damaging. The DoS attacks are of many types:

- An attacker can overload the target with the massive amount of junk data that consume the network bandwidth and resources, for example, UDP floods, and ICMP floods.
- An attacker can make use of a particular character (blank space) that associated with several networking protocols to overload target resource, for example, fragment packet attack, SYN floods, ping of death, etc.
- An attacker can make HTTP request in the massive amount so that the server cannot handle it, for example, Hypertext Transfer Protocol-Distributed Denial of Service (HTTP-DDoS) attack, and extensible Markup Language Distributed Denial of Service (XML-DDoS) attack.
- The making of a new virtual machine is not a challenging assignment, and thus, creating rogue virtual machines and allocating enormous spaces for them can lead to a denial of service attack for cloud service providers when they opt to generate a new virtual machine on the cloud. This kind of attack is generally known as virtual machine sprawling.

- **Meltdown and Specter:** Meltdown and Specter exploit dangerous vulnerabilities in latest processors. These hardware vulnerabilities allow programs to steal data which is currently processed on the computer. While programs are generally not permitted to read data from other programs, a malicious program can exploit Meltdown and Specter to get hold of secrets stored in the memory of other running programs. Specter and Meltdown both open up possibilities for dangerous attacks. For instance, JavaScript code on a website could use Specterto trick a web browser into exposing user and password information. Attackers could exploit Meltdown to observe data owned by other users and even other virtual machine hosted on the same physical hardware, which is potentially harmful to cloud computing hosts.

12.4.2 DATA SECURITY OVER THE CLOUD

These classes of attacks include unauthorized alteration of sensitive data or performing configuration changes to enable a sniffing attack via a specific device, etc. These attacks are centered on both the cloud service providers and cloud users. Some essential points those are responsible for developing vulnerability for data security over the cloud:

- **Data Integrity:** This in the cloud computing system means preserving information integrity. The data should not be lost or altered by unauthorized users.
- **Data Confidentiality:** It is essential for users to store their private or confidential data in the cloud.
- **Data Availability:** The general thought about data availability means when accidents such as data corrupted, hard disk damage, IDC fire, and network failures occur, the extent that user's data can be used or recovered and data also available on time when it demanded.
- **Distributive Data Storage:** It is also a promising approach in the cloud computing environment.
- **Data Privacy:** It means when users visit the sensitive data in the cloud computing environment, the cloud services can prevent the potential adversary from inferring the user's behavior by the user's visit model (not direct data leakage). The issues related to security of data privacy in the cloud computing are the intrusion, the integrity of data and availability of service in the cloud.
- **Data Breach:** A cloud environment has multiple users and organizations, whose data are in the same place. Any breach of this cloud environment would expose all users' data. In the multi-tenancy, customers using different applications on virtual machines that use the same database, if any corruption of data, going to affect others tenants who are using the same database. Keeping a backup of data offline may reduce the risk of data loss, but will increase the risk of data exposure.
- **Risk Unknown Profile:** Exposure to log details and data, kind of people who will be people sharing infrastructure, network intrusion logs, redirection strategies and other log information overlooked before relying on the vendor.
- **Loss of Data:** If data affected due to deletion, alteration, unlinking a record and storing of data on an unreliable medium, is another serious threat. It leads to the loss of essential data, reputation, and the trust of cloud customers. Sometimes the loss of data may cause severe legal and policy compliance issues.

S3 is an online, high-speed, economical, scalable, web-based interface for the data backup and archiving of data files in the AWS cloud. S3 supports data transfer over SSL protocol, and the data file becomes encrypted automatically once it is uploaded to the AWS cloud. The data owner has complete control beyond their data. Amazon RDS is a relational database service over the AWS cloud with the feature of an access control mechanism.

12.4.3 CLOUD CHANNEL SECURITY

Some key points that are related to developing vulnerability for cloud channel:

- **Increased Attack Surface:** Cloud has an increased attack surface since for an organization, the data is stored outside the organization with an unknown entity being in charge. This entity is typically doing even all the computations on the stored data. Also, all the communications during, before, and after storing data in the cloud take place over the Internet, thereby making the cloud more vulnerable through the communication links which are public/insecure.
- **Insecure API:** Application User Interface, software, and other interfaces are shared among the users of the cloud. Security in sharing such resources is dependent upon the security policies those software and APIs. Software and APIs which are going to be distributed upon the cloud should have strong security in every aspect it authentication or encryption to avoid any malicious attacks.
- **Crypto Mining:** With the growth of cryptocurrency, there is also a rise in crypto mining. Attackers will target systems with the internet of hijacking processing power in secrecy and steal resources to control as much power as possible to mine cryptocurrency. An attacker attempts to compromise the cloud system by placing a malicious virtual machine near a target cloud server system and then launching a side-channel attack. Side-channel attacks have emerged as a kind of active security threat that is targeting system implementation of cryptographic algorithms. Evaluating cryptographic systems resilience to side-channel attacks is therefore essential for secure system design. Side-channel attacks use two

steps to attack VM co-residence and placement, i.e., an attacker can often place his or her instance on the same physical machine as a target instance and VM extraction, i.e., the ability of a malicious instance to utilize side channels to learn information about co-resident instances. It can be effortless to gain secret information from a device so security against side-channel attack in cloud computing should be provided.

Security is the primary issue in cloud computing. The cloud service providers implement top security criteria and enterprise certifications, but, storing cloud users' data and import data files from cloud continually convey a risk. Amazon's Elastic Compute Cloud (EC2) offers a specialty known as the security group, which is equivalent to a traditional network firewall, in which we have to specify the protocols, network ports and secure IP addresses ranges that are permitted to reach EC2 instances. Each EC2 instance can be specified more than one security group, each of which routes the relevant traffic to each instance. The security groups can be customized based on specific subnets or IP address which restricts admittance to EC2 instances.

12.4.4 SECURING CLOUD COMPUTING SYSTEM MANAGEMENT

Some key points that are related to developing vulnerability for cloud system management:

- **Loss of Control on Data:** Cloud promotes outsourcing. The industries are thriving in this outsource business model. According to this model, the responsibilities of acquiring resources and using these resources be it storage or computing requirements, are delegated to contracted third-party service providers, with the customer concentrating on more delicate parts of the work. That is one of the best-known strategies that have rendered cloud popularity.
- **Lack of Trust:** In the cloud, the customer infrastructure, data, and compute all located off-premise and are managed by a third-party entity. That is an appearance of the cloud that is special. It is vital for the consumer to be able to trust a provider. However, trust is a concept rather than a measurable element since it is more subjective

than objective. Measuring trust must involve several factors that make a decision-making and evaluating a decision harder.

- **Difficulty in Auditing:** The cloud user does not have direct access to any physical resources since everything in the cloud has virtualized. Hence it is difficult to have a monitoring mechanism. Hence auditing can be done only by the service provider, and we have to ask how trustworthy that is.
- **Malware Attack:** The cloud environment is less visible and more exposed so that a cloud network is prone to malware attacks. Several times it happens so that the cloud provider may not produce the details of how they grant access to software and other system functionality, how they trace a user and how they add their policies. In cloud malware injection attack an attacker tries to inject malicious service or virtual machine into the cloud. To fulfill this purpose, the attacker creates its malicious service implementation module (SaaS or PaaS) or virtual machine instance (IaaS) and tries to add it to the Cloud system.
- **Account/Service Hacking:** Attacks like fraud, phishing, and software exploitation are various types of account or service hacking techniques. Usually, such exercise is carried out by stealing credentials (like illegally obtaining access to someone's account). Account hacking can cause critical destruction to integrity and reputation.
- **Misconfigurations:** These are known as the security measures, and access controls are present but were either poorly implemented or not enabled. Misconfigurations will proceed to be a primary source of security points.
- **Backdoor:** It is one of significant security threat in a cloud computing environment. If anyone received sensitive information like hosting files of the virtual machine or any other sensitive pieces of information through the back door. Using this sensitive information the hackers will be able to allow illegal access to the whole system indirectly.
- **Malicious Insiders:** These are the persons who are authorized to handle the data such as system administrator, database administrators or employees of the data center that offering cloud services, partners, and contractors who have access to the data. Those people can steal or corrupt the data whether they are getting paid by other companies or to just hurt a company (Table 12.3).

TABLE 12.3 Cloud Security Areas

S. No.	Cloud Security Area	Role in the Cloud Security
1	Cloud Security behind the virtualization	Management of isolation of cloud tenants, In SaaS model chance to break data integrity, data confidentiality, In PaaS model data integrity, data availability, in IaaS model, securing unauthorized and illegal access of customers data, Securing resource attack like botnet and denial of service, prevention of virtual machines
2	Data Security over the cloud	Management of data integrity, data confidentiality, data availability, data privacy, loss of data
3	Cloud Channel Security	Monitor attack surface, insecure API, implementation of data decryption techniques during data transaction and storing
4	Securing Cloud Computing system management	Management of control of data, prevention of malware attack, conducting a security audit, trust management, malicious insider

12.5 CLOUD SECURITY MODELING

A security model is a conceptual design for specifying and enforcing security policies. A security model may be described as a sequence of authentication modeling, authorization modeling, access control modeling, threat modeling (Figure 12.4).

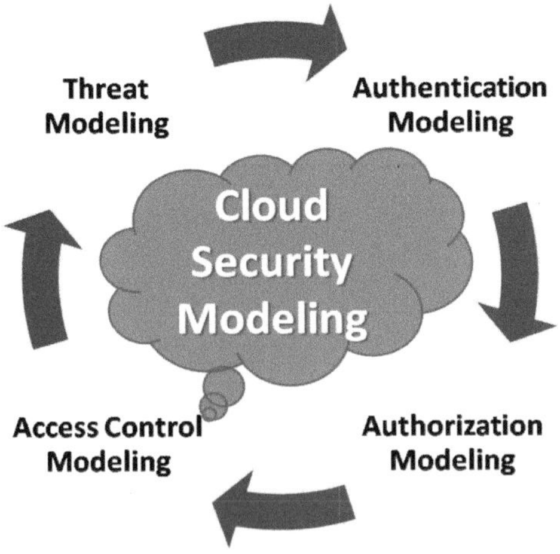

FIGURE 12.4 Cloud security modeling

12.5.1 THE AUTHENTICATION MODELING

The object of authentication is to assure the identity of the entity involved in the cloud system. Password, Biometric, and Public Key Infrastructure (PKI) are some authentication methods using currently. Problem with the cloud system, it is required multiple authentication processes, that means every service has separate authentication for a single cloud user. Authentication is a significant constraint on cloud security, so it is necessary to model the authentication process. The authentication consists of three entities:

1. Certificate authority (CA);
2. End-entity; and
3. Identity verifier.

The identity verifier is the entity that determines as to whether or not attested identity is correct. The CA certifies the end-entities by issuing public-key certificates for them. The identity verifier trusts the CA as the authority for a given certification for the third entity. If that CA does not issue an end entity's certificate, then the identity verifier must locate a certification path of certificates from that of the entity to one issued by the CA. In the present scenario, cloud providers provide multi-factor authentication (MFA) like OTP, captcha, etc. AWS also introduced a MFA mechanism for the authentication of cloud users using to enter a unique and random authentication code generated from an authenticated device or text message on the mobile when user access AWS resources. If entered MFA code is accurate, then the user can access AWS resources otherwise not.

12.5.2 THE AUTHORIZATION MODELING

The cloud security is based on a two-step process one is authentication, which ensures that the entity requesting access to the cloud is what or who it claims to be, and second authorization to allow access only to those cloud resources which are appropriate to entity's identity. Authentication and authorization are both critical enablers of the cloud computing system. That is because the cloud resource available over the internet anyone accesses these anytime and anywhere. The authorization consists privilege management matrix consists of two entities: the privilege holder, and the privilege verifier. The privilege holder is the entity that holds a particular privilege and asserts its privileges for a particular context of use. The privilege verifier is the entity that determines as to whether or not asserted privileges are sufficient for the given context of use. Authorization models of cloud system are the development of privilege management matrix.

12.5.3 THE ACCESS CONTROL MODELING

Access control is a security technique that can be used to systematize who, what, and where can view or use resources in a cloud computing environment. Access control systems perform authorization identification, authentication, access approval, and accountability of entities through login credentials including passwords, personal identification numbers (PINs), biometric scans, and physical or electronic keys. There are four significant types of access control are:

- **Mandatory Access Control (MAC):** MAC is a security strategy that binds the individual resource owners to have to grant or deny access to the resource. MAC criteria are defined by the cloud provider and are unable to be changed by cloud users.
- **Discretionary Access Control (DAC):** The DAC is another type of access control and based on the Discretionary Access Control List (DACL). A DACL identifies the trustees that are allowed or denied access to a securable object.
- **Role-Based Access Control (RBAC):** The RBAC is an approach to restricting system access to authorized users. RBAC is a policy unbiased access control mechanism defined around roles and privileges.
- **Rule-Based Access Control:** This is a method, where access is allowed or denied to resource objects based on a collection of rules established by the cloud provider.

12.5.4 THE THREAT MODELING

Anything that has the potential to make dangerous to a computer system in the context of computer security it is known as the threat. A threat is something destructive entity that may or may not happen, but has the potential to cause severe damage. Threats can lead to attacks on software, hardware, networks, and more. The threat modeling is an approach for analyzing the security of an application. It is a structured approach that enables to identify, quantify, and address the security risks associated with the computer system. The threat modeling is not an approach to reviewing code, but it does complement the security code review manner. The core to threat modeling is to determine where the most effort should be applied to keep a system secure. Threat modeling is an iterative process to identify security threat, assign priority of threat, estimate events of threat activation and prepare advisory documents of threat (Table 12.4).

TABLE 12.4 Cloud Security Model Objectives

S. No.	Security Model	Objectives	Security Aspects
1	Authentication model	Assure the identity of the entity	Unauthorized access
2	Authorization model	Which cloud resources are appropriate to the entity	Illegal use
3	Access control model	Who, what, and where can view or use resources	unauthorized modification, Illegal use, personal inspection, personal disclosure
4	Threat model	Systematic security review	Illegal recording or destruction of information, personal inspection, personal disclosure, Unauthorized access, illegal use, unauthorized modification

12.6 CLOUD SECURITY STANDARDS

The security standards are techniques usually outlined in published materials that attempt to protect the cyber environment of a user or organization. The primary intention is to reduce the risks and prevention from cyberattacks. These published materials consist of collections of policies, tools, security concepts, security safeguards, security guidelines, risk management procedures, actions, training, best practices, assurance, and technologies. Some security standards related to cloud environment as following -

- **SAS-70 (The Statement on Auditing Standards No. 70):** This is a commonly adopted security standard among cloud service providers with the adoption rate of almost 67%. That was developed by the American Institute of Certified Public Accountants (AICPA). The focus is on a CSP's infrastructure, policies, and procedures. It ensures that a CSP follows best practices ensuring the security of clients' data.
- **ISO-27001:** This standard was published in 2005 for an Information Security Management System (ISMS). Adoption of this standard is about 33%. It is generally accepted to be a holistic information security mechanism since it examines in great detail how an organization manages information security management. ISO 27001 is Standard for protecting privacy in the cloud.
- **Safe Harbor:** Typically this standard is used by organizations in the U.S. and European Union.
- **NIST:** National Institute of Standards and Technology is a body that regulates many things in the cloud, and they have proposed a set of standards known as the NIST standards.
- **HIPAA:** The U.S. Health Insurance Portability and Accountability Act (HIPAA) is a popular standard among CSPs. It is followed by roughly 16 percents of cloud service providers.
- **FISMA:** This is the Federal Information Security Management Act used by U.S. federal government processes or programs to measure security for federal IT systems.

KEYWORDS

- **channel security**
- **cloud security**
- **cloud security model**
- **cloud storage security**
- **security attacks**
- **system management**
- **virtualization security**

REFERENCES

Al-Saiyd, N. A., & Sail, N., (2013). Data integrity in cloud computing security. *Journal of Theoretical and Applied Information Technology*, *58*(3).

Barhate, S. M., & Dhore, M. P., (2016). User authentication issues in cloud computing. *IOSR Journal of Computer Engineering (IOSR-JCE),* pp. 30–35. www.iosrjournals.org (Accessed on 28 August 2019).

Cloud Standards Customer Council, (2017). *Security for Cloud Computing*. Ten steps to ensure success, version 3.0.

Gawande, M. R., & Kapse, A. S., (2014). Analysis of data confidentiality techniques in cloud computing. *International Journal of Computer Science and Mobile Computing*, *3*(3), 169–175.

Hussain, Z., & Gummadi, A., (2013). *Vulnerabilities in Cloud Computing*. George Mason University, www.chinacloud.cn/upload/2013–05/13050613111874.pdf (Accessed on 28 August 2019).

Mebrahtu, A., & Srinivasulu, B., (2017). Key architectural models, security issues, and solutions on cloud computing. *International Journal of Computer Science Trends and Technology (IJCST), 5*(1), 88–94.

Rajput, R. S., & Pant, A., (2018). Optimal resource management in the cloud environment-a review. *International Journal of Converging Technologies and Management (IJCTM), 4*(1), 12–24.

Ramachandran, R., & Ramachandran, A., (2012). Rapid and proactive approach to exploration of vulnerabilities in cloud-based operating systems. *International Journal of Computer Applications*, *42*(3).

Sabahi, F., (2011). Cloud computing reliability, availability and serviceability (RAS): Issues and challenges. *International Journal on Advances in ICT for Emerging Regions, 4*(2), 12–23.

Vani, B., & Priya, C. M. R., (2015). Availability in cloud computing. *International Journal of Innovative Research in Information Security (IJIRIS), 4*(2), p. 2015.

Yunchuan, S., Junsheng, Z., Yongping, X., & Guangyu, Z., (2014). *International Journal of Distributed Sensor Networks 2014*, Article ID 190903, http://dx.doi.org/10.1155/2014/190903.

Zou, J., (2016). *Accountability in Cloud Services*. Doctor of Philosophy Thesis, Faculty of Science and Engineering at Macquarie University.

CHAPTER 13

Unpacking Cybersecurity Trends

VANNIE NAIDOO[1] and RAHUL VERMA[2]

[1]*University of KwaZulu – Natal, South Africa,*
E-mail: Naidoova@ukzn.ac.za

[2]*Bhai Parmanand Institute of Business Studies, India*

ABSTRACT

With technological advancements taking place at such a rapid rate in the world today, the world as we know has changed. However, how safe are we? This is a crucial question that comes to mind when one has to explore cybersecurity. Cybercriminals are prowling the internet as we speak, looking for the vulnerabilities that they can use to wage attacks against us. In this digital age security and privacy of our data should be a major security issue that all organizations take care of. Cybersecurity is vital in the Fourth Industrial Revolution, which can cause major technology disruption and affect lives in ways we never thought possible. This chapter will explore crucial issues associated with cyber trends namely; cybersecurity, cybercriminals, cybercrimes, cyberethics, cybersecurity techniques, social media and cybersecurity and challenges faced in implementing cybersecurity.

13.1 INTRODUCTION

Cybersecurity affects all citizens of the world as technology permeates all levels of society. Any country and its people can come under a cyberattack from the unseeing and evil terrorist who have in mind to create mass destruction for that country and its people. Cybersecurity is, therefore, a crucial issue facing a country. A country has to be able to protect itself from cybercriminals who want to destroy the county's business sector, hospitals, infrastructure, or government. A Company can also come under attack by cybercriminals. A company, therefore, has to have in place the appropriate level of cybersecurity to safeguard its products and resources against cyberattacks perpetrated by its competitors. Cybersecurity is crucial in keeping a company, government, or individual safe. The lack of cybersecurity can expose governments, companies, and individuals to cybercriminals. To protect itself, a company or government must be aware of possible threats that they can come up against.

13.1.1 CYBER CRIME

Computer oriented crime or, cybercrime is a crime that includes a network and a computer (Moore, 2005). The computer might have been utilized in the commission of a crime, or it might be the target (Warren and Jay, 2002). There are broad definitions on cybercrime, never the less it speaks about a crime perpetrated against a person, organization, or government. In the discussion that enfolds thoughts on cybercrime will be explored. Cybercrimes could be characterized as: Offences that are committed against a person or group of persons with a criminal thought process to intentionally damage the repute of the victim or cause mental or physical loss, or harm, to the victim indirectly or directly, by means of modern telecommunication systems such as internet (systems including but not restricted to groups, notice boards, e-mails, and Chat rooms) and mobile phones (MMS/SMS/Bluetooth) (Halder and Jaishankar, 2011).

Cybercrime might intimidate an individual or a nation's financial health and security (Steve, 2016). Issues encompassing these sorts of crimes have become high-profile, especially those encompassing copyright infringement, hacking, unwarranted mass-surveillance, child grooming, child pornography, and sextortion (Halder and Jaishankar, 2011). There are also issues of privacy when private data is disclosed or intercepted, legally or otherwise. Debarati Halder and Jaishankar further characterize cybercrime from the viewpoint of sexual orientation and defined 'cybercrime against women' as crimes focused on against women with a rationale to deliberately harm the victim physically and psychologically, using modern telecommunication networks such as mobile phones and internet. Globally, non-state and governmental actors both engage in cybercrimes, including financial theft, espionage, and other cross-border crimes (Halder and Jaishankar, 2011). Cybercrimes crossing global borders and relating the activities of at least one nation-state is at times alluded to as cyber warfare.

A report (sponsored by McAfee) gauges that the yearly harm to the worldwide economy is at $445 billion (Reuters, 2014); however, a Microsoft report appears that such study-based estimates are "hopelessly imperfect" and overstate the genuine losses by orders of greatness (Microsoft, 2011). Roughly $1.5 billion was misplaced in 2012 to online debit and credit card extortion in the US (North Denver News, 2015). In 2016, a survey by Juniper Research evaluated that the costs of cybercrime can be as high as 2.1 trillion by 2019 (Juniper Research, 2016).

At the global level, it is on record that cybercrime is a growth industry. The returns are great, and the risks are low since these criminals can commit these crimes from remote location and remain undetected. It is expected that the probable annual cost to the worldwide economy from cybercrime is more than $ 40 Billion, which reached $ 12 Trillion in 2012. A traditional estimate would be $ 375 Billion in losses, whereas the greatest can be as much as $ 575 Billion (CSIS, 2014). Krone (2005) indicates that the Council of Europe's Cybercrime Treaty uses the term "Cybercrime" to refer to offenses ranging from criminal activity against data to content and copyright infringement.

Sharma and Alam (2016) argue that cybercrimes could be in any form from user privacy violation to hate speeches or any other crimes in which a computer or electronic device and a network has been involved. These crimes know no boundaries, time, and location. Even they can be conducted by anyone irrespective of any gender or age group. These crimes not only affect an individual but also the whole community. Today, blogs, Twitter, Facebook, Instagram, community websites, and other web-based platforms allow users to reach out to millions of people at the click of a mouse.

According to Stanwick and Stanwick (2009), the e-mails or websites used in phishing schemes look like legitimate business e-mails or websites, but they are not. Cyber criminals can use these phishing schemes to commit cyber crimes like identity theft, wire fraud, credit card fraud, bank fraud, and computer fraud. Lininger and Dean (2005) add that Cybercrime is motivated by fraud, typified by the bogus e-mails sent by 'phishers' that aim to steal personal information. The tools driving their attacks and fuelling the black market are crimeware. Examples of crimeware are bots, Trojan horses, and spyware.

In the next part of this discussion, different types of cyber-related crimes will be discussed. They are as follows:

- **Financial Fraud Crimes:** Computer fraud is any unscrupulous deception of reality, expecting to let another refrain from doing or to do something which causes loss. A range of internet scams, numerous based on social engineering and phishing, target businesses and consumers. Other forms of fraud may be encouraged using computer systems, including extortion, identity theft, carding, bank fraud, and theft of classified information.
- **Cyber Terrorism:** This, in general, can be characterized as an act of terrorism committed through the usage of computer resources or cyberspace (Parker, 1983). A cyber terrorist is somebody who coerces or intimidates an organization or a government to progress his/her social or political targets by propelling a computer-based attack against computers, networks, or the data stored on them.
- **Cyber Extortion:** It happens when a computer system, e-mail server, or website is threatened with or subjected to the rehashed refusal of service or other attacks by malicious hackers. These hackers demand cash in return for promising to offer "protection" and to halt the attacks. An illustration of cyber extortion was the assault on Sony Pictures of 2014 (Mohanta, 2014).
- **Cyber Warfare:** This includes the battlespace use and targeting of networks, online control systems, and computers in warfare. It includes both hostile and cautious operations relating to the risk of sabotage, espionage, and cyberattacks.
- **computer as a Target:** These crimes are committed by a chosen group of criminals. In Contrast with crimes using the computer as a tool, these crimes require the specialized information of the culprits. As such, as innovation advances, so as well does the nature of the crime. These crimes are generally latest, having been in subsistence for only as long as computers have — which clarifies how ill-equipped world and the society in common is towards fighting these crimes.
- **computer as a Tool:** When the person is the major target of cybercrime, the computer can be referred as the tool instead of the

target. These crimes generally include less specialized skill. Human shortcomings are by and large misused. The harm dealt is to a great extent intangible and psychological, making legitimate action against the variants more troublesome. These are the crimes which have been existed for centuries within the offline world.

13.1.2 CYBERSECURITY

IT security, computer security or cybersecurity is the assurance of computer systems from the damage and theft to their information, software or hardware, over, and above from misdirection or disruption of the services they offer. Cybersecurity incorporates controlling physical access to the hardware, plus shielding against damage that might come by network access, data, and code injection (Daniel et al., 2017). Moreover, due to negligence by administrators, whether accidental or intentional, IT security is vulnerable to being deceived into veering off from secure process through various methods (Rouse, n.d.). The field is of increasing significance due to the growing dependence on the Internet (The Australian, 2013) and computer systems, wireless networks such as Wi-Fi and Bluetooth, the development of "smart" gadgets, including tiny devices, televisions, and smart phones as part of the IoT (Internet of Things).

Experts working in the cybersecurity field can be known by a few of the following terms (Menking, 2017):

- **White Hat Hacker:** Moreover known as an "ethical hacker" or "penetration tester." They are proficient programmers that break into frameworks and utilize abuses to get to target systems for reasons relating to hardening the security of a target or prevention of crime.
- **Black Hat Hacker:** A criminal who breaks into frameworks and compromises security against the law.
- **Grey Hat Hacker:** Somebody who conducts black hat hacks for white hat programmer reasons.

In the world today organizations and governments should be in charge of ensuring the privacy, security, and safety of its information. If information is not secure and privacy is not guaranteed by the organization or government, then if this information gets into the wrong hands it can create chaos and destruction that can result in a shortfall of millions of dollars. To prevent this, cybersecurity is becoming vital for any organization or government. Cyber criminals are becoming very techno savvy and can defraud organizations or governments of millions of dollars by the touch of a button. Another very hateful crime that cybersecurity relates to is when terrorists and hateful people post horrible hate-speak, prejudice, and abusive messages on social media. If proper cybersecurity is in place such cyber-criminal activity can be minimized by penalizing and locking up cyber criminals who want to steal, harm, and terrorize people, governments or organizations online.

There are various definitions of cybersecurity which will be highlighted. Lewis (2006) contends that cybersecurity involves the defending of computer systems and the data they contain from malicious disruption or damage and from penetration. Cybersecurity includes diminishing the chance of malicious attack to networks, computers, and software. This incorporates devices utilized to identify break-ins, halt viruses, block malicious access, uphold authentication, empower encrypted communications, and on and on (Amoroso, 2006).

As per Canongia and Mandarino (2014), cybersecurity is the art of guaranteeing the continuity and existence of the information society of a country, protecting, and guaranteeing, in cyberspace, its critical infrastructure, assets, and information. The USA has designed policies and laws that they have put in place to protect their country and citizens from cyber crime attacks. A study conducted on information security by Price Waterhouse Cooper's, outlines the fact that cyber crime has developed to an extent that brings over 117,000 attacks per day (Price Waterhouse Coopers and Info Security, 2014).

According to Sharma and Alam (2016) the European Commission has played a key role in the development of European Cyber Crime Centre as termed as EC3, which started operations in January 2013. The EC3 acts as the central point in the fight against cyber crime within the Union, pooling European cyber crime skills to back Member State's cyber crime investigations and giving a collective voice of European cyber crime investigators across the judiciary and law enforcement. This collaboration amongst European government can be instrumental in keeping cyber criminal activities to a minimum.

Hale (2002) argues that cyber-crime has in fact reached epidemic proportions. In a survey conducted, more than 90% of the respondent government agencies and corporations reported computer security breaches at one time or the other. It is common place for hackers and disgruntled employees to commit numerous cyber crimes whereas others are committed by crooks utilizing the Web to execute identity theft, auction fraud, and other scams. The financial institutions get constantly hard hit as identity thefts have allegedly cost them $2.4 billion in expenses and losses in 2000 alone.

13.2 MAIN FOCUS OF THE CHAPTER

13.2.1 CYBERSECURITY TECHNIQUES

Proper cybersecurity is vital in this day and age when anyone can be under attack by cyber criminals who remain unseen, yet cause so much damage, harm, and confusion to internet users. In the discussion that follows cyber-security techniques will be put forward and discussed.

Singleton and Singleton (2004) identified several measures that can be used to protect a business from internet attacks. They are as follows:

- Conducting and an effective risk assessment.
- Embarking on a comprehensive review of the company's procedures and policies relating to security.
- Having a sound business recovery plan that is described in the company's polices and procedure manuals.
- Developing an incremental response plan.
- Having proper education, training, and seminars on cybersecurity issues for key managers and employees.
- Ensuring that the company or organization has comprehensive insurance coverage if an attack occurs.

According to Stanwick and Stanwick (2015) customers need to protect themselves as follows:

- Customers need to watch for watchwords and phases.
- Customers must always protect their private information.
- They must not download electronic files sent by strangers.

- When busying items online, customers must pay with credit card.
- Customers must never share their password with anyone.
- Customers when using the internet must pay attention.
- Customers must file a complaint if they feel their privacy was violated or if they were prone to cyber crime.

Reddy and Reddy (2014) puts forth the following essential cybersecurity techniques to adhere to:

- **Access Control and Password Security:** The concept of user name and password has been fundamental way of protecting our information. This might be one of the prime measures with respect to cybersecurity.
- **Authentication of Data:** The documents that we receive must always be authenticated before downloading it. Authenticating of these records is often done by the anti-virus program present within the gadgets. Thus a good anti-virus software is also essential to protect the devices from viruses.
- **Malware Scanners:** This is program that ordinarily scans all the documents and files present within the system for harmful viruses or malicious code. Trojan horses, worms, and viruses are illustrations of malicious software that are frequently grouped together and alluded to as malware.
- **Firewalls:** It is a piece of equipment or software program that helps to screen out worms, viruses, and hackers that attempt to reach your computer over the web. Every bit of messages leaving or entering the web go all the way through the firewall there, which analyzes every message and blocks those that don't meet up the required security criterion. Consequently firewalls play a critical part in identifying the malware.
- **Anti-Virus Software:** It is a computer program that identifies, anticipates, and takes action to remove or disarm malicious software programs, such as worms and viruses. Most anti-virus programs incorporate an auto-update feature that empowers the program to download profiles of latest viruses so that it can check for the new viruses as soon as they are found. An anti-virus program is a must and fundamental need for each system.

13.2.2 COMPUTER EMERGENCY RESPONSE TEAM (CERT)-INDIA: AN ILLUSTRATION

Most nations have their own CERT (Computer Emergency Response Team) to guard network security. A few provisions for cybersecurity have been integrated into rules framed under the IT (Information Technology) Act, 2000. The NCCP (National Cybersecurity Policy), 2013 is a policy structure by MeitY (Ministry of Electronics and Information Technology) which points to look after the private and public infrastructure from cyberattacks, and protect information, such as sovereign data, personal information (of web users), and banking and financial information. CERT-India is the nodal agency which screens the cyber threats in the nation. The post of National Cybersecurity Coordinator has also been made in the Prime Minister's Office (PMO). The Indian Companies Act, 2013 has also presented cybersecurity and cyber law commitments on the part of Indian directors. A few provisions for cybersecurity have been integrated into rules framed under the IT Act, 2000 update in 2013 (NCSSS, 2017).

13.2.3 CHALLENGES FACED IN IMPLEMENTING CYBERSECURITY

- People are so free in sharing their private information with an audience of millions of online users on social media cites that this becomes a major security challenge.
- The cost of cybersecurity systems to protect a company is expensive and since the cyber criminals are becoming so good at hacking the cybersecurity systems, continuous systems upgrades can also be a costly affair for a company or a private user.
- People are sometimes so easily charmed by social engineers online that they can innocently give up their private information to a stranger without even suspecting this person.
- Children who use computers and smart phones need parental monitoring so that predators do not hunt their children and hurt them. This is not easy when homes today are run by single parents or parents who are busy working that they don't always have the time to monitor their kid's computer use.
- No proper leadership and collaboration amongst governments to combat cyber terrorism. Governments of the world should come

together and share information and come up with a common law or policy on cyber terrorism. This is not possible as many countries are at war or competing with each other for a global market share that they want to work alone for fear they will be beaten by other nations if they were to share their policy or plans on cyberattacks or cyber terrorism.

- The need for training and education on cybersecurity is pivotal. However, cyber criminals can also use this information to their benefit.
- Developing a code of ethics for computer users that organizations can implement within the workplace to safeguard the organization against cyberattacks.
- Awareness of cyberattacks and cybersecurity in the media should be present so that people are aware of what is happening in the world. This is often difficult when media is controlled by the state.
- Prejudice and hate is a human trait that is often very difficult to overcome. Making people respectful and tolerant is not simple and what people say on social media is often the prejudice they feel for others.
- Some countries are very lenient on criminals therefore cyber criminals still prowl cyberspace and continue their crime.
- Social media cites must be continuously monitored for hate speech, religious bigotry, racism, or terrorism and this are often not the case. There need to be proper social media etiquettes in place to prevent people from hate speech, bigotry, or terrorist remarks online.

13.3 SOLUTIONS AND RECOMMENDATIONS

13.3.1 CYBER ETHICS

Cyberethics is the philosophic learning of morals relating to computers, including user conduct and what computers are planned to do, and how this influences society and individuals. For a long time, different governments have enacted rules, whereas organizations have defined procedures about cyber ethics.

A code of conduct is essential, and people who use computers need to adhere to it. There is a wrong and right, and the ethical paradigm clearly outlines what practice is ethical and what is not. In the discussion below cyber ethics will be discussed in detail. In cybersecurity, cyber ethics is crucial. As per Jones and George (2009), ethics are the inward directing benefits, values, and moral principles that individuals utilize to analyze and interpret a situation and at that point choose the appropriate way to behave or what is right.˙ Irwin (2002) adds that ethics is a system of accepted beliefs and principles of conduct characteristically based on moral objectives that administer the behavior of groups and individuals within the organization.

Another important ethical debate is privacy of customer information. Some companies sell or trade customer information. A client can take a company to court if they leak their sensitive private personal information. In the USA, for example, there are laws in place to protect a customer's personal information.

According to The Computer Ethics Institute (1992), the 10 commandments of computer ethics is as follows:

- **Commandment 1:** Don't utilize the computer in ways which might hurt other people.
- **Commandment 2:** Don't utilize computer technology to cause intrusion in other user's work.
- **Commandment 3:** Don't spy on another persons' computer data.
- **Commandment 4:** Don't utilize computer technology to steal data.
- **Commandment 5:** Don't contribute to the spread of deception utilizing computer technology.
- **Commandment 6:** Abstain from copying software or buying pirated copies. Pay for program unless it is free.
- **Commandment 7:** Don't utilize someone else's computer assets unless authorized to.
- **Commandment 8:** It is morally wrong to claim rights on a work which is the productivity of someone else's intellect.
- **Commandment 9:** Before making program, think approximately the social influence it can have.
- **Commandment 10:** In utilizing computers for communication, be courteous and respectful with the individual members.

Reddy and Reddy (2014) outline essential cyber ethics that can be adhered to. They are as follows:

- Do use the internet to communicate and interact with other people. Instant messaging and E-mail make it simple to remain in touch with friends and family members, communicate with work colleagues, and share information and ideas with individuals over town or halfway around the world.
- Don't be a bully on the internet. Do not call people names, send embarrassing pictures of them, lie about them, or do anything else to try to hurt them.
- Internet is considered as world's largest library with information on any topic in any subject area, so using this information in a correct and legal way is always essential.
- Don't operate others accounts by using their passwords.
- Never attempt to send any sort of malware to other's systems and make them corrupt.
- Never share your individual data to anybody as there's a great chance of others abusing it and at last you'd end up in a trouble.
- When you're online never pretend to the other individual, and never attempt to form fake accounts on somebody else because it would land you as well as the other individual into trouble.
- Always download videos or games only if they are permissible and adhere to copyrighted information.

Bush et al., (2000) identified through a survey of marketers the following ethical issues facing Internet users:

- Security transactions;
- Illegal activity (fraud, hijacking, etc.);
- Privacy;
- Honesty/truthfulness;
- Judged by same standards as other mediums;
- Pornography;
- Product warranty;
- Plagiarism;
- Targeting children;
- Unsolicited e-mail;
- False advertising.

A good and responsible code of ethics should be the cornerstone of any organization today, be it a large corporate, a non-profit NGO or a small family-owned business. When it comes to the use of computers and ICT's in social and business constructs, it is essential that there is cyber ethics in place that people and organizations adhere to. Ethics go hand in hand in instilling responsible management and good business practice, and in the age of computers, this is pivotal.

13.3.2 SOCIAL MEDIA AND CYBERSECURITY

Social media are computer-mediated innovations that encourage the creation and sharing of career interests, ideas, information, and other forms of expression through virtual networks and communities. The assortment of built-in and stand-alone social media services right now assessable presents challenges of definition; however, there are a few common features (Obar and Wildman, 2015; Kaplan and Haenlein, 2010; Boyd and Ellison, 2007):

- Social media are intuitively Web 2.0 Internet-based applications.
- User-generated substance, such as text comments or posts, digital videos or photos, and information produced through all online interactions, is the soul of social media.
- Users make service-specific profiles for the app or website that are maintained and designed by the social media organization.
- Social media encourage the improvement of online social systems by interfacing a users' profile with those of other groups or individuals.

People love to connect to one another; this is what makes us human. Social media websites allow us to speak to family and friends and other users online. However, the downside of this is that social media can become a hunting ground for cyber criminals to gather sensitive information on users and phish or hack them. Cyber criminals can be charming social engineers that can use excellent people skills to gain private and sensitive information from a user on social media without the user even suspecting them. Rose (2011) contends that as long as social media sites share data with other social media sites, location-sharing is permitted and individuals are getting to be comfortable with unveiling such data, at that point anybody will promptly be able to induce sensitive information. Combine this with excessively excited users, who intentioned or not, share as much personal data and developers who can access private data, then social media presents a very serious security risk.

Cyber-criminals attack popular social media sites. Many home-based internet users and corporate employees who use social media networks can be at risk. Studies have uncovered that social networking can open a back door into corporate ICT stages, putting individuals and businesses at risk of identity theft, information compromise, and other malevolent attacks. Cyber-criminals are seeking out for sites that have numerous users with destitute security mindfulness to infect, and social networking sites are a fabulous place to chase (Bazelon et al., 2006). Huber et al., (2009) adds that social media sites such as Facebook provide the ability for a user to maintain his or her own web page and share content with their personal connections. By breaching the faith a user has with his/her online network, hackers are able to plant malware into friend's content and cause yet more persons to fall victim to the malicious link.

Social media cites can also attract young and impressionable children and adults into cult worship. Some children often meet religious fanatics on Facebook that fool them into becoming possible terrorists or religious soldiers. Social media websites can be especially harmful to young children who are preyed on by pedophiles. Social media cites can get people together to wage war and cause kayos and anarchy within a country. In today's world anyone of us can be prey to Cyber criminals who can gain personal information about us on social media websites and use it to incite war or terrorism.

According to Gharibi and Shaabi (2012) the following strategies are recommended to protect people on social media:

- **Building Awareness the Information Disclosure:** Users must beware and be exceptionally cognizant with respect to the revealing of their individual data in profiles in social websites.
- **Encouraging Awareness-Raising and Educational Campaigns:** Governments need to offer and provide educational classes about security issues and awareness-rising.
- **Modifying the Existing Legislation:** Existing legislation should be adjusted related to the new attacks and frauds, and new technology.

- **Empowering the Authentication:** Access authentication and control must be exceptionally solid so that cyber crimes done by spammers, hackers, and other cyber criminals can be minimized as much as possible.
- **Using the Most Powerful Anti-Virus Tools:** Users must utilize the foremost effective anti-virus tools with normal updates and must keep the appropriate default setting, so that the anti-virus tools can work more effectively.
- **Providing Suitable Security Tools:** Here, we grant proposal to the security software providers and are that they ought to offer few extraordinary tools for clients that empower them to remove their accounts and to control and manage the diverse security and privacy issues.

13.4 FUTURE RESEARCH DIRECTIONS

Cybersecurity is an area that is new and debates surrounding the issues are often very contentious. Research on government laws and policy in developed and undeveloped countries on cybersecurity needs to be under way so that cyber crimes can be minimized. Training and education on cybersecurity should be investigated so that universities can draw up new curriculum to introduce such theories and practices. The mind of the cyber criminal should be investigated, as understanding cyber criminals and what makes them tick can assist law enforcement to catch such people.

13.5 CONCLUSION

Anyone who uses the internet or computer is susceptible to attack by cyber criminals. There needs to be proper legislation or policy in place at government and company level to safeguard the information of its citizens and customers. Anyone can be under attack since we live in a technologically advanced society. Proper steps need to be taken to safeguard our information. Therein lays the challenge since there is a lack of collaboration amongst governments in dealing with cyberattacks and cybersecurity issues.

KEYWORDS

- **cyber crimes**
- **cyber criminals**
- **cyber ethics**
- **cybersecurity**
- **cyberspace**
- **social media**

REFERENCES

Amoroso, E., (2006). *Cybersecurity*. New Jersey: Silicon Press. Available at: https://www.amazon.com/Cyber-Security-Edward-Amoroso/dp/0929306384

Bazelon, D. L., Choi, Y. J., & Conaty, J. F., (2006). Computer crimes. *American Criminal Law Review, 43*(1), 259.

Boyd, D. M., & Ellison, N. B., (2007). Social network sites: Definition, history, and scholarship. *Journal of Computer-Mediated Communication, 13*(1), 210–230. doi: 10.1111/j.1083–6101.2007.00393.x.

Bush, V., Venable, B., & Bush, A., (2000). Ethics and marketing on the internet: Practitioner's perceptions of societal, industry and company concerns. *Journal of Business Ethics, 23*(1), 237–248.

Canongia, C., & Mandarino, R., (2014). Cybersecurity: The new challenge of the information society. In: Hershey, P. A., (ed.), *Crisis Management: Concepts, Methodologies, Tools and Applications* (pp. 60–80). IGI Global.

Center for Strategic and International Studies (CSIS), (2014). Net losses-estimating the global cost of cyber crime. *Economic Impact of Cyber Crime II*, pp. 1–24. Available at: http://www.mcafee.com/us/resources/reports/rp-economic-impact cyber crime2.pdf (Accessed on 28 August 2019).

Daniel, S., Rabih, B., & Julie, W., (2017). "Towards a more representative definition of cybersecurity." *Journal of Digital Forensics, Security and Law, 12*(2). ISSN 1558–7215. Archived from the original on 28 December 2017.

Florêncio, D., & Herley, C. (2013). Sex, lies and cyber-crime surveys. In Economics of information security and privacy III (pp. 35-53). Springer, New York, NY.

Gharibi, W., & Shaabi, M., (2012). Cyber threats in social networking websites. *International Journal of Distributed and Parallel Systems (IJDPS), 3*(1), 119–126.

Halder, D., & Jaishankar, K., (2011). *Cyber Crime and the Victimization of Women: Laws, Rights, and Regulations*. Hershey, PA, USA: IGI Global. ISBN 978–1–60960–830–9.

Hale, C., (2002). *Cyber Crime: Facts & Figures Concerning the Global Dilemma, Crime and Justice International, 5*(1), 18–65.

Huber, M., Kowalskiy, S., Nohlbergz, M., & Tjoa, S., (2009). Towards automating social engineering using social networking sites. CSE '09: *Proceedings of the 12th IEEE*

International Conference on Computational Science and Engineering, 3, 117–124. doi:10.1109/CSE.2009.205

Irwin, L., (2002). Ethics in organizations: A chaos perspective. *Journal of Organizational Change Management, 15*(4), 359.

Jones, G. R., & George, J. M., (2015). *Essentials of Contemporary Management* (6th edn.). McGraw-Hill: New York.

Juniper Research, (2016). *Cybercrime will Cost Businesses Over $2 Trillion by 2019 (Press Release)*.

Kaplan, A. M., & Haenlein, M., (2010). Users of the world, unite! The challenges and opportunities of social media (PDF). *Business Horizons, 53*(1), 61. doi:10.1016/j.bushor.2009.09.003.

Krone, T., (2005). *High Tech Crime Brief*. Australian Institute of Criminology: Canberra. Available at : https://researchprofiles.canberra.edu.au/en/publications/child-exploitation-high-tech-crime-brief

Lewis, J. A., (2006). *Cybersecurity and Critical Infrastructure Protection*. Washington, DC: Center for Strategic and International Studies. http://csis.org/publication/cybersecurity-and-criticalinfrastructure-protection (Accessed on 28 August 2019).

Lininger, R., et al. (2005). *Phishing: Cutting the Identity Theft Line*, ISBN: 0-7645-8498-7, Indiana: Wiley Publishing, Inc.

Menking, C., (2017). Hacker. In: Paul J., (ed.), *Encyclopedia of Cyber Warfare* (pp. 126–128). Springer, ABC-CLIO. Gale virtual reference library. http://link.galegroup.com/apps/doc/CX7353200101/GVRL?u=mcc_pv&sid=GVRL&xid=62916c40 (Accessed on 28 August 2019).

Mohanta, A., (2014). "*Latest Sony Pictures Breach: A Deadly Cyber Extortion*." Retrieved from: https://forums.juniper.net/t5/Engineering-Simplicity/Announcing-Juniper-s-Intent-to-Acquire-Cyphort/ba-p/312174

Moore, R., (2005). "*Cyber Crime: Investigating High-Technology Computer Crime*." Cleveland, Mississippi: Anderson Publishing.

NCSSS, (2017). "*National Cyber Safety and Security Standards (NCSSS)-Home*." www.ncdrc.res.in (Accessed on 28 August 2019).

North Denver News, (2015). *Cybercrime — What are the Costs to Victims-North Denver News*.

Obar, J. A., & Wildman, S., (2015). Social media definition and the governance challenge: An introduction to the special issue. *Telecommunications Policy, 39*(9), 745–750. doi: 10.1016/j.telpol.2015.07.014. SSRN 2647377 Freely accessible.

Parker, D., (1983). *Fighting Computer Crime*. U.S.: Charles Scribner's Sons. Available at : https://dl.acm.org/citation.cfm?id=538685

Price Waterhouse Coopers & Info Security, (2014). *Information Security Breaches Survey*. Available at: http://www.pwc.co.uk (Accessed on 28 August 2019).

Reddy, G. N., & Reddy, G. J. U., (2014). A study of cybersecurity challenges and its emerging trends on latest technologies. *International Journal of Engineering and Technology, 4*(1), 48–51.

Reuters, (2014). *Cyber Crime Costs Global Economy $445 Billion a Year: Report*.

Rose, C. (2011). The Security Implications Of Ubiquitous Social Media. *International Journal of Management & Information Systems* (IJMIS), 15(1). https://doi.org/10.19030/ijmis.v15i1.1593

Rouse, M., (n.d.). "*Social Engineering Definition.*" TechTarget. Archived from the original on 5 January 2018.
Sharma, I., & Alam, M. A., (2016). Privacy and freedom issues in cyberspace with reference to cyber law. *International Journal of Computer Applications, 145*(3), 11–18.
Singleton, T., & Singleton, A., (2004). Cyber terrorism; are you at risk? *Journal of Corporate Accounting and Finance, 15*(5), 3–15.
Steve, M., (2016). "*Cyber Crime Costs Projected to Reach $2 Trillion by 2019.*" Forbes. Available at: http://www.forbes.com/sites/stevemorgan/2016/01/17/cyber-crime-costs-projected-to-reach-2-trillion-by-2019/#16be59a43bb0
The Australian, (2013). "Reliance spells end of road for ICT amateurs." 7 May 2013. The Australian.
TheComputerEthicsInstitute,(1992).*The10CommandmentsofComputerEthics.CEIOnline.* Retrieved from www.brook.edu/its/cei/overview/Ten_Commanments_of_Computer_Ethics.htm (Accessed on 28 August 2019).
Warren, G. K., & Jay, G. H., (2002). *Computer Forensics: Incident Response Essentials* (p. 392). Addison-Wesley. ISBN 0–201–70719–5.

Index

D

E

F

M

N

O

P

Q

R